10672227

ANNEXE DE LA BIBLIOTHÈQUE

uOttawa

LIBRARY ANNEX

PROGRESS IN CLINICAL AND BIOLOGICAL RESEARCH

Series Editors

Nathan Back
George J. Brewer
Vincent P. Eijsvoogel
Robert Grover

Kurt Hirschhorn
Seymour S. Kety
Sidney Udenfriend
Jonathan W. Uhr

RECENT TITLES

See pages following the index for previous titles in this series.

INDUSTRIAL HAZARDS OF PLASTICS AND SYNTHETIC ELASTOMERS

HS

INDUSTRIAL HAZARDS OF PLASTICS AND SYNTHETIC ELASTOMERS

Proceedings of the International Symposium on Occupational
Hazards Related to Plastics and Synthetic Elastomers,
Espoo, Finland, November 22–27, 1982

Editors

JORMA JÄRVISALO
PIRKKO PFÄFFLI
HARRI VAINIO
Institute of Occupational Health
Helsinki, Finland

ALAN R. LISS, INC. • NEW YORK

Address all Inquiries to the Publisher
Alan R. Liss, Inc., 150 Fifth Avenue, New York, NY 10011

Copyright © 1984 Alan R. Liss, Inc.

Printed in the United States of America.

Under the conditions stated below the owner of copyright for this book hereby grants permission to users to make photocopy reproductions of any part or all of its contents for personal or internal organizational use, or for personal or internal use of specific clients. This consent is given on the condition that the copier pay the stated per-copy fee through the Copyright Clearance Center, Incorporated, 21 Congress Street, Salem, MA 01970, as listed in the most current issue of "Permissions to Photocopy" (Publisher's Fee List, distributed by CCC, Inc.), for copying beyond that permitted by sections 107 or 108 of the US Copyright Law. This consent does not extend to other kinds of copying, such as copying for general distribution, for advertising or promotional purposes, for creating new collective works, or for resale.

Library of Congress Cataloging in Publication Data

International Symposium on Occupational Hazards Related
   to Plastics and Synthetic Elastomers (1982 :Espoo,
   Finland)
   Industrial hazards of plastics and synthetic elastomers.

   (Progress in clinical and biological research;
v. 141)
   Includes bibliographical references and index.
   1. Plastics—Toxicology—Congresses. 2. Elastomers—
Toxicology—Congresses. 3. Industrial toxicology—
Congresses.   I.   Järvisalo, Jorma. II. Pfäffli, Pirkko.
III. Vainio, H. (Harri), 1947–   . IV. Title.
V. Series.
RA1242.P66I58   1982   615.9'51   83-24848
ISBN 0-8451-0141-2

RA
1242
.P66
I58
1982

JUN 2 2 1984

Contents

Contributors

Antero Aitio, Institute of Occupational Health, Helsinki, Finland **[287]**

G. Belvedere, Istituto di Recerche Farmacologiche "Mario Negri", Milan, Italy **[215]**

E. Enwald, Nokia Local Administration Medical Department, Nokia Corporation, Nokia, Finland **[397]**

Väinö Erä, Department of Wood and Polymer Chemistry, University of Helsinki, Helsinki, Finland **[11]**

Lawrence Fishbein, National Center for Toxicological Research, Food and Drug Administration, Department of Health and Human Services, Jefferson, AR 72079 **[19,113,239]**

K. Hemminki, Institute of Occupational Health, Helsinki, Finland **[79,373]**

T. Hemminki, Institute of Occupational Health, Helsinki, Finland **[79]**

Paul K. Henneberger, Environmental Sciences Laboratory, Mount Sinai School of Medicine, City University of New York, New York, NY 10029 **[65,155]**

Ariel Hoff, Department of Polymer Technology, Royal Institute of Technology, Stockholm, Sweden **[91,299]**

Bo Holmberg, Research Department, Unit of Occupational Toxicology, National Board of Occupational Safety and Health, Solna, Sweden **[99,319]**

J.E. Huff, National Toxicology Program, National Institute of Environmental Health Sciences, Research Triangle Park, NC 27709 **[43,137,227,347,421]**

M. Jääskeläinen, Technical University of Helsinki, Helsinki, Finland **[279]**

Jorma Järvisalo, Institute of Occupational Health, Helsinki, Finland **[xiii,287]**

Niels Kjaergaard Jørgensen, Danish Labor Inspection Services, Copenhagen, Denmark **[385]**

Pentti Kalliokoski, University of Kuopio, Kuopio, Finland **[193,279]**

W.M. Kluwe, National Toxicology Program, National Institute of Environmental Health Sciences, Research Triangle Park, NC 27709 **[137]**

T. Koistinen, University of Kuopio, Kuopio, Finland **[279]**

M.-L. Lindbohm, Institute of Occupational Health, Helsinki, Finland **[79]**

Jukka M. Martinmaa, Department of Wood and Polymer Chemistry, University of Helsinki, Helsinki, Finland **[3,365]**

J.A. Moore, National Toxicology Program, National Institute of Environmental Health Sciences, Research Triangle Park, NC 27709 **[43]**

William J. Nicholson, Environmental Sciences Laboratory, Mount Sinai School of Medicine, City University of New York, New York, NY 10029 **[65,155,263]**

H. Norppa, Institute of Occupational Health, Helsinki, Finland **[215]**

Pirkko Pfäffli, Institute of Occupational Health, Helsinki, Finland **[203]**

The number in brackets after each author's name is the opening page number of that author's article.

Claes Ramel, Wallenberg Laboratory, Division of Toxicological Genetics, University of Stockholm, Stockholm, Sweden **[407]**

Agneta Rannug, Wallenberg Laboratory, Division of Toxicological Genetics, University of Stockholm, Stockholm, Sweden **[407]**

Ulf Rannug, Wallenberg Laboratory, Division of Toxicological Genetics, University of Stockholm, Stockholm, Sweden **[407]**

Christina Rosenberg, Institute of Occupational Health, Helsinki, Finland **[337]**

Herbert Seidman, American Cancer Society, New York, NY 10001 **[155]**

A.J.M. Slovak, Fisons Occupational Health Service, Loughborough, England **[309,313]**

Ole Svane, Danish Labor Inspection Services, Copenhagen, Denmark **[385]**

S. Tarkowski, Regional Office for Europe, World Health Organization, Copenhagen, Denmark **[177]**

Diane Tarr, Environmental Sciences Laboratory, Mount Sinai School of Medicine, City University of New York, New York, NY 10029 **[65,263]**

H. Vainio, Institute of Occupational Health, Helsinki, Finland **[79,215,373]**

PREFACE

Both the production and the uses of the synthetic polymers known as plastics have increased continuously in recent decades. As a result a constantly increasing number of people are exposed, both occupationally and otherwise, to monomers, additives, and degradation products of various plastics.

Despite general recognition that occupational exposure to various (toxic) compounds in the production and uses of different plastics may make work conditions hazardous, only few textbooks, monographs or other concise material are available which professional people such as occupational safety and health care personnel, industrial hygienists, and representatives of government bodies may consult to gain further insight into the industrial hygienic, toxicologic or occupational medical aspects of the production and uses of some specific plastics products.

Due to the evident need for an up-to-date evaluation of both the chemical hazards in the plastics industry and the measures by which they can be prevented the Nordic Institute of Advanced Occupational Environment Studies arranged an international symposium on the issue, which was held in Espoo, Finland, on 22-26 November, 1982. Several experts in the field acted as lecturers at the symposium. The participants were from both the Nordic countries and also some other European countries.

This book is a relatively short account of the proceedings of the symposium. For publication reasons the various aspects dealt with had to be covered in short form. However, during the course, it was accepted that there are many aspects of the hazards in the plastics industry that have not been discussed to any extent in the published literature. Thus there is a clear need for additional literature on toxicology, industrial hygiene, and hazard prevention in the plastics industry.

The lecturers and participants of the symposium also took part in evening seminars organized on the following four issues: industrial hygiene; occupational medicine and health care; toxicology; and studies of preventive measures. The reports of the various groups clearly indicated the special need for knowledge and expertise on

industrial hygiene in the use and production of plastics.
Another aspect stressed was the vast use of additives in
the plastics industry. However, there is almost no
literature about the toxicology or hazards of most of the
additives now used.

Many of the communications in this book have no
equivalents in previously published works. Many identify
special needs for further research. As more literature on
developments related to industrial toxicology and hygiene
of plastics is being published continually, hopefully
there will soon be other gatherings of experts where the
present volume can be updated.

Jorma Järvisalo
Principal Editor

CHAPTER I. SYNTHETIC POLYMERS AND CURRENT FEATURES OF OCCUPATIONAL PROBLEMS

Industrial Hazards of Plastics and Synthetic Elastomers, pages 3-10
© **1984 Alan R. Liss, Inc., 150 Fifth Ave., New York, NY 10011**

SYNTHETIC POLYMERS: MAIN CLASSES OF PLASTICS AND THEIR
CURRENT USES

Jukka M. Martinmaa

Department of Wood and Polymer Chemistry

University of Helsinki, Meritullinkatu 1 A
SF-00170 Helsinki 17, Finland

INTRODUCTION

   Plastics are deeply rooted in almost every type of human
activity. Application areas are many and varied: architec-
tural, packaging, transprtation, electronics, agriculture,
recreation, aircraft and aerospace, medical, etc. In all its
applications plastics have replaced other materials by its
own properties, such as low price, light weight, and easy
processability. They can be metallized, colored, or transparent. They provide design freedom, and they can be mass
produced.

WHAT ARE PLASTICS?

   Plastics are a group of man-made materials having at
least two features in common: 1) They are built up of a polymer component and an additive component. These components
must be compatible to get a mixture having chemical and
physical properties to meet the specific end-use requirements.
Only a slight change in chemical formulation will result in
a variation of a major property. 2) They also must be capable
of being formed.

WHAT ARE POLYMERS?

   Polymers are materials - mostly organic - characterized
by the big size of their molecules. They are prepared, either
synthetically by man or by living organisms, from simple

chemicals called monomers. Among commodities they are present
for example in fibers, enamels, lacquers, adhesives, wax, and
paints.

Polymers used in plastics are mostly prepared synthetically and they are considered to be inert and nontoxic. Due
to the low reactivity always connected with a large size of
a molecule, polymers can be held the most safe part of plastics. In fact, the chemical problems and hazards arising
with plastics are almost exclusively related to the ingredients or degradation products with low molecular size and,
therefore, higher reactivity. New demands presented by
modern life cause that polymeric materials become increasingly
more advanced and safe.

TYPES OF PLASTICS

Actually, there are several ways to classify polymeric
materials. For traditional and practical reasons plastics
are usually divided into three major categories: thermoplastic
types, thermosetting types, and elastomers. This classifica-
tion, however, refers to the chemical and physical changes
taking place during processes used to produce plastic articles. From a consumer's point of view: thermoplastics are
generally fusible at elevated temperatures, thermosets are
not. Elastomers are considered apart from other polymeric
materials because of their special properties.

The number of distinct, commercially important plastics
is today about 50. In addition, many of these comprise
several tens, some even thousands, of different qualities.
Because technical and other properties of plastics and polymeric materials are strongly formulation-dependent, the
variation possibilities are countless. The situation is
even more complicated if we think, that the structure and
other variables of an individual base polymer can be changed
in wide ranges.

THERMOPLASTICS

From the total consumption of petroleum products in the
world only about 8 per cent is used to produce plastic materials. Of the total amount of plastics consumption about
75 per cent consists of the so called basic plastics which

are: polyethylene (PE), polyvinyl chloride (PVC), polystyrene (PS), and polypropylene(PP). All these materials
belong to the thermoplastic type.

Polyethylene (LDPE, LLDPE, HDPE)

The repeating unit of PE is the following: $-CH_2-CH_2-$.
Polyethylene has long been the first in volume among plastics
known already about 50 years. It is produced by polymerization of ethylene, $CH_2=CH_2$, at high or low pressures aided
by catalysts an initiators.

According to the density, polyethylenes are grouped into
three main categories: low density PE (LDPE), linear LDPE
(LLDPE), and high density PE (HDPE). All of these types are
lighter than water and belong to the most inexpensive plastics. Producers commonly offer for sale more than 50 grades
of PE having different technical properties.

Films and sheets for packaging uses are the most widespread forms of PE plastics. Because LDPE is soft and flexible, transparent, and nontoxic due to the absence of plasticizers, it is utilized in food packages. In addition, shopping bags, sacks, and shrinkfilms are among the most popular
applications of LDPE.

Due to its better mechanical strength LLDPE is the current star plastics of the film manufacturing industry. Most
films contain blends of LLDPE and conventional LDPE. Because
LD polyethylenes have an outstanding chemical and frost resistance, hoses, coatings of electric cables and wires, many
sorts of housewares, such as jars, containers, deep-freeze
boxes and cases are made of them.

Injection molded products from HDPE also have diverse
shapes and a host of uses. This plastic is the stiffer PE,
best known in containers, mainly bottles with tight lids,
but also shopping bags, pipes and conduits. A relatively
long product life is characteristic for all PE plastics.

Polypropylene (PP)

Polypropylene has the repeating unit: $-CH_2-CH(CH_3)-$.
PP is a close relative to HDPE, but it is slightly harder

and tougher. It is commercially manufactured by polymerization of propylene, $CH_2=CH-CH_3$, catalyzed by metallic salts and alkyls. Of the four largest-volume thermoplastics PP is the lightest.

In addition to filament applications, such as home furnishings, nonwoven products, and carpets, PP is generally used as pipes and oriented films.

Polyvinyl chloride (PVC)

Vinyl chloride, $CH_2=CH-Cl$, is a gaseous monomer polymerized by suspension, emulsion, solution, or bulk processes. The repeating unit of the polymer is $-CH_2-CHCl-$.

Because PVC is one of the most inexpensive thermoplastics, it is second in consumption after PE. The presence of chlorine in the hydrocarbon backbone creates rigidity and toughness to the polymer and makes it self-extinguishing. For the same reason, however, PVC liberates hydrogen chloride (HCl) when exposed to elevated temperatures. To prevent this stabilizers are added to the polymer.

The toughness of hard-PVC makes application in sewage, agricultural, and drinking-water pipes, furniture, window profiles, dishes, and packages of various shapes.

Plasticizers are added to PVC mainly to impart flexibility to the finished products and to improve flow and processability of the melt. A typical composition might be PVC with about 30% plasticizer, 5% stabilizer, and some filler or pigment. Usually more than one plasticizer is used when other properties in addition to flexibility are required in the end product. Plasticized PVC is very popular in applications such as artificial skin, foamed wallpapers, laminated table cloths, packaging sheets an films, floorings, toys, garden hoses, wire coatings, shower curtains, etc.

Polyvinylidene chloride (PVDC)

PVDC is closely related to PVC. Its chlorine content is twice that of PVC. The monomer structure is $CH_2=CCl_2$, and the repeating unit $-CH_2-CCl_2-$.

Due to its transparency, flexibility, and low gas permeability, PVDC is used as films and laminates especially in aroma packages. Because it has a good resistivity against fats, PVDC films and laminates are widely used in cheese packages. In addition, PVDC is used in deep-freeze and shrinkfilm applications. It also is an outstanding material for fibers used in textiles of transport vehicles, furnitures, mats, and nets.

Polystyrene Plastics (PS)

Polystyrene (PS) is a hard and transparent plastic. It is manufactured by polymerization of styrene, $CH_2=CH-C_6H_5$, using a peroxide initiator. The repeating unit $-CH_2-CH(C_6H_5)-$ is typical for a vinyl polymer.

In the fields of packaging and insulation a very important application is the foamed or expanded PS (EPS). The excellent heat isolation properties of EPS arise from the great amount of air it contains. Pentane is the usual gassing agent employed.

Because the thermal stability and the impact strength of PS are poor, modified PS-plastics with a co- or terpolymer structure have been developed. Examples of these are the styrene-butadiene (SB), styrene-acrylonitrile (SAN), and acrylonitrile-butadiene-styrene (ABS) plastics. The good shock-resistance properties of these materials are utilized, for example, in housewares, toys, electronic and electrical appliances, recreational articles, inner parts of a refrigerator, handles, bags, and pipes.

PS-products can usually be identified by their metallic sound when dropped on a hard surface. This special property has found its application in baby rattles. PS-products are widely used in food packages and disposable dishes: spoons, forks, and knives.

Polymethylmethacrylate (PMMA)

The most important plastic in the group of acrylics is polymethylmethacrylate with a repeating unit as follows: $-CH_2-C(CH_3)(COOCH_3)-$. Because of its excellent transparency with a glassy surface, roof windows, housewares, watch glasses, bags, basins, bathtubes, etc. are made of PMMA. It does not, however, resist scratches, boiling water or organic solvents.

Polyamides (PA)

The most common polyamides are PA-66 and PA-6. The first of them is manufactured by condensation polymerization of adipic acid, $HOOC-(CH_2)_4-COOH$, and hexamethylene diamine, $H_2N-(CH_2)_6-NH_2$. The resulting polymer has a linear structure with the repeating unit $-OC-(CH_2)_4CONH-(CH_2)_6-NH-$. PA-6 is polymerized from caprolactam and water.

Both polyamides are tough, strong, and chemically re-sistant. PA-66 is an excellent fiber polymer. Most of its production goes to textile industry for home furnishings and carpets. Both PAs are widely used in gear wheels for machin-ery and laboratory equipments. The transparency of PA films makes them very useful for packaging purposes. Hospital wares made of PA plastics have a good stability at steril-ization temperatures. Combined films or laminates are used, for example, in vacuum packages of meat.

Polycarbonates (PC)

PC is one of the most important engineering plastics having a clearly higher price than many of its competitors. It is a very transparent, tough and physiologically inert material. A polycarbonate plastic, characterized by the $-O-CO-O-$ group, can be made from phosgene, $OCCl_2$, and bis-phenol A (4,4'-dihydroxy-diphenyl-2,2'-propane). It has the structure: $-O-(C_6H_4)-C(CH_3)_2-(C_6H_4)-O-CO-$.

The exceptionally high impact resistance is utilized among other things in safety helmets, bullet-proof windows, machine guards, shields, doors, bottles, and lamp globes.

THERMOSETS

Phenol-formaldehyde Resins

Phenol-formaldehyde (PF) resins are manufactured by polycondensation of phenol, C_6H_5-OH, and formaldehyde, $CH=O$.

One-stage products, resoles, are cured by heating. Two-stage resins, novolacs, are cured by heating with hexamethylene tetramine. Phenolics are considered the most versatile materials among polymers. This is partly due to the low price of the raw materials and partly to the diversified ways phenolic resins can be utilized.

The most part of PF resin production goes into adhesive and bonding applications: plywood, fibrous and granulated wood, binders for moulding compounds and foundry core binders for metal casting, glass-fiber laminates, etc.

Unsaturated Polyesters (UP)

UP resins are mainly produced by polyesterification of maleic and phthalic anhydrides with propylene glycol. These resins are mixed in certain proportions with readily polymerizable liquid monomers such as styrene or allyl esters. To initiate the crosslinking (curing) reaction, organic peroxides, such as benzoyl peroxide, are added to the mixture. The curing time is reduced by accelerators, usually cobalt naphthenate or tertiary amines.

Major end uses of UP resins are found in reinforced building constructions, marine castings, transport vehicles (cars, trucks, and buses), and appliances.

Epoxy (EP) Resins

Epoxies are synthetized from bisphenol A with epichlorohydrin followed by crosslinking with hardeners. These high-performance resins have a very broad base of applications. Coatings, laminates and composites, moldings, floorings, and adhesives are the major end uses. Almost half of all epoxy uses goes to coatings.

Polyurethanes (PUR)

Polyurethanes are typically formed through the reaction of a diisocyanate and a glycol. Urethane polymers occur in many forms: coatings, adhesives, elastomers, castings, foams, and fibers. Flexible PU foams are used for mattresses, cushions, dashboards, and packages.

Aminoplasts (UF, MF)

Urea, NH_2CONH_2, and melamine, 2,4,6-triamino-1,3,5-triazine, when reacted with formaldehyde give thermosetting resins called aminoplasts.

Although both resins are quite similar in appearance, water-clear syrups or white powders, the heat-cured melamine-formaldehyde (MF) resins have superior water resistance when compared to that of cured urea-formaldehyde (UF) resins. Both aminoplasts are relatively unaffected by common organic solvents, oils, and greases.

MF and UF thermosets are widely used as laminating and bonding materials in the wood, furniture, and allied industries. They also are utilized to improve the wet strength of paper and crease-resistance of textiles. Moulded closures made of UF resins are extensively used in containers of cosmetic products. UF foams have found application as insulants in refrigerators and between walls of houses. Other typical uses of UF resins are: air-dryer and mixer housings, clock cases, lavatory seats, and buttons.

Table-ware, molded from powders filled with alpha-cellulose, are very common articles made of MF resins. Examples are plates, cups, and ladles. In addition, high-quality decorative laminates are made of MF. MF resins are often used in conjunction with fillers and reinforcements such as glass-mat and cloth, silica, cotton fabric, asbestos, and certain synthetic fibers.

REFERENCES

Milby RV (1973). "Plastics Technology." New York: McGraw-Hill.

Miles DC, Briston JH (1965). "Polymer Technology." London: Temple.

Industrial Hazards of Plastics and Synthetic Elastomers, pages 11–17
© 1984 Alan R. Liss, Inc., 150 Fifth Ave., New York, NY 10011

POLYMER PROCESSING

Väinö Erä

University of Helsinki,
Department of Wood and Polymer Chemistry
Meritullink.1 A, 00170 Helsinki 17, Finl

INTRODUCTION

Processing refers to the method of fabricating or converting the liquid or solid form into a product. Thermoplastic polymers can be processed by means of heat and pressure by molding into desired products which preserve their thermo-plastic properties after processing. Thermosetting polymers undergo chemical reactions while processing and cannot be softened by heat and pressure after molding.

1. COMPRESSION MOLDING

Compression molding is one of the oldest known methods of processing polymers. It is used mainly for thermo-setting plastics although it can be used to process thermoplastics e.g. for making records from vinyl plastics.

1.1 METHODS

In the process of compression molding the heated mold is placed between platens in a press, by which the required pressure is applied onto the mold. The molding operation consists of the following stages:

1. A known weight of molding powder is placed into an open-mold cavity. The mold consists of two halves, a male and a female part, which are normally pre-heated.

2) The mold is closed and low pressure is applied from a compression press consisting of a ram in a cylinder operated hydraulically. The molding powder is compacted and heated by convection from the walls of the heated

molds. Then the material becomes plastic and starts to flow.

3) When the optimal degree of flow is reached, the halves of the mold are opened to relieve the air, water vapour and other gases.

4) Thereafter the mold is closed under high pressure, whereas the material will flow within the mould cavity and assume the required shape prior to the actual hardening stage.

5) Excess material flows between the surfaces of the mold, hardens, and "flash" is formed.

6) The mold is kept closed until the material has hardened, after which it is opened and the moulding is ejected by ejection mechanism.

1.2 TECHNIQUE

The most common molding materials, which are processed by compression molding are phenolic-, melamine- and urea resins. In hardening of the resins the polycondensation takes place and water formed in the reaction is vaporized and liberated by "venting". The materials are delivered in the form of molding powders.

The molding powder can be initially compressed into the shape of the molding before being placed in the mold or can be used in the form of pellets. Pelleting is a preforming process and consists of compressing the molding powder into pellets of approximately known weight.

Preheating of the molding material provides a method for molding relatively thick sections without porosity. The most efficient method is the radio-frequence technique by which times of heat convection and hardening can be considerably reduced. The modern screw plasticizing provides an efficient method, where granulated molding material is metered and preheated in the same operation.

There are many variables which should be taken into consideration in the compression molding. For the optimum conditions the factors such as pressure, pressing time, preheating time, and mold temperature should be experimentally determined.

1.3 PRESSES

Nowadays hydraulic presses are mainly used for molding thermosets and their capacity range from 15-10000 Mp. They are classified to upstroke and downstroke types depending on the pressing direction of the movable ram. Platens for molds are bolted in the press.

The drive for the presses is supplied by hydraulic pumps. Additional control devices regulate the pressure, closing of the mold, temperature of the mold and the hardening time of the molding. Modern presses are of semi- or full automatic types. Compression molding of small pieces is carried out in full automatic presses, where metering of molding material and ejection of molding takes place automatically. Thus one operator may take care of several presses simultaneously and the wage costs of molding are low.

2. INJECTION MOLDING

Most thermoplastic are molded by the process of injection molding. Here the polymer is preheated in a cylindrical chamber to a temperature at which it will flow and then is forced into a closed mold cavity by means of high pressures applied hydraulically through a plunger. Injection molding temperatures are higher than those for compression molding, rising above 250°C for many materials. An outstanding feature of injection molding is the speed with which finished articles can by produced. Cycle times of 10-30 sec are common.

2.1 METHODS

Injection molding is a cyclical process. A complete cycle consists of the following operations:

1) The mold is closed and a locking force is applied

2) The plunger moves forward, carrying a fresh charge of material into the heating zone of the cylinder. The heated material flows to the mold cavity through the nozzle.

3) Pressure is maintained on the plunger for the period during which the material in the mold cools and contracts.

4) The plunger is retracted.

5) The mold is opened and the molding is ejected.

2.2 TECHNIQUE

Injection molding machines are characterized by their shot capacity, plasticizing capacity, rate of injection, injection pressure and mold locking force.

Shot capacity is given in terms of the maximum weight which can be injected per shot. Injection molding machines are designed which can be applied to the production of articles ranging in weight of a few grams up to hundred kilogram.

The plasticizing capacity is

defined as kg/h of a
particular material which
can be brought to the
temperature required for
molding.

Injection rate is a factor
determining the output of a
machine and it is expressed
as the volume of material
discharged per second
through the nozzle during
an injection stroke. This
obviously depends on
pressure, temperature and
the material used.

The injection pressure of
an injection moulding
machine is the pressure
exerted by the face of the
plunger and can vary from
about 2000 lb/in^2 to 25000
lb/in^2. The clamp pressure
is an important factor in
determining the maximum
projected area that can be
molded on a particular
machine. The molds are
normally closed by means of
hydraulic rams.

The range of injection
molding machines is an
extremely wide one. Modern
machines incorporate a
single screw for melting the
material and are equipped
with automatic control
devices including the
programmer unit. The latest
design developed by Electro-
lux Co in Sweden is equipped
with microprocessor, steer-
ing units and robots. In
their plant of 65 injection

molding machines 30 robots
take care of handling of the
ejected articles including
the piling up operations. The
pneumatic transportation of
molding material to the hopper
of the injection molding machine
combined with automated control
and handling devices makes the
process reliable and economic
for producing multitude of
articles for household,
industry and other areas.

3. EXTRUSION

In contrast to the above
compression and injection
molding techniques the extrusion
process is a continuous method
for producing a wide variety
of shapes including rods, tubing,
hose, sheeting, film, paper
coatings and cable coverings
from thermoplastic materials
such as PVC, PE, PS, PA etc..
In this method the polymer is
conveyed continuously along a
screw through regions of high
temperature and pressure
where it is melted and compacted
and forced through a die shaped
to give the final object.

3.1 METHODS

The extrusion operation can be
divided into the following
stages:

1) Plasticization of the
   granular raw material

2) Metering of the plasticized

mass through a die which forms it to the desired shape

3) Solidification into the desired shape and size

4) Winding into reels or cutting into units

Processes 1 and 2 are carried out in the extruder, while 3 and 4 are ancillary processes. The extruder consists of an Archimedian screw which revolves inside a heated cylinder. The thermoplastic granules are fed through a hopper at one end, and carried forward along the cylinder by the action of screw. As the granules move along the screw, they are melted by contact with the heated walls of the barrel and by the generation of frictional heat in the viscous melt. The final action of the screw is to force the melted polymer through a die which determines its final form.

3.2   TECHNIQUE

The most important component of any extruder is the screw. Screws are characterized by their length/diameter ratios (l/d ratios) and their compression ratios, namely the ratio of the volume of one flight of the screw at the hopper end to the volume of one flight at the

die end. L/D ratios most normally used for single-screw extruders vary from 15:1 to 30:1 while compression ratios are usually in the range 2:1 to 4:1. The screw is usually divided into three sections, namely, feed, compression and metering. The feed section merely conveys the material from under the hopper mouth along the compression section where the gradually diminishing depth of thread causes volume compression of the melting granules and consequent removal of trapped air which if forced back through the feed section. Improved mixing results, together with the generation of frictional heat, which leads to a more uniform temperature distribution in the molten extrudate. The function of the final section of the screw is to meter the molten polymer through the die at a steady rate and to iron out pulsations.

Heat is supplied to the granules by external electrical heaters or by frictional heat generated internally by the shearing and compression action of the screw. The frictional heating is appreciable and in modern machines it can supply most of the heat energy required for steady running.

Extrusion is a very versatile process and the machines of special design are constructed to produce blown films, flat films, sheets, wire and cable covering, pipes, profiles and paper coating. In the blown film extrusion the molten polymer from the extruder is expanded into a bubble of the required diameter by blowing air through the center. The film bubble is cooled by blowing air on to it from a cooling device. The film is led through the pressure rolls and reeled. In wire and cable covering process the polymer melt is forced through the extruder and then out through the die which forms the coating around the wire. For the extrusion of pipe a specially designed die is used which control the dimension of the inner and outer diameters. Paper coating is carried out by the direct extrusion process and low-density polyethylene is the commonest polymer used for this process, the basic process consisting of extruding a thin film of polyethylene and pressing it on to the substrate without the use of an adhesive.

4. OTHER PROCESSING METHODS

Extension to the extrusion process is blow molding. In this technique a section of tubing is extruded into an open mold. By means of compressed air the plastic is then blown into the configuration of the mold. This technique is widely used for the manufacture of bottles.

Calendering is a process for the continuous manufacture of sheet. Granular resin is passed between pairs of highly polished heated rolls under high pressure. Calendering requires precise control of roll temperature, pressure, and speed of rotation. An embossed design can be produced on the surface by means of engraved calender roll.

In casting process, a liquid material is poured into a mold and solidified by physical (e.g. cooling) or chemical (polymerization) means, and the solid object is removed from the mold. Casting method can be applied for thermosetting, and thermoplastic resins in making films, sheets, rods and tubings.

In coating method, used mainly for paper and fabrics, the plastic may be utilised as a melt, solution, latex, paste or lacquer. It may be applied to the substrate by spreading with a knife, brushing, using a roller, calendering, casting or extrusion.

The production of plastic foams is accomplished by generating a gas in a fluid polymer, usually at an elevated temperature. Thermoplastics are foamed by incorporating either a blowing agent which decomposes to a gas at elevated temperature, or an inert gas. The plastic foams are widely used as an insulating material in the building industry as well as cushioning material in furniture.

Lamination process consists of the following steps, a) the impregnation of sheets (wood, paper, fabric) with a liquid or dissolved thermosetting resin b) assembly of the individual sheets and c) compression and curing. This process may be used for making glass fibre reinforced plastic articles such as boats, pipes and a wide variety of articles of different shapes. Paper laminates impregnated with phenolic resins are used for durable table coverings.

Processing of polymer to fibre form is accompleshed by spinning. In melt spinning the molten polymer is pumped under high pressure through a plate containing a large number of small holes. The solidified fibers are brought together to form a thread and drawn to orient the fibers.

BOOKS

1) Miles, D.C., Briston, J.H. Polymer Technology, Temple Press Books, London 1965

2) Milby, R.V., Plastics Technology, McGraw Hill, Inc., New York, 1973

3) Domininghaus, H., Einführung in die Technologie der Kunststoffe, Farbwerke Hoechst AG, Frankfurt

Industrial Hazards of Plastics and Synthetic Elastomers, pages 19–42
© 1984 Alan R. Liss, Inc., 150 Fifth Ave., New York, NY 10011

ADDITIVES IN SYNTHETIC POLYMERS: AN OVERVIEW

Lawrence Fishbein

Department of Health and Human Services, Food
and Drug Administration, National Center for
Toxicological Research, Jefferson, AR (USA)

INTRODUCTION

Plastics additives are products that are combined with
the basic resins and polymers as extenders, or to modify
their properties, or to facilitate their processing, or to
achieve special color and finish (DuBois and John, 1981).
The plastics industry, which includes the production of
man-made fibers and synthetic rubbers, began a period of
rapid growth in the late 1930's and is now one of the most
important branches of the chemical industry and uses a wider
variety of chemicals than any other technology (Vouk, 1976).

The principal objective of this overview on additives is
to review the salient features of the 14 major classes of
additives, (e.g., plasticizers, flame retardants, heat stabilizers, antioxidants, ultraviolet light absorbers, blowing
agents, initiators, lubricants and flow control agents,
antistatic agents, curing agents, colorants, fillers and
reinforcements, solvents and optical brighteners) from a
viewpoint of their scope of usage and a focus on their
potential toxicity based on structural considerations.   It
is important to stress that there are nearly 2500 individual
chemicals or mixtures that are utilized in the above 14
major classes of additives.   Withey (1977) has also noted
that excluding the polymer resins themselves there are at
least 7×10^{19} different possible combinations of plastics
additives available.   Additional additives that have been
introduced in the last 5 years should swell the possible
combinations to even greater astronomical levels.

PRODUCTION OF PLASTICS AND ADDITIVES AND FUTURE TRENDS

Since the scope of additive use and future trends is based primarily on plastics production it is initially instructive to briefly consider aspects of current and projected production. The total U.S. production in 1981 of thermoplastic resins (e.g., polyethylene (low and high density), polypropylene, polyvinyl chloride, polystyrene acrylonitrile-butadiene-styrene, styrene-acrylonitrile, styrene-butadiene copolymers alcohol and other vinyl resins) totalled approximately 28 billion pounds; that of the thermosetting resins (epoxy, polyester, urea, melamine and phenolic resins) totalled approximately 5 billion pounds; and that of synthetic rubber (e.g, styrene-butadiene rubber, polybutadiene, nitrile, ethylene-propylene, polychloroprene, butyl, polyisoprene) totalled approximately 1.7 billion pounds (Anon, 1982a). The world plastics capacity in 1981 for the 10 largest volume plastics, e.g., polyethylene (low density), polyvinyl chloride, polyethylene (high density), polypropylene, polystyrene, polyethylene (linear low density), acrylonitrile-butadiene-styrene, isocyanates, polyols, unsaturated polyesters and other (including acrylics, amino resins, cellulosics, fluoropolymers, phenolics, polyacetols, polycarbonates, polyphenylene oxide and styrene-acrylonitrile) totalled approximately 140 billion pounds (63.7 million metric tons) in 1981 and was projected to increase by 17% by the mid-1980's to almost 165 billion pounds (74.8 million metric tons) (Anon, 1982b). It is projected that conventional low density polyethylene (LDPE) (not including the new linear grades) will grow 20% in worldwide capacity in the first half of the 1980's to 38.0 billion pounds which will rank it as the largest volume plastic ahead of PVC which is predicted will grow only 8.9% in capacity during this period to 35.3 billion pounds (Anon, 1982b). Table 1 lists the 10 largest volume plastic resins in terms of their production capacity in 1981, and projected mid 1980's as well as their rankings during these periods and percent change. Among the 10 leading industrial countries, the U.S. had about 40 billion pounds of plastics capacity in 1981 (approximately one-third of world capacity). Japan was second with more than 15 billion pounds and the Federal Republic of West Germany was third with more than 11 billion pounds. The U.S. has also the most plastics companies, e.g., 101 (Anon, 1982b).

TABLE 1

WORLD PLASTICS CAPACITY OF 10 LARGEST VOLUME RESINS

IN 1981 AND MID-1980'S PROJECTIONS

| Rank | | | Capacity, millions of lb/year | | |
Mid 1980s	1981	Resin	Mid 1980s	1981	%Change
1	2	Polyethylene, low-density	38,047	31,702	20.0
2	1	Polyvinyl chloride	35,339	32,430	8.9
3	3	Polyethylene, high density	22,895	18,071	26.7
4	5	Polypropylene	17,809	15,631	13.9
5	4	Polystyrene	17,626	16,491	6.9
6	10	Polyethylene, linear low density	8,081	2,669	202.8
7	6	Acrylonitrile-Butadiene-Styrene	4,662	4,374	6.6
8	8	Isocyanates[a]	4,270	3,634	17.4
9	7	Polyols[a]	4,255	4,246	0.2
10	9	Unsaturated Polyesters	3,538	3,516	0.6
		Other[b]	8,006	7,359	8.8
		Total	164,528	140,123	17.4

[a]Used as indicators for derivative polyurethanes.
[b]Includes acrylics, amine resins, cellulosics, fluoropolymers, phenolics, polyacetols, polycarbonates, polyphenylene oxide, styrene-acrylonitrile.

Additives consumed by the plastics industry comprise approximately twenty-four chemical types that perform 14 major functions (as noted above). The plastics additive categories can be divided into three groupings: processing aids/catalysts, end-use performance additives and colorants. Projections as to the rate of growth of plastics additives for the period 1980 through 1985 differ somewhat. According to one report (Storck, 1981), the total market for plastics additives in the United States in 1980 amounted to 2.6 billion pounds. The demand for additives is projected to grow at a rate of 4 to 5%/year for the next five years which corresponds to an average growth rate of about 6% for plastics. In 1980, the largest volume plastics additive were the plasticizers which accounted for 58% of the total 2.6 billion pounds, and in decreasing order (%) the remaining high volume additives were: colorants, 12; blowing agents, 9; flame retardants, 8; heat stabilizers, 3; lubricants, 3 and others (including organic peroxides, antioxidants, antistats, ultraviolet radiation absorbers, catalysts, curing agents, impact modifiers, etc.) (Figure 1) (Storck, 1981).

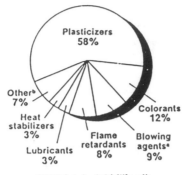

1980 Total = 2.6 billion lb

a Includes both physical and chemical blowing
agents. b Includes organic peroxides, antioxidants,
antistats, ultraviolet radiation absorbers, catalysts,
curing agents, impact modifiers, and others.

FIGURE 1.   Largest Volume Plastics Additives
in 1980 (Storck, 1981)

The major reasons for the generally slower growth rate
of additives compared to their end-use plastics is that
growth of PVC which consumes by far the largest bulk of the
additives is expected to be slower than for many of the
other plastics and that the amount of additives used per 100
pounds of resin is dropping as producers attempt to develop
more efficient materials. For example, although 0.19 pounds
of chemical additives was used per pound of resin in 1979,
this use will decline to 0.18 pounds in 1985, and concomit-
tantly the use of non-chemical fillers and extenders and
resin forcing fibers as additives for resins will increase
in this period (Storck, 1982).

Another recent report (SRI, 1982) listed the total 1980
market for plastics additives in 1980 in the U.S., Western
Europe and Japan as in excess of 815 thousand metric tons.
Consumption of plastics additives in the U.S. in 1980 was
estimated at 488 thousand metric tons with the consumption
of all types of plastics additives expected to grow at an
average rate of 6% per year for 1980 through 1985.   The
fastest growing plastics additives (e.g., those expected to
grow at 7% or more/year) include catalysts, impact modifiers

and UV stabilizers. Their growth rates are expected as a consequence of the forecast higher level of consumption of unsaturated polyester which will impact the market for peroxide catalysts and rigid polyvinyl chloride (used in production of pipe and siding), spurring demand for impact modifiers. The expected growth in U.V. stabilizers is due to an anticipated growth in polyolefin consumption (e.g., almost 80% of all U.V. stabilizer production is consumed in polyolefins). Additionally, there is an increasing demand for hindered amines which are expected to replace older, less efficient U.V. stabilizers (SRI, 1982).

In 1980, Western European consumption of plastics additives amounted to 309 thousand metric tons and the overall demand for these additives during the next five years is expected to grow at a rate of about 4%/year through 1985. The highest growth rate is expected in flame retardants (ca 8%/year). Other plastics additives with above-average growth in annual demand are chemical blowing agents, impact modifiers, antioxidants, and U.V. stabilizers, all forecast to grow at 6%/year. In the U.S. polyvinyl chloride is the largest market for plastics additives both in terms of volume and value (SRI, 1982). The current market size in Japan for antistatics, chemical blowing agents, and brominated flame retardants amounts to about 20,200 metric tons. The Japanese plastics additives business has the largest number of manufacturers, about 200 companies. There are 17 major producers in the U.S., Western Europe, and Japan of broad lines of plastics additives while many of the other companies in the market offer narrow product lines frequently limited to only one functional type of additive (SRI, 1982).

These figures give some measure of the magnitude of the potential environmental health problems related to the growth of the polymer industry. The annual per capita consumption of plastics varies from country to country, and for the 9 countries of largest per capita consumption (kg) ranging from Finland, 92; Federal Republic of Germany, 89; Sweden, 79; Austria, 71; U.S., 64; Switzerland, 60; France, 48; U.K., 44; and Japan, 41 (Nowea Presse-Informationen, 1979).

It is possible to divide the potential health hazards of plastics initially into two broad categories, e.g., those hazards attributed to the monomers per se (e.g., vinyl chloride, acrylonitrile, vinylidene chloride) and hazards associated largely with the various additives used to improve the technological processes and the quality of these products, for which very much less is known of their individual toxicological properties. Additional broad areas of potential health hazards include: direct toxicity from oral ingestion, respirable particulate inhalation, and the toxicity of unreacted chemicals (as well as additives); dermatological problems; acro-osteolysis; toxicity of thermal decomposition products as well as the toxicology of plastics in medical application.

Aspects of the chemical and health hazards in the plastics industry (Eckardt and Hindin, 1973; IARC, 1979, 1982a,b,c; Vainio et al, 1980) and mutagenic, carcinogenic and teratogenic hazards arising from human exposure to plastics additives (Withey, 1977) have been previously reviewed.

ADDITIVES

Plasticizers

Plasticizers constitute a broad range of chemically and thermally stable products of a variety of chemical classes, that are added to improve the flexibility, softness and processibility of plastics. Their principal use is in thermoplastic resins. PVC, because of its unique molecular structure, is especially suited for beneficial modification of plasticizers and reflects the largest use area of organic plasticizers (Sears and Touchette, 1982). Plasticizers frequently constitute more than 50% by weight of the finished product (DuBois and John, 1981; Sears and Touchette, 1982). Plasticization of PVC consumes about 80% of plasticizer production (Sears and Touchette, 1982).

Approximately 450 plasticizers are commercially available, comprising moderately high molecular weight liquids (or occasionally low melting solids). Most are esters of carboxylic acids (e.g., phthalic, isophthalic, adipic, azelaic, benzoic, citric, abeitic, trimellitic, oleic, palmitic, sebacic, stearic, myristic, tartaric acids) or phosphoric acid. Additional chemical classes include: epoxy, ether, glycerol, polyglycols, biphenyl, lactam, paraffin, halogenated hydrocarbons, polyester, sucrose and styrene derivatives. Of the 450 available plasticizers, perhaps 100 are of significant commercial value. Annual plasticizer production has grown from 500 thousand metric tons in 1965 to 947 thousand metric tons in 1978 (Sears and Touchette, 1982). The U.S. production of the more common plasticizers in 1978 (in thousands of metric tons) was as follows: phthalates, 572; phosphates, 53; adipates, 31; polymerics, 24; trimellitates, 15; others (including epoxys, citrates, stearates, benzoates and glycol derivatives), 252.

The rapid growth of the plasticizer industry has resulted from its symbiotic relationship with the polyvinyl chloride industry (Sears and Touchette, 1982). The use of plasticizers has grown to an annual worldwide use of over 2.5 billion pounds. The annual production of plasticizers in the U.S. in 1980 was about 1.5 billion pounds, down 9.7% from 1979. It is expected that plasticizer demand will recover to 1979 levels sometime in 1982 and grow about 4 to 5% per year on average (Storck, 1981). Esters of phthalic acid account for almost two-thirds of the volume of plasticizers produced and more than 80% of these plasticizers are used for the plasticization of flexible PVC which market is growing much slower than rigid PVC. Although there are about 100 phthalates that have been employed as plasticizers, about 14-15 phthalates account for over 90% of commercial phthalate production (e.g., approximately 1.3 billion pounds/year) (Anon, 1981a). The major phthalates utilized are di(2-ethylhexyl)phthalate (DEHP) (dioctyl phthalate) (DOP) which is produced at a rate of 300 to 400 million pounds/year (Anon, 1981b). A recent NCI report that DEHP given in the diet was carcinogenic for rats and mice may lead to the diminished use of DEHP or to a possible ban (Anon, 1981a,b). Additional high volume general-purpose phthalate plasticizers include: di-isononyl phthalate (DINP) and di-isodecyl phthalate (DIDP) whose current U.S. demand is about 175 million pounds/year each (Anon, 1981b). Of the specialty plasticizers, butyl benzyl phthalate (approximately 150 million pounds/year in U.S.) is used most commonly. The major low temperature plasticizers include di-2-ethyl hexyl esters of adipic, azelaic and sebaic acids and tri-2-ethyl hexyl phosphate (Sears and Touchette, 1982). Table 2 lists the consumption of plasticizers by type for the years 1980 to 1982 (Anon, 1982c).

TABLE 2

Consumption of Major Types of Plasticizers in
Plastics, 1980-1982 (Anon, 1982c)

Plasticizer	Consumption, 1000 metric tons		
	1980	1981	1982
Adipates	28	28	23
Azelates	5	5	5
Epoxidized esters	55	57	58
Glutarates*	--	--	31
Phosphates*	--	--	41
Phthalates	455	450	400
Polyesters	22	21	21
Trimellitates	13	13	13
Others	97	100	32
TOTAL	675	684	624

*Included in "Others" in 1980 and 1981.

Flame Retardants

Flame retardancy is required for high performance thermoplastic resins because of their use in electrical and high
temperature applications. In 1974 flame retardants ranked
second to plasticizers with a volume of 384 million pounds
(Liepins and Pearce, 1976), while in 1980 in the U.S. they
ranked as the third largest volume additive with approximately 208 million pounds. The average annual growth rate
had been approximately 2% for the previous 5 years. Before
that, government and other standards had almost ensured that
growth in fire retardants was high, about 10% per year. The
current restricted growth in the use of flame retardants is
that major uses are in two of the lower-growth plastics,
e.g., PVC and urethane (Storck, 1981). The uses, chemistry
and toxicity of flame retardants for plastics has been
reviewed by Pearce and Liepins (1975) and Liepins and Pearce
(1976). Flame retardation is achieved in four ways: 1) by
using flame retardants as additives prior to polymer processing; 2) by using them as a finish or a surface coating;
3) by using flame retardants as co-monomers in the polymerization or grafting and 4) by producing inherently flame-
resistant structures. The major chemical classes of agents
that have been employed as flame retardants include:

a) chlorine-containing aliphatic, cycloaliphatic and aromatic compounds (e.g., chlorinated paraffins, hexachlorocyclopentadiene tetrachlorophthalic anhydride, chlorinated naphthalenes, tetrachlorobisphenol, dimethylchlorendate); b) phosphate esters (e.g., tricresyl phosphate, cresyl diphenyl phosphate, triphenyl phosphate, tris (isopropylphenyl)phosphate and tris(2-ethylhexyl)phosphate; c) bromine-containing aliphatic, cycloaliphatic, aromatic and ionic agents (e.g., 2,3-dibromopropanol, dibromoneopentyl glycol, tribromoneopentyl alcohol, dibromobutenediol, hexabromocyclododecane, decabromodiphenyl ether, dibromopolyphenylene oxide, hexabromobenzene, hexabromobiphenyl, decabromobiphenyl oxide, tetrabromophthalic anhydride, bis(3,3-dibromopiopylether) of tetrabromobisphenol A; d) chlorine and phosphorus and bromine and phosphorus agents (e.g., tris(2,3-dichloropropyl)-phosphate, tris(2-chloroethyl)phosphate, chlorinated polyphosphates, bis(2-chloroethyl)vinyl phosphate, tris(2,3-dibromopropyl)phosphate, tris(4-bromophenyl phosphate and tris(2,4,6-tribromophenyl)phosphate; e) inorganic agents (e.g., antimony oxide, aluminum oxide trihydrate, zinc borate, ammonium orthophosphate, ammonium sulfamate (Liepins and Pearce, 1976; Pearce and Liepins, 1975). Table 3 lists the consumption of flame retardants used in plastics for the years 1981 and 1982 (Anon, 1982c).

The largest volume end-use and the percentage of fire-retarded plastics used in 1973 in the U.S. are as follows: building and construction, 10.0; electrical/electronic, 22.5; transportation, 20.0; furnishings, 15.0; packagings, 1.0; housewares and appliances, 1-2; and all other 1.0% (Liepins and Pearce, 1976).

Heat Stabilizers

Plastics, particularly chlorine-containing polymers such as poly(vinyl chloride), vinyl chloride co-polymers and poly(vinylidene chloride) are susceptible to thermal decomposition when exposed to high temperatures or prolonged heating. For example, PVC undergoes complete dehydrochlorination at $300^{\circ}C$, and partial dehydrochlorination at lower temperatures (Boettner et al, 1982). Approximately 83 million pounds of heat stabilizers were produced in the U.S. in 1980 (3% of the total additives production) and their growth rate is projected as 6 to 6.5% for the next few years (Storck, 1981). The major chemical classes of stabilizers

TABLE 3

Consumption of Flame Retardants Used in
Plastics, 1981-1982 (Anon, 1982c)

Type	Consumption, 1000 metric tons	
	1981[a]	1982
Additives		
Alumina hydrates	80	76
Antimony oxides	15	13
Boron compounds	3	3
Bromine compounds	14	11
Chlorinated paraffins and cycloaliphatics[b]	16	14
Phosphate esters		
Nonhalogenated	18	16
Halogenated	11	9
Others[c]	8	6
Total	165	148
Reactive Systems		
Epoxy intermediates	5	5
Polycarbonate intermediates	2	2
Polyester intermediates	5	4
Urethane intermediates		
Flexible foam	1	1
Rigid foam	9	8
Others	4	3
Total	26	23
Grand Total	191	171

[a]Revised in light of new data supplied by the Fire Retardant Chemicals Assn.
[b]Does not include elastomer and coating applications.
[c]Includes molybdenum, zinc, and metal oxides.

include: 1) organotins (e.g., di-N-octyl tin mercaptide; dilauryl and dibutyl tin maleate); 2) metal salts (e.g., inorganic salts of barium, cadmium, zinc and lead, usually as phosphites, carboxylates and phenates; 3) epoxides (e.g., epoxidized oils and esters (alkyl epoxy stearate, epoxidized soybean oil, epoxidized linseed oil) and 4) pentavalent phosphorus compounds (e.g., cotyl diphenyl phosphate; sodium capryl phosphate, hexamethyl phosphortriamide. Analogously to many other additives, the largest use for heat stabilizers is in PVC, e.g., barium-cadmium stabilizers for flexible PVC and organotin stabilizers for rigid PVC where the stabilizer content can be up to 10%. There is a faster rate of growth for organotin stabilizers since the market for rigid PVC is growing faster than that of flexible PVC (Storck, 1981). Specific compounds which are the most widely used organotins for stabilizing poly vinyl chloride

(PVC) and chlorinated polyvinyl chloride (CPVC) potable water pipe include methyl-, butyl- and octyl tin esters particularly of lauric, maleic and thioglycolic acids. Organotin stabilizers are used in the range of 0.3 to 1.5 parts per 100 parts resin for PVC pipe and fittings and in the range of 1.5 to 3.5 parts per 100 parts resin for CPVC pipe and fittings (Boettner et al, 1982).

Lead stabilizers account for approximately 60% of total stabilizer consumption. Commonly used lead stabilizers include dibasic lead phthalate, lead chloro-silicate, basic lead carbonates and lead silicate (Fischbein et al, 1982). Table 4 lists the consumption of heat stabilizers for 1980 to 1982 (Anon, 1982c).

TABLE 4

Consumption of Heat Stabilizers
1980-1982 (Anon, 1982c)

Stabilizer	Consumption, metric tons		
	1980	1981	1982
Barium-cadmium	15,250	15,700	14,000
Tin	10,125	10,650	9,600
Lead	9,950	10,450	9,400
Calcium-zinc	1,970	2,050	1,970
Antimony	430	450	410
Total	37,725	39,300	35,380

Antioxidants

Although not considered strictly as heat stabilizers, antioxidants can also be considered to some extent in a consideration of heat stabilization since the degradation process in PVC is somewhat an oxidation process under heat. Hence, antioxidants reduce the need for heat stabilizers in PVC processing (Storck, 1981). Oxidative degradation of polymers during the manufacturing process or during their useful lifetimes is a major industrial concern. The oxidation of polymers can be inhibited by two major ways. The first way involves the scavenging of free radicals by the use of primary antioxidants (hindered phenols) hence preventing propagation and coupling of radicals (which can arise from both the brain reactions of polymerization, from free radical initiators themselves, and in a finished plastic via the activation of reaction centers by atmospheric oxygen or by thermal, mechanical and photochemical energy (Seymour, 1975; Withey, 1977). The second way involves the utilization of synergists or peroxide decomposers which are often used in conjunction with hindered phenols (e.g., 2,6-di-tert.butyl-4-methylphenol and 2,2'-methylene bis(4-methyl-6-tert.butyl phenol). Synergists may be sulfides (RSR) such as dilauryl thiopropionate, or phosphites [(RO)$_3$P] such as tris-nonylphenol phosphite (Seymour, 1975; Withey, 1977). Examples of primary antioxidant and synergist chemical classes that have been utilized include: 1) alkylated phenols and polyphenols (butylated hydroxytoluenes; 4-tert.-butyl catechol; 2) esters (dilauryl thiodipropionate; distearylthiodipropionate; ditridecyl thiodipropionate; 3) organic phosphates and phosphates (didecyl phosphite; phenyldidecyl phosphite; phenyl neopentyl phosphate; 4) hydroquinones (hydroquinone; 2,5-di-tert.amylhydroquinone). Additionally, a large number of phenol condensation products, amines hydrazides and triazoles (proprietary compounds of undisclosed composition) have also been employed as antioxidants (Vainio et al, 1980; Withey, 1977). Antioxidants are generally employed in very low concentrations (e.g., as low as 0.1 pound per 100 pounds of resin).

The antioxidant demand in the United States in 1980 was about 32 million pounds with 30% of this market being utilized in the production of polypropylene. It is projected that the demand for antioxidants will increase about 4 to 4.5% per year in the U.S. for the next few years with the market growth of antioxidants paralleling those for plastics, especially polyolefins (Storck, 1981). Table 5 lists the antioxidant consumption by plastics for the years 1980 to 1982 (Anon, 1982c).

Ultraviolet-Light Absorbers

Radiation from the sun or fluorescent lighting (in the range 280 to 440 nm) is sufficient to break carbon-carbon, carbon-chlorine and carbon-hydrogen bonds and is responsible for the rapid degradation of most plastics. Chlorine-containing polymers such as poly(vinylchloride), vinyl chloride co-polymers and poly(vinyldiene chloride) are particularly susceptible to U.V. decomposition. For example, the U.V.

TABLE 5

Antioxidants Consumption by Plastics
1980-1982 (Anon, 1982c)

Plastics	Consumption, metric tons		
	1980	1981	1982
ABS	4,300	4,500	3,700
Polyethylene	2,900	3,100	2,900
Polypropylene	4,000	4,300	4,000
Polystyrene	2,000	2,100	1,900
Other[a]	1,400	1,500	1,500
Total	14,600	15,500	14,000

[a]Urethane polyols, unsaturated polyesters, epoxides engineering resins.

decomposition of PVC is autocatalytic, e.g., the first molecule of HCl that is released catalyzes additional decomposition in a "zipper" reaction producing chromophoric groups. Hydrocarbon polymers containing tertiary carbon atoms such as polystyrene and polypropylene are also susceptible to U.V. as well as thermal degradation (Seymour, 1975). High energy photons (hv) can be either filtered out or screened by pigments or U.V. absorbers (Carbon black, calcium carbonate and titanium dioxide) are often used as pigments in PVC).

The most widely used UV absorbers belong to six distinct chemical classes: 1) benzophenones (e.g., 2,4-dihydroxybenzophenones, 2-hydroxy-4-methoxybenzophenones); 2) benzotriazoles (e.g., 2-hydroxyphenylbenzotriazole, 2-(2-hydroxy-5-methylphenyl)benzotriazole, 2-(2-hydroxy-3,5-di-tert.butyl phenyl)benzotriazoles); 3) salicylates (e.g., strontium salicylate, phenyl salicylate, carboxyphenyl salicylate); 4) acrylates (e.g., 2-cyanodiphenylacrylates); 5) organonickel derivatives (e.g., nickel bisoctylphenylsulfides, [2,2-thiobis(4-T-octylphenolato)]-N-butylamine nickel and 6) hindered amines. The most widely used U.V. absorbers are 2-hydroxybenzophenones, 2-hydroxyphenylbenzotirazoles and

2-cyanodiphenyl acrylate (Seymour, 1975). Table 6 lists the plastics use of UV stabilizers for 1980 to 1982 (Anon, 1982c).

TABLE 6

Consumption of UV Stabilizers in Major Plastics, 1980-1982 (1982c)

Material	Consumption, metric tons		
	1980	1981	1982
Acrylic	68	69	67
Cellulosics	18	19	18
Polycarbonate	139	148	145
Polyester	65	69	68
Polyolefins	1,643	1,740	1,635
Polystyrene	58	63	56
Polyvinyl chloride	65	71	64
Coatings	60	63	60
Other	33	35	32
Total	2,149	2,277	2,145

Foaming (Blowing) Agents

Cellular or "foamed" plastic products are widely employed as insulating and flotation materials (e.g., as hot-beverage containers) as well as in an increasing number and range of other applications. To achieve this physical form, two processes are commonly employed, viz., a gas stream is introduced into the plastic melt or a compound which decomposes into a gas at the molding temperatures is introduced into the formulation (Withey, 1977). Physical foaming agents which have been employed in the former case include: nitrogen, carbon dioxide, heptane, pentane, methylene chloride, dichlorodifluoromethane and hydrogen peroxide with nitrogen being the blowing agent of choice. It should be noted that when thermally labile compounds are employed, decomposition products other than the generated gas remain with and may be part of the ultimate plastic formulation. Thermally labile compounds which have been employed fall into three main classes: 1) azo compounds (e.g., azodicarbonamide, diazoaminobenzene and azobisisobutyronitrile); 2) sulfonyl hydrazides (e.g., benzene sulfonyl hydrazide, diphenylsulfon-3,3-disulfonyl hydrazide) and 3) N-nitroso compounds (e.g., N,N'-dimethyl-N,N'-dinitrosoterephthalamide, N,N'-dimethyl-N,N'-dinitrosoterephthalamide and N,N-dinitrosopentamethylene tetramine). In all, chemical blowing agents constitute a fairly small volume additive (14.9 million pounds) with the largest use being in PVC. However, there has been an increasing utilization of methylene chloride in place of fluorocarbon-11 as a physical blowing agent of choice in urethane foam production (Anon, 1982c). Approximately 14.9 million pounds of blowing agents were employed in the U.S. in 1980. Industry forecasts for the chemical blowing agents market is for 6 to 7% per year growth for the next few years. The largest end-use market for these additives is in packaging (Storck, 1981). Table 7 lists the consumption of chemical blowing agents (CBA) by type for 1980 to 1982 (Anon, 1982c).

TABLE 7

Consumption of Chemical Blowing Agents in
Plastics, 1980-1982 (Anon, 1982c)

Blowing Agent	Consumption, metric tons		
	1980	1981	1982
Azodicarbonamide	3,860	4,250	3,995
Modified azodicarbonamide[a]	560	620	650
OBSH[b]	230	230	230
HTBAs[c]	110	120	105
Total	4,760	5,220	4,980

[a]Nonplating, dispersible, calcium-stearate-modified, and rigid-vinyl types.
[b]Oxybis-benzenesulfonylhydrazide
[c]High-temperature blowing agents for use with engineering resins needing decomposition temperatures of 500°F or higher.

Initiators

Most commercial synthetic polymers are produced by a chain reaction polymerization process (e.g., addition polymerization). Polymerization initiators that have been employed include: protonic acids (H_2SO_4), Lewis acids

(AlCl$_3$), organometallic compounds, e.g., butyl lithium, a combination of an organometalic compound such as triethyl aluminum and a salt of a transition metal, such as titanium chloride, called the Ziegler-Natta catalyst, additionally, unstable compounds which are readily decomposed to produce free radicals are employed to initiate the polymerization of monomers (Syemour, 1975). There are six major chemical classes of free radical initiators: 1) diacyl peroxides (e.g., benzoyl peroxide, acetyl peroxide, carprylyl peroxide, lauroyl peroxide, decanoyl peroxide); 2) alkyl peroxides (e.g., dicumul peroxide, di-tert butyl peroxide); 3) peroxy esters (e.g., tert.butyl peroxy acetate, tert.butyl peroxybenzoate, tert.butyl peroxy acetate); 4) hydroperoxides (e.g., tert.butyl hydroperoxide, p-methane-hydroperoxide, cumene hydroperoxide); 5) ketone peroxides (e.g., acetyl acetone peroxide, methyl ethyl ketone peroxide, cyclohexanone peroxide) and 6) peroxy carbonates and dicarbonates (e.g., diisopropyl peroxydicarbonate, tert.butyl peroxyisopropyl carbonate) (Sheppard and Mageli, 1982).

As a consequence of their propensity to undergo homolysis, organic peroxides commonly are used as initiators for many free-radical reactions primarily to initiate the copolymerization of vinyl and diene monomers, (e.g., ethylene, vinyl chloride, styrene, acrylic acid and esters, methacrylic acid and esters, vinyl acetate, acrylonitrile and butadiene) and to cure or cross-link resins, e.g., unsaturated polyester-styrene blends, thermoplastics such as polyethylene; elastomers (e.g., ethylene-propylene co-polymers and terpolymers and ethylene-vinyl acetate co-polymer) and rubbers such as silicone rubber, and styrene-butadiene rubber. There are more than 65 commercially available organic peroxides in over 100 formulations (e.g., liquids, solids, pastes, powders, solutions and dispersions).

In 1978 the consumption of organic peroxides in the U.S. amounted to 13,263 metric tons and according to chemical type was as follows: diacyl peroxides (benzoyl, 3,595; decanoyl, 59; and lauroyl peroxide, 690); dialkyl peroxides, 2,242; methyl ethyl ketone peroxide, 3,250; peroxy esters, 2,378 and other peoxides, 1,049) (Shepard and Mageli, 1982). Ninety percent of the uses of commercial organic peroxides are related to the polymer industry. In 1978, 40% of organic peroxides were used for vinyl monomer polymerization, and 35%, 15% and 10% were used for unsaturated polyester curing, cross-linking and miscellaneous applications respectively (Shepard and Mageli, 1982). The consumption of organic peroxides of major types is shown in Table 8 (Anon, 1982c).

TABLE 8

Consumption of Major Types of Organic Peroxides
in Plastics, 1980-1982 (Anon, 1982c)

Type	Consumption, metric tons		
	1980	1981	1982
Diacyl peroxides			
Benzoyl	3,800	4,200	3,800
Decanoyl	75	135	120
Lauroyl	725	700	630
Dialkyl peroxides	2,200	2,200	1,900
MEK peroxides	3,450	4,150	3,600
Peresters	3,900	4,200	3,700
Others (including peroxyketals)	1,200	1,300	1,200
Total	15,350	16,855	14,950

Lubricants and Flow Control Agents

Lubricants reduce adhesion and/or viscosity of the plastic formulations during processing; hence permitting more facile molding or extrusion. There are over 200 lubricants that have been employed with the largest number being non-specific formulations of waxes, polymers, or metallo-organic acid salts such as calcium, zinc and lead stearates; additionally, high molecular weight esters, polyethylene and paraffin waxes are used. These are added in small quantities to resins such as PVC and acrylonitrile-butadiene-styrene (ABS) to facilitate the processing of these materials. It is assumed that these additives reduce the coefficient of friction between plastic molecules and act as slip agents. These lubricants and other mold release agents such as talc are often dusted on mold surfaces to prevent adhesion of the plastic to the mold. Additionally, inorganic silicates or fumed silica are often added to plastics such as PVC to reduce surface tension while firmly divided polyfluorocarbon particles in the 10 micron range are often added to plastics

to assure their release from metallic surfaces during processing as well as providing plastics with a low coefficient
of friction (Seymour, 1975; Withey, 1977). In the United
States in 1980, lubricants accounted for 3% (78 million
pounds) of the total 2.6 billion pounds produced (Storck,
1981). The consumption of additive lubricants by type for
1980 to 1982 is shown in Table 9 (Anon, 1982c).

TABLE 9

Consumption of Major Types of Lubricants
in Plastics, 1980-1982 (Anon, 1982c)

Lubricants	Consumption, metric tons		
	1980	1981	1982
Fatty acid amides	8,410	9,090	8,640
Fatty acid esters	4,090	4,770	4,770
Metallic stearates	14,550	15,360	14,550
Paraffin waxes	7,500	7,950	7,270
Polyethylene waxes	2,510	2,640	2,550
Total	37,060	39,810	37,780

Antistatic Agents

Since plastics are electrical insulators, they tend to
hold any charge that builds up on them accidently (Deanin,
1975). Antistatic agents (antistats) which are usually
hygroscopic agents that attract and incorporate a small
amount of moisture to the plastic surface may be incorporated in the plastic formulation be applied externally to the
plastic surface to dissipate electrostatic charges. There
is a major and increasing utilization of antistats in electronic hardware as well as flexible PVC applications, e.g.,
hospitable disposables and packaging, mine belting and
directing, and other important film and sheet areas (Anon,
1982c). Almost all of the approximately 180 compounds which
are employed as antistats are listed as proprietary names of
undisclosed composition. The five major chemical classes
that constitute the majority of antistatic agents are: 1)
amines (RNH_2); 2) quaternary ammonium salts (R_4N^+,X^-); 3)
organic phosphates (R_3PO_4); 4) esters of poly(ethylene
glycol) [$RCO_2CH(O(CH_2)_2O)_nCH_2OCR$] and 5) metallic salts,

e.g., stannous chloride. When used internally, 0.1 to 1.0% of the antistat is added to the plastics while concentrations as high as 2% of antistat are applied to the plastic surface (Seymour, 1975; Withey, 1977). The volume of antistatic agents used in plastics in 1981 and 1982 was 2620 and 2800 metric tons respectively (Anon, 1982c).

Curing Agents

The usefulness of a number of plastics such as unsaturated polyester, epoxy and phenolic resins is limited unless there linear polymer chains are crosslinked or cured. In addition, the heat resistance of thermosetting resins such as phenolic resins is dependent on the addition of different crosslinking agents. Epoxy resins obtained by the condensation of bisphenol A and epichlorohydrin are cured using polyamines such as diethylene triamine at room temperature or dicarboxylic acid anhydrides such as phthalic anhydride at elevated temperature. Urea and melamine resins produced by the condensation of formaldehyde with urea or melamine are crosslinked in the presence of phosphoric acid. Hexamethylene tetramine is used for the curing of Novolak A stage phenolic resins formed via the interaction of phenol and formaldehyde in the presence of sulfuric acid (Seymour, 1975). Other compounds which have been used as epoxy resin curing agents are anhydrides such as hexahydrophthalic-, tetrahydrophthalic-, pyromellitic- and methyl hexahydrophthalic anhydrides. Peroxides have been used for curing of glass reinforced polyesters for specialty elastomers such as EPR and silicone rubber and for crosslinking of polyethylene as in wire and cable insulation. Benzoyl peroxide and methyl ethyl ketone peroxide are most commonly employed with a variety of other peroxides used in small amounts (Deanin, 1975).

Colorants (Dyes and Pigments)

While pigment-free products are often used as transparent sheets (e.g., poly(methyl methacrylate), cellulose acetate, and polycarbonates), many plastics are treated with dyes and pigments to impart color as well as other visual qualities (e.g., hue, opacity, fluorescence, irridescence and pearlescence). A large number of colorants have been employed including: 1) soluble dyes (e.g., based on

anthraquinone, nigrosines, and indulines and aniline black);
2) organic pigments (e.g., quinacridone, madder lake,
naphthol red, peryline, monazo red, phthalocyanine blue); 3)
inorganic pigments (e.g., titanium pigments, blue cobalt,
yellow nickel titanate, lithophone, ultram arene red,
molybdate-orange, chrome orange, ceramic yellows) and 4)
special compounds (e.g., metallic oxide browns, copper
plastic grades, bismethoxychloride, titanium dioxide-mica
composite, fluorescents and optical brighteners) (Seymour,
1975; Withey, 1977). Most colorants are inorganic pigments
with titanium dioxide being the most commonly used, and iron
oxides the second most common. More specialized pigments
based on cadmium, chromium and molybdenum are undergoing
increasingly critical scrutiny for toxicity especially in
both FDA and toy applications (Deanin, 1975). In the United
States in 1980, 312 million pounds, or 12% of the additives
produced were colorants (Storck, 1981). Table 10 illus-
trates the pigment and dye consumption in plastics for 1980
to 1982 (Anon, 1982c).

TABLE 10

Consumption of Major Types of Pigments and Dyes
in Plastics, 1980-1982 (Anon, 1982c)

Colorant	Consumption, metric tons		
	1980	1981	1982
Inorganics			
Titanium dioxides	94,000	102,000	100,000
Iron oxides	3,400	3,600	3,500
Cadmiums	2,000	2,200	2,000
Chrome yellows	2,100	2,300	2,100
Molybdate oranges	1,500	1,600	1,400
Others	1,200	1,300	1,200
Total	104,200	113,000	110,200
Organics			
Carbon blacks	28,000	30,000	29,000
Phthalo blues	1,700	1,750	1,700
Phthalo greens	920	950	940
Organic reds	1,290	1,300	1,400
Organic yellows	230	240	260
Others	540	580	600
Total	32,680	34,820	33,900
Dyes			
Nigrosines	1,500	1,550	1,400
Gil solubles	640	660	640
Anthraquinones	200	220	210
Others	230	250	230
Total	2,570	2,680	2,480
Grand Total	139,450	150,500	146,580

Fillers and Reinforcements

Fibrous or particulate reinforcing agents are widely used in plastics to induce less creep, greater rigidity, improved hardness and heat resistance. Fillers have been used to reduce cost as well as modify physical properties such as mechanical strength. The types of fillers that have been employed in polymers include: 1) silica products (e.g., minerals such as quartz, sand, diatomaceous earth; and synthetic amorphous silica such as silica aero gel, wet process silica and fumed colloidal silica); 2) silicates (e.g., minerals such as kaolin, mica, talc, asbestos, wollastonite, and synthetic products such as calcium silicate and aluminum silicate); 3) glass (flakes, hollow glass spheres, solid glass spheres, granules); 4) calcium carbonate; 5) metallic oxides (e.g., zinc, alumina, magnesium, titanin, and beryllium oxides); 6) metal powders (aluminum, bronze, lead, zinc, stainless steel); 7) carbon black (channel and furnace blacks, pyrolyzed products); 8) cellulosic fibers (wood and shell flour) and 9) miscellaneous inorganic compounds.

There are six major types of fibrous reinforcements for polymers, viz., 1) cellulose fibers; 2) synthetic fibers [polyamide (Nylon), polyester (Dacron), polyacrylonitrile (Dynel, Orlon), poly(vinylalcohol)]; 3) carbon fibers; 4) asbestos fibers; 5) fibrous glass; 6) metallic fibers (Seymour, 1975; Withey, 1977).

Solvents

Approximately fifty solvents, all of high volatility and primarily of low molecular weight, are employed at a number of stages in the processing of some plastics. For example, the initial solution of two plastic surfaces in a suitable solvent and subsequent evaporation of the latter is frequently used to bond two plastic components. The major chemical classes used include: 1) alcohols (methyl-, sex.-butyl-, butyl-, methylisobutyl alcohols); 2) esters (ethyl acetate, amyl acetate, and ethylene glycol monobutyl ether acetate); 3) glycol ethers (ethylene glycol, monoethyl ether, diethylene glycol monobutyl ether); 4) ketones (acetone, methyl ethyl ketone, diisobutyl ketone); 5) nitroparaffins (nitriethane, sec.nitropropane) (Withey, 1977); 6) glycidyl ethers (n- and tert.butylglycidyl ethers, phenylglycidyl ether).   The glycidyl ethers are used as reactive diluents in epoxy-resin systems (Potter, 1970; Stein et al, 1979).

Optical Brighteners

    Optical brighteners are also referred to as fluorescent whitening agents (FWA's) or fluorescent brightening agents. In 1975, worldwide consumption of FWA's was estimated at over 45,000 metric tons of commercial product of which approximately 15,000 tons were used in the paper industry, 500 tons in plastics, 20,000 tons in washing powders, and 10,000 tons in the textile industry.   FWA's used in poly-vinyl chloride and polystyrene include derivatives of the following classes: 1) 2-(stilben-4-yl)-naphtho triazoles; 2) 4,4'-bis(stryl)biphenyls; 3) bis(benzoxazol-2-yl) and 4) 3-phenyl-7-(azol-2-yl)coumarins and those for acrylo-nitrile-butadiene-styrene plastics include FWA's related to 2), 3) and 4) derivatives above (Zweidler and Hefti, 1978).

SUMMARY

    The 14 major classes of additives in plastics, e.g., plastizers, flame retardants, heat stabilizers, antioxi-dants, UV absorbers, foaming agents, initiators, lubricants, antistatic agents, curing agents, colorants, fillers and reinforcements, solvents, and optical brighteners were reviewed from a primary consideration of their structural chemical classes, areas of application as well as their production volumes and future use trends.   The number of possible combinations of plastics additives that can be employed are staggering.   In most cases, little is known of their toxicological properties and in many cases their identity is not commonly known.   It should also be noted that many new additives of new generic types in each of the above classes are being developed and introduced into the market yearly as well as new technology in polymer produc-tion, e.g., increasing use of additive polymers for PVC.

REFERENCES

Anon (1971).   Flame retardant growth flares up.   Chem Eng News 18:14.

Anon (1975). For chemicals and additives, a new splash in R and D. More Plastics 52:41.

Anon (1981a). A wary eye on plasticizer testing. Chem Week May 27, p. 12.

Anon (1981b). Phthalates are alive but being watched. Chem Week June 24, p. 18.

Anon (1982a). Production by the U.S. chemical industry. Chem Eng News June 14. p. 37.

Anon (1982b). World plastics capacity due up 17% by mid-80's. Chem Eng News May 3, p. 24.

Anon (1982c). Chemicals and additives. Modern Plastics 59:55.

Boettner EA, Ball GL, Hollingsworth Z, Aquino R (1982). Organic and organotin compounds leached from PVC and CPVC pipe. Project Summary EPA-600/SI-81-062. Health Effects Research Laboratory, US Environmental Protection Agency, Cincinnati, OH, February, pp. 1-9.

Deanin RD (1975). Additives in plastics. Env Hlth Persp 11:35.

DuBois, JH, John FW (1981). "Plastics", 6th ed., New York: Van Nostrand-Reinhold, p. 19.

Eckardt RE, Hindin R (1973). The health hazards of plastics. J Occup Med 15:808.

Fischbein A, Thornton JC, Berube L, Villa F, Selikoff, IJ (1982). Lead exposure reduction in workers using stabilizers in PVC manufacture: Effects of a new encapsulated stabilizer. Am Ind Hyg Assoc J 43:652.

IARC (1979). IARC Monographs on the Evaluation of the Carcinogenic Risk of Chemicals to Humans, Vol. 19, Some Monomers, Plastics and Synthetic Elastomers and Acrolein. International Agency for Research on Cancer, Lyon, France, pp. 495.

IARC (1982a). Di(2-ethylhexyl)phthalate. In: IARC Monographs on the Evaluation of the Carcinogenic Risk of Chemicals to Humans, Vol. 29, Some Industrial Chemicals and Dyestuffs. International Agency for Research on Cancer, Lyon, France, pp. 270-294.

IARC (1982b). Butylbenzyl phthalate. In: IARC Monographs on the Evaluation of the Carcinogenic Risk of Chemicals to Humans, Vol. 29, Some Industrial Chemicals and Dyestuffs. International Agency for Research on Cancer, Lyon, France, pp. 194-201.

IARC (1982c). Di(2-Ethylhexyl)adipate. In: IARC Monographs on the Evaluation of the Carcinogenic Risk of Chemicals to Humans, Vol. 29, Some Industrial Chemicals and Dyestuffs. International Agency for Research on Cancer, Lyon, France, pp. 258-267.

Liepins R, Pearce EM (1976). Chemistry and toxicity of
flame retardants for plastics. Env Hlth Persp 17:55.
Modern Plastics Encyclopedia 1975-1976 (1976). Vol. 52, No.
10A, pp. 657, 661, 705. McGraw-Hill, New York.
NCI (1982). Carcinogenesis bioassay of di(2-ethylhexyl)-
phthalate in F344 rats and B6C3F1 amice (feed study).
Tech. Rep. Series No. 217, DHEW Publication No. (NIH)
82-1773, Washington, DC.
Nowea Presse-Informationen (1979). International Trade Fair
Plastics and Rubber, Düsseldorf, October 10-17.
Pearce EM, Liepins R (1975). Flame retardants. Env Hlth
Persp 11:59.
Potter WG (1970). Epoxide resins. New York: Springer-
Verlag, pp. 109-115.
Sears JK, Touchette NW (1982). Plasticizers. In: Grayson M
(ed) Kirk-Othmer Encyclopedia of Chemical Technology, New
York: Wiley, p. 111.
Seymour WB (1975). Stabilizers, plasticizers and other
additives. In: Modern Plastics Technology, Reston, VA:
Reston Publ., p. 64.
Sheppard CS, Mageli OL (1982). Organic peroxide. In:
Grayson M (ed) Kirk-Othmer Encyclopedia of Chemical
Technology, New York: Wiley, p. 27-90.
SRI (1981). CEH Report Abstract Plasticizers. Chemical
Industries Division Newsletter, Stanford Research
International, Menlo Park, CA, March-April, p. 4.
SRI (1982). CEH Report Abstract Plastics Additives Chemical
Industries, Division Newsletter, Stanford Research
International, Menlo Park, CA, March-April, pp. 4-5.
Stein HP, Leidel NA, Lane JM (1979). NIOSH Current
Intelligence Bulletin, Glycidylethers. Am Ind Hyg 40:A36.
Storck WJ (1981). Plastics additives heard for turnaround.
Chem Eng News July 13:9.
Vainio H, Pfäffli P, Zitting A (1980). Chemical hazards in
the plastics industry. J Toxicol Env Hlth 6:1179.
Vouk VB (1976). Introductory statement. Environ Hlth Persp
17:1.
Withey JR (1977). Mutagenic, carcinogenic and teratogenic
hazards arising from human exposure to plastics additives.
In: Hiatt HH, Watson JD, Winsten JA (eds): Origins of
Human Cancer Book A, Incidence of Cancer in Humans, New
York: Cold Spring Harbor Laboratory, p. 219.
Zweidler R, Hefti H (1982) Brighteners, fluorescent. In:
Grayson M (ed) Kirk-Othmer Encyclopedia of Chemical
Technology, New York: Wiley, pp. 215-224.

Industrial Hazards of Plastics and Synthetic Elastomers, pages 43–64
© 1984 Alan R. Liss, Inc., 150 Fifth Ave., New York, NY 10011

CARCINOGENESIS STUDIES DESIGN AND EXPERIMENTAL
DATA INTERPRETATION/EVALUATION AT THE
NATIONAL TOXICOLOGY PROGRAM

J.E. Huff and J.A. Moore

National Toxicology Program
National Institute of Environmental Health Sciences
Research Triangle Park, NC USA

In the absence of adequate data on humans,
it is reasonable, for practical purposes,
to regard chemicals for which there is sufficient
evidence of carcinogenicity in animals as if
they presented a carcinogenic risk to humans.

IARC (1983)

CHEMICALS AND CANCER -- To enter the topical aspects of
evaluating chemicals for cancer causing potential the
overruling and persuasive bases for doing chronic (long-
term, life-term, 2-year) studies using rodents stems from
the clear observation that chemicals cause cancer (quod erat
demonstrandum). Fortunately most do not. Certain chemicals
cause cancer in humans (Althouse, et al., 1979; Althouse, et
al., 1980; IARC, 1982; NTP 1983a); and certain chemicals
cause cancer in animals (Chu, et al., 1981; IARC, 1972-1983;
NCI, 1976-1980; NTP, 1981-1983). All chemicals known to
induce cancer in humans cause cancer in laboratory animals
(recent data show arsenic and benzene -- classic exceptions
-- as causing carcinogenic responses in animals). Ergo,
reducing exposure to chemicals known or suspected to cause
cancer in humans or animals will reduce chemically-induced
cancer in humans. This public health stance of preventive
oncology stands as one of the primary goals of the National
Toxicology Program (NTP): identify with certainty those
chemicals most likely to be hazardous to humans. From a
preventive health view, the NTP underscores the concept that
chemicals found to cause cancer in animals must be considered being capable of causing cancer in humans

(Freireich, et al., 1966; Huff, 1982; Huff, et al., 1983; IARC, 1972-1983; IARC, 1982; NAS, 1977; NTP, 1983a; OTA, 1981; Rall, 1979a; Rall, 1979b); and thus, those chemicals so identified in well-conducted experiments should be controlled accordingly (IARC, 1972-1983; Huff, 1982; Huff and Moore, 1982; Huff et al., 1983).

Clearly the characteristics of chemical carcinogenesis in animals and in humans appear to be identical, even though a single chemical may produce different cancers in different species. Using animal experiments as surrogates render data necessary to better predict or hypothesize actual human experience from exposure to the same chemical. Often animal data showing a carcinogenic response to a chemical precede human case reports or epidemiological findings. For instance, Tomatis (1979) identified seven such chemicals -- aflatoxin, 4-aminobiphenyl, bis(chloromethyl)ether, diethyl-stilbestrol, melphalan, mustard gas, vinyl chloride; if this advance warning would have been heeded, considerable suf-fering could have been avoided and appropriate protective measures might have been initiated sooner.

The NTP strives to accomplish this by evaluating those large volume chemicals known to have a high index of human expo-sure. This scientific appraisal process comprises an integrated toxicological characterization approach: chemi-cal disposition (absorption, distribution, metabolism, excretion), genetic toxicology, fertility and reproductive assessment, systemic toxicology (14-day and 90-120 day exposures), specific studies as needed (immunological, biochemical, and inhalation toxicology), clinical pathology where applicable (hematology, urinalysis, endocrine func-tion, and clinical chemistry), and long-term (2-year) car-cinogenesis studies.

This paper does not attempt a treatise on what currently is known among the scientific community about carcinogenesis mechanisms. These have been regarded elsewhere (see for example, Boyland, 1980; Farber, 1982, Miller, 1978; Miller and Miller, 1981; Weinstein, 1981). Rather the contents explain the approaches taken by the NTP to design and interpret carcinogenesis studies. Topics highlighted include prechronic and two-year studies, use of tolerable dosage regimens, chemical pathology, statistical methodolo-gies, Technical Report series process, and evidence of car-cinogenicity.

TOXICOLOGY AND CARCINOGENESIS STUDIES -- Chemicals tested in the NTP carcinogenesis program are chosen primarily on the basis of human exposure, available (or lack of) toxicology data, level of production, and chemical structure. Selection per se is not an indicator of a chemical's carcinogenic potential. The "standard" two-year carcinogenesis bioassay remains as the most definitive method for detecting chemical carcinogens in animals. The standard protocol as developed by the NCI (and sometimes still used by the NTP) typically uses two rodent species (usually Fischer 344 rats and B6C3F$_1$ mice), both sexes, and administration of multiple dose levels (concurrent controls, low dose, and high dose) of a chemical to groups of 50 animals, beginning at weaning and ending after two years (Sontag, Page, and Saffiotti, 1976). These experiments are designed primarily to determine whether selected chemicals produce cancer in animals.

Under the NTP, the carcinogenesis procedures have been and continue to be changed to meet the objective of a broadened toxicologic characterization of chemicals, and further, to lead or stay abreast of advancing scientific developments. Prior to NTP involvement, the prechronic phases -- which include single dose (acute), 14-day repeated dose, and 90- to 120-day repeated dose studies --were conducted to determine gross toxicity and general target organ effects at different dose levels as a primary basis for setting appropriate doses for the two-year studies. Now, the NTP has begun to gather routinely other information related to target organ effects: chemical disposition, fertility and reproduction, urinalysis, clinical chemistry, and hematology also are obtained from the prechronic studies -- especially the 90-day study; certain other specific studies as applicable are included in the chronic two-year studies as well. Once those parameters that may be altered through exposure to the tested chemicals are identified, then suspect chemicals are referred to specific organ system groups for more detailed study of the functional, biochemical, and morphologic effects of the test compounds. Also, wider analysis of the quantitative and comparative absorption, distribution, metabolism, and excretion patterns may be desired. Most chemicals started on test since 1981 had an expanded design including other select studies. Significantly most chemicals selected for carcinogenesis studies will be profiled for chemical disposition patterns. The goal is to ensure that all major toxic effects will be identified for each chemical being considered for long-term toxicology and carcinogenesis studies.

Genetic Toxicology -- Prior to commencing the actual long-term carcinogenesis studies, all chemicals undergo genetic toxicology testing in at least five in vitro short-term assays: i) gene mutations in bacteria -- Salmonella typhimurium/microsome; ii) gene mutations in mammalian cells -- mouse lymphoma (L5178Y, thymidine kinase); iii) chromosome damage in mammalian cells -- cytogenetic damage and sister chromatid exchange (in vitro, CHO); iv) a mammalian cell transformation assay --(BALB/c-3T3); and v) a direct measure of DNA damage/repair (which does not necessarily result in mutation or transformation) -- unscheduled DNA synthesis (rat hepatocytes).

Liver Model -- The NTP is evaluating short term in vivo rodent liver carcinogenesis models to help clarify the nature of carcinogenic responses associated with two year carcinogenesis studies in rodents. A major objective is to assess the ability of selected chemicals to act as initiators, promoters, or complete carcinogens. Initial emphasis will be directed toward further model development including assessment of chemical dosimetry. The research will attempt to quantitatively assess response through the use of preneoplastic markers and correlate the results with histomorphologic tumor endpoints. Selection of chemicals will focus on those known to induce liver tumors in rodents, taking into account their genetic toxicity (see for example Peraino et al., 1981; Pitot and Sirica, 1980).

Design -- These data, together with other prechronic results, are used by the experimental design groups for preparing appropriate study protocols and are used by staff for assisting in establishing priorities for chemicals queued for long-term carcinogenesis studies. A key decision that must be made at this juncture between the completion of the prechronic phase and the beginning of the chronic study squares directly on whether indeed the lifetime bioassay should be done at all.

With this composite information base, the doses for the chronic study are selected. The high dose, termed the estimated maximum tolerable dose (EMTD), represents the highest dose of a chemical or substance given during a chronic study that can be predicted not to alter the treated animals' normal longevity from toxic effects other than carcinogenicity. The low dose ordinarily equals one-half EMTD. Other empirical factors include weight gain/food consumption data; for instance, a decrease in weight gain near 10-20% (not associated with a neoplastic response) is often used as a general indication that the EMTD was selected properly.

Thus, while the lifetime animal bioassay remains the best procedure for determining the carcinogenic potential of chemicals, NTP does not ordinarily use a standardized design. Rather the design is adapted to the special testing needs identified for the particular chemical. The NTP tailors its testing protocols to the particular chemicals, based on the results from the prechronic testing phases, on available literature, and on structure-activity relations. These new protocols permit better, more specific information to be generated for the tested compounds, which increases the effectiveness of the tests for potential human risk estimations. Such protocols also will be useful as guidelines for testing undertaken by other agencies and by industry. As examples, the NTP has investigated, initiated, and continues to pursue actively other design methodologies -- increased the number of dose levels, begun "unbalanced" distribution of animals among dose groups, uses interim kills, has modified histopathology requirements, and so on.

To accomplish these scientific advances, for instance, the NTP in March 1982 (NTP, 1982) presented a series of biomathematical simulations aimed at improving the basic experimental design of the two-year studies to provide information useful for low-dose extrapolation while retaining appropriate "power" for detecting carcinogenic effects (Portier and Hoel, 1983; 1983b). Three and four-dose designs were examined. The optimal design involved four groups (control plus low, medium, and high doses). The maximum tolerable dose (MTD) would be the high dose, the middle dose one-half the MTD, and the low dose 20-30% of the MTD. The most appropriate designs are being implemented for long-term carcinogenicity studies.

CONCEPT OF MAXIMUM DOSAGE IN CARCINOGENICITY (Kluwe, Haseman, and Huff, 1983) -- The practice of using high doses (or estimated maximum tolerable doses) in the carcinogenesis studies continues to receive excessive criticism, albeit considered by most to be essential for optimizing the capability of the experimental model to determine with feasible assuredness whether a chemical does or does not induce a carcinogenic response in animals. Metabolic or physiologic overload has been cited as a key factor against using high doses, the argument being that the body deviates from normal metabolic processes and preferentially detours chemicals to secondary or tertiary pathways; thus labeling any carcinogenic response spurious. Others, however, maintain an adamantine posture that because most carcinogens require metabolism to an active form any alteration in absorption, metabolism, distribution, and excretion pathways renders a chemical less, not more, capable of inducing a carcinogenic response. On balance both likely pertain. The National Academy of Sciences (1977), in addition to declaring the principle that "effects in animals, properly qualified, are applicable to man", clearly endorses the further principle that "the exposure of experimental animals to toxic agents in high doses is a necessary and valid method of discovering possible carcinogenic hazards in man".

The simple fact that a relatively small number of experimental animals (generally 50 per sex and dose) is to be used to predict response in a much larger human population nescessitates that the experimental conditions be maximized to detect a potential carcinogenic response. Thus, it is reasonable and valid in animal carcinogenicity studies to utilize doses in excess of expected human exposures (NAS, 1977; IARC, 1980; OSHA, 1980; Food Safety Council, 1980; OTA, 1981). To use excessively toxic doses is equally unreasonable, however, as repeated tissue injury may of itself induce neoplasia (bladder stones and bladder papillomas), and early deaths may preclude the development and detection of chemically induced tumors occurring late in life (OSHA, 1980; OTA, 1981).

With these criteria in mind, Sontag, Page, and Saffiotti (1976) suggested that the highest dose used for carcinogenicity testing in animals be the dose that could be predicted not to alter normal lifespan. Importantly, this dose must, in practice, be predicted on the basis of prechronic toxicity testing, and has more appropriately been labeled as an "estimated" maximal tolerable dose (EMTD) (OSHA, 1980). The concept of EMTD and its role in carcinogenicity testing has been reviewed thoroughly in recent forums (Food Safety Council, 1980; OSHA, 1980; IARC, 1980).

A major risk in exceeding an EMTD is that shortened lifespan may preclude the appearance of tumors developing late in life, while data that clearly indicate an abnormal physiological state or recurrent acute toxic lesion in test animals can cast doubt on the etiology of any increased tumor incidences.

Sontag, Page, and Saffiotti (1976) proposed that a 10% differential weight gain between control and chemical exposed animals in a 90-day study would be a general predictor of an MTD for the two year study. Although this general strategy has been endorsed by other groups as well (Food Safety Council, 1980; IARC, 1980), there is no credible body of evidence to support the 10% weight-gain differential as a predictor of animal survival in the chronic study. Moreover, stronger arguments could be made for predicting chronic toxic response on the basis of other criteria, such as cumulative toxic potential, dose-dependent pharmacokinetics, clinical signs of toxicity, and the presence or absence of histopathological lesions at various doses in prechronic studies. Some have suggested that carcinogenic effects in dosed groups where more than a 10% weight-gain differential was present should be ignored, on the belief that a so-called MTD was exceeded. These authors have apparently confused the weight differential in the prechronic studies as a predictor of MTD, as originally proposed by Sontag, Page, and Saffiotti (1976), with actual measurements of an MTD during the two year study.

Some further suggest that data from any study where a theoretical MTD has been exceeded should be rejected from analysis for carcinogenic potential. The consensus amongst practicing toxicologists, in contrast, is that clear distinctions must be made between positive and negative studies in which the MTD has been exceeded. "Negative" results where an MTD has been exceeded may be of little biological significance, since early deaths can preclude tumor development or detection. Clear "positive" results even at doses in excess of an MTD, however, can be evidence of carcinogenic potential in the absence of convincing evidence that the developed tumors were not chemical induced (OSHA, 1980).

CHEMICAL PATHOLOGY -- Modified Pathology Protocol -- The traditional procedure requires that 42 sections from 32 tissues be examined microscopically from all animals that die during the carcinogenesis studies and from those at the end of the 104 week treatment period. Analyses of the results from previous NCI and NTP bioassays indicate that significant reductions in the number of tissues examined could be implemented without compromising the ability to detect chemically induced tumors. Further, the quality of the toxicologic pathology will be markedly improved through examination of some animals at a time period earlier than 24 months, when normal ageing lesions often interfere with the detection and interpretation of chemical related lesions. The current pathology procedures in the two year studies are detailed by McConnell (1983a; 1983b) and are summarized below:

o Necropsy Examination -- All animals from all dose groups which die or are killed during and at the end of the experiment receive a complete necropsy examination.

o Fifteen Month Evaluation -- Ten animals/dose/sex/species will be killed and evaluated at 15 months: i) Organ Weights - liver, kidney, brain; ii) Clinical Pathology - includes complete blood count and selected serum chemistry; iii) Complete Necropsy of all animals; iv) Complete Histopathology (32 tissues) of the high dose and control groups, extending to lower dose groups based on findings in the high dose group.

o Twenty-Four Month Examination -- All animals in the high dose and control groups receive an "essential" histopathologic examination (see below). Histopathology extends to lower doses based on the findings from these observations.

-- Essential Histopathology -- Sixteen (female) or eighteen (male) organs or tissues (21-23 sections) from the remaining animals from the high dose and control groups are prepared and examined microscopically:

Adrenal gland - left and right	Ovary/uterus
Brain - 3 sections	Pancreas
Heart	Pituitary glands
Kidney - left and right	Prostate/seminal vesicles
Liver - left and right anterior lobe	Spleen
Lung	Stomach
Lymph node - submandibular	Testis/epididymis
	Thyroid and parathyroid glands
	Urinary bladder

Additional tissues examined microscopically include: nasal cavity and turbinates (inhalation studies, 3 sections), skin (dermal study), all gross lesions detected at necropsy, and target organs identified in the prechronic studies, in the 15-month evaluation, or in animals which died or were sacrificed during the two year bioassay.

-- Extended Histopathology -- Once the 16 or 18 organs are examined microscopically and pathologic effects are identified, tissues from all animals in lower doses are examined using the following convention: all gross lesions detected at necropsy; organs in which neoplasms were found significantly increased (P<0.10) compared to controls; organs in which rare tumors were found even if the comparison with controls does not achieve the P<0.10 level; and organs in which toxic lesions were observed.

o Pathology Quality Assessment (Maronpot and Boorman, 1982) -- The quality assessment (QA) for the pathology portion of the carcinogenesis program has been modified and expanded. Following an initial complete pathology examination by the performing laboratory, the pathology summary tables, the microscopic slides, the original individual animal data records (IADR), and the individual animal pathology tables (IAP) are sent to an independent QA pathology laboratory for assessment and validation. All records and tables are compared, tissues counted, and slides evaluated for quality of histotechnique. Using the pathology summary tables, the QA pathologist identifies target tissues. These in turn are verified by the NTP pathologists. All tumor diagnoses and target tissues are then reviewed by a pathologist expert in rodent diagnostics.

A second set of slides is organized for the NTP Pathology Working Group (PWG). This set includes all target tissues and discrepancies (if any) in tumor diagnoses between the original and reviewing pathologists. For each study a PWG composed of four to five pathologists (two to three not part of the Program) reviews the QA report, slides, IADRs, and tumor summary tables and reaches a consensus of all diagnostic discrepancies.

A third set of slides consisting only of diagnostic discrepancies between the original and PWG pathologists is returned to the original pathologist along with the PWG

comments for review and update of diagnosis. Where significant differences in diagnosis remain after the review, both
diagnoses are included in the final report. In rare cases,
where numerous significant differences exist, the entire
study is reread by another contract pathologist. In this
infrequent instance, the study is again subjected to the QA
and PWG.

The QA and PWG procedures, as now used for the 2-year carcinogenesis studies, are being redrafted for use in the
90-day prechronic studies to assure more uniformity of both
neoplastic and nonneoplastic diagnoses.

STATISTICAL DATA ANALYSIS METHODS (Haseman, 1983a) --
Statistical analyses of data from two-year carcinogenesis
studies are concerned primarily with survival and comparisons of tumor incidence.

o Survival Analysis -- Survival curves are calculated and
plotted for the controls and each dose group using the product limit method of Kaplan and Meier (1958). Comparison of
these curves reveal whether or not the test chemical under
study significantly altered survival. If survival differences do exist, these must be taken into account in the statistical analyses of tumor incidence.

Observed differences in survival are tested by Cox's (1972)
life table method. The Tarone (1975) modification of the
Cox method is used to determine whether mortality increases
monotonically as a function of dose.

o Analysis of Tumor Incidence -- Analysis of tumor incidence data is to some extent dependent upon whether or not
a particular tumor was the cause of death, or an incidental
finding at necropsy (Peto, et al., 1980). Consequently, for
the statistical analysis of tumor incidence data, NTP
employs two methods of adjusting for intercurrent mortality.
Each uses the classical methods for combining contingency
tables developed by Mantel and Haenszel (1959). Tests of
significance include pairwise comparisons of individual dose
groups with controls, and tests for overall dose-response
trends. Most investigators prefer one-tailed tests for
evaluating tumor incidence data, since the primary objective
of these studies is the detection of carcinogenic effects (a
one-tailed alternative).

The first method of analysis assumes that all tumors of a given type observed in animals dying before the end of the study are fatal: directly or indirectly cause the death of the animal. Proportions of tumor bearing animals in the dosed and control groups are compared at each point in time at which an animal dies with a tumor of interest. The denominators of these proportions are the total number of animals at risk in each group. These results, including the data from animals killed at the end of the study, are then combined by Mantel-Haenszel (1959) methods to obtain an overall P-value. This method of adjusting for intercurrent mortality is essentially the life table method.

The second method of analysis assumes that all tumors of a given type observed in animals dying before the end of the study are "incidental": observed at necropsy in animals dying of an unrelated cause. According to this approach, the proportions of animals found to have tumors in dosed and control groups are compared in each of five time intervals (generally 0-52 weeks, 53-78 weeks, 79-92 weeks, week 93 to the week before the terminal kill period, and the terminal kill period). The denominators of these proportions are the numbers of animals actually necropsied during the time interval. The individual time interval comparisons are then combined to obtain a single overall result.

Fisher's exact test for pairwise comparisons (Gart, et al., 1979) and the Cochran-Armitage linear trend test for dose-response trends (Armitage, 1971) are also used for the statistical analysis of primary tumors. These tests are based on the overall proportion of tumor-bearing animals.

For studies in which there is little effect of compound administration on survival, the results of the three alternative analyses will generally be similar. When differing results are obtained by the three methods, the final interpretation of the data will depend on the extent to which the tumor under consideration is regarded as being the cause of death.

When interpreting the tumor incidence analyses, a number of components must be considered in addition to the results of the statistical tests. While no formal statistical decision rules are used in the evaluation, a $P<0.05$ increase (or decrease) in the incidence of a neoplasm indicates a possible chemically related effect. Although the statistical significance of an observed increase in tumor incidence is one of the most important pieces of evidence used in the evaluation process, these other factors must be considered as well: (1) whether the effect was dose-related, (2) the mortality patterns of dosed and control animals, (3) whether the effect was supported by related non-neoplastic lesions or by similar evidence in both sexes or species, (4) whether the effect occurred in a target organ, (5) the biological "meaningfulness" of the effect, and (6) the historical rate of the particular tumor. NTP is also cognizant of the issue of false positivity that might be introduced by testing at multiple organ sites (Fears, et al., 1977; Haseman, 1977; Haseman, Huff, and Moore, 1983).

o False Positives -- One hypothetical issue frequently being trumpeted by critics of carcinogenesis studies in laboratory animals (Salsburg, 1977; Salsburg, 1983) centers on the occurrence or frequency of "false positive" results (Type I Errors). Specifically, concern has been expressed that certain "innocuous" or "non-toxic" or "harmless" chemicals may be labeled carcinogenic when in fact they may not be carcinogenic. Unfortunately, the studies that purport to show elevated false positive rates in carcinogenesis studies base their calculations on statistical decision rules that are never used in practice (for example, Salsburg's (1977) rule which declares a compound to be carcinogenic if there is any statistically significant increase in any tumor at any site in any dosed sex/species group by any of a battery of statistical tests). Hence, these investigations are of limited value and greatly over-estimate the true false-positive rate.

Many of the reports addressing this issue have assumed that any statistically significant increase in tumor incidence automatically implies that the test chemical is regarded as carcinogenic. This is not the case. As noted by the International Agency for Research on Cancer (and adhered to by the NTP) "P-values are objective facts, but unless a P-value is very extreme, the proper use of it in the light of other information to decide whether or not the test agent really is carcinogenic involves subjective judgment" (IARC, 1983). Worded somewhat differently, "a statistically significant difference is not necessarily an important difference, and a difference that is not statistically significant may be an important factor" (Carver, 1978).

In an attempt to better define the false positive rate Haseman (1983b) has examined 25 recent feeding studies and compared the statistical significance of observed tumor increases with the interpretation placed on the carcinogenic effect of the chemical under study. Although no rigid decision rule was (or should be) used in the evaluation of these studies, a simple statistical decision procedure was established that closely approximated the final interpretation of these carcinogenesis results. Using this more realistic procedure as well as the updated historical control tumor frequencies, the actual overall false positive rate in NTP carcinogenesis studies appears to be no greater than 7-8%. Importantly this figure represents an estimated upper bound on the true false-positive rate.

In summary, re-examination of recent carcinogenesis data has revealed little evidence to support the hypothesis that these studies suffer from inflated false-positive rates. Moreover, even though the NTP remains concerned that any false positive studies may occur, an even greater public health concern (and one rarely voiced by those critical of the carcinogenesis studies) centers on false negatives: those chemicals rendering no evidence of carcinogenicity in laboratory rodents when these are in fact positive and thus represent a silent hazard to humans.

TOXICOLOGY AND CARCINOGENESIS BIOASSAY TECHNICAL REPORTS -- Data from the toxicology characterization studies and the chronic exposure experiments are collected into comprehensive and evaluative technical reports. These compilations receive consideral staff attention during generation, initial data analyses, and draft report preparation; further intense focus is devoted to the draft report by the NTP staff as an iterative review process. After the draft receives internal approval, copies are sent for external peer review to the NTP Board of Scientific Counselors' Technical Reports Review Subcommittee (Peer Review Panel). This ad hoc panel is comprised of non-government scientists.

o The Review Process -- Meetings to perform peer review of NTP draft technical reports are held approximately three times yearly in Research Triangle Park, NC. A list of the reports to be reviewed, reviewer assignments, review forms, and other information about a particular meeting are sent to Peer Review Panel members at least one month prior to the meeting date. At the same time notices about the meeting are published in the Federal Register and in the NTP Technical Bulletin. These draft technical reports are also made available to anyone upon request.

Members of the Peer Review Panel receive copies of all reports to be reviewed. Two and sometimes three reviewers are assigned for each technical report and each Panel member gives an oral critique at these open-to-the-public meetings accompanied by written comments. Panel members are asked to provide the NTP with a critical review of each report in advance of the meeting. Deficiencies in design, conduct, or interpretation of the study should be identified, and errors or omissions in the draft report should be stated. Routinely, comments and questions are requested from individuals or groups attending these open meetings. Further, panel members are requested to read all reports scheduled for a particular meeting and to contribute their opinions and personal dissertation during the discussion period on each report. The recommendations of the reviewers and summary comments recorded at the meeting are incorporated where appropriate and relevant in the final revision of the report.

Following the meeting, draft summary minutes for each technical report review are prepared and are sent to the reviewers for any necessary corrections and alterations. The edited minutes are then made available for distribution to any interested party. Likewise, immediately after the peer review, reports are readied for publication.

When the draft reports are considered by staff and the peer review panel to be scientifically acceptable and soundly based, the technical reports undergo a final technical and style edit prior to printing and distribution.

CARCINOGENESIS RESULTS INTERPRETATION -- The interpretation of carcinogenesis results has been the subject of considerable attention. Negative results, in which the test animals do not have a greater incidence of cancer than control animals, do not necessarily mean that a test chemical is not a carcinogen, inasmuch as the experiments are conducted under a limited set of conditions. Positive results demonstrate that a test chemical is carcinogenic for animals under the conditions of the test and indicate that exposure to the chemical has the potential for hazard to humans.

-- <u>Evidence of Carcinogenicity</u> -- Five categories of inter-pretative conclusions have been adopted for use in the NTP Technical Reports series on toxicology and carcinogenesis studies to specifically emphasize consistency and the con-cept of actual evidence of carcinogenicity. For each defi-nitive study result (male rats, female rats, male mice, female mice) one of the following categories will be selected to describe the findings. These categories refer to the strength of the experimental evidence and not to either potency or mechanism.

o <u>Clear Evidence of Carcinogenicity</u> is demonstrated by stu-dies that are interpreted as showing a chemically-related increased incidence of malignant neoplasms, studies that exhibit a substantially increased incidence of benign neoplasms, or studies that exhibit an increased incidence of a combination of malignant and benign neoplasms where each increases with dose.

o <u>Some Evidence of Carcinogenicity</u> is demonstrated by stud-ies that are interpreted as showing a chemically-related increased incidence of benign neoplasms, studies that exhi-bit marginal increases of neoplasms in several organs/tissues, or studies that exhibit a slight increase of uncommon malignant or benign neoplasms.

o <u>Equivocal Evidence of Carcinogenicity</u> is demonstrated by studies that are interpreted as showing a chemically-related marginal increase of neoplasms.

o <u>No Evidence of Carcinogenicity</u> is demonstrated by studies that are interpreted as showing no chemically-related increases of malignant or benign neoplasms.

o <u>Inadequate Study of Carcinogenicity</u> demonstrates that because of major qualitative or quantitative limitations the studies cannot be interpreted as valid for showing either the presence or absence of a carcinogenic effect.

Use of these categories takes advantage of not only cate-gorizing <u>the evidence</u> yet records <u>the evidence</u> (or effect) as well. This allows some leeway for the reader of the Technical Reports to form a personal interpretation about the data.

As an example of the use of these categories of evidence, evaluations from an earlier draft and the Peer review approved version (NTP, 1983) of benzyl acetate are given:

-- Benzyl acetate was administered in corn oil by gavage to F344/N rats (0, 250, or 500 mg/Kg body weight) and to B6C3F$_1$, mice (0, 500, or 1000 mg/Kg b.w.) five times per week for 103 weeks. Under these conditions, benzyl acetate caused an increased incidence of acinar cell adenomas of the exocrine pancreas in male F344/N rats; the gavage vehicle may have been a contributing factor. No evidence of carcinogenicity was found for female F344/N rats. For male and female B6C3F$_1$, mice there was some evidence of carcinogenicity, in that benzyl acetate caused increased incidences of hepatocellular adenomas.

Contrast this version with the statement in an earlier DRAFT:

-- Under the conditions of these studies, benzyl acetate should be considered carcinogenic for male and female B6C3F$_1$, mice, causing increased incidences of hepatocellular adenomas. Benzyl acetate also caused an increased incidence of acinar cell adenomas of the pancreas in male F344/N rats; the vehicle may have been a contributing factor. For female F344/N rats there was no evidence of carcinogenicity.

Additionally, the following concepts (as patterned from the International Agency for Research on Cancer Monographs) have been adopted by the NTP to give further clarification of these issues (IARC, 1983):

o The term chemical carcinogenesis generally means the induction by chemicals of neoplasms not usually observed, the earlier induction by chemicals of neoplasms that are commonly observed, or the induction by chemicals of more neoplasms that are usually found. Different mechanisms may be involved in these three situations. Etymologically, the term carcinogenesis means the induction of cancer, that is, of malignant neoplasms; however, the commonly accepted meaning is the induction of various types of neoplasms or of a combination of malignant and benign neoplasms. In the NTP Technical Reports the words tumor and neoplasm are used interchangeably.

AFTERWORD -- History remains clear: all known human carcinogens cause cancer in animals. Should society not demand
therefore that until absolutely and unequivocally proven
otherwise unnecessary exposure to all convincing animal carcinogens must be eliminated or substantially reduced?
Dreadfully the 1000 cancer deaths per day in the United
States alone testify that better and more complete preventive measures are justified. Even if one adopts the conser-
vative and lower estimate of cancers definitely caused by
chemical exposure this represents 20,000 human deaths mini-
mum per year attributed directly to chemicals per se.
Further, other chemically-induced subtle or clinically manifest toxicities must be viewed with equal concern.

Given the 6,000,000 unique chemicals cataloged in the
Chemical Abstracts system, given the 70,000 chemicals in
regular use, given the 1,000 to 3,000 chemicals introduced
annually into the market place signal a formidable toxicological task ahead.

REFERENCES

1. Althouse, R., Huff, J.E., Tomatis, L., and Wilbourn,
J.D. (1979). Chemicals and Industrial Processes Associated
with Cancer in Humans. (IARC Monographs, Volumes 1 to 20),
IARC Monographs Supplement 1, 71 pages, International Agency
for Research on Cancer, Lyon, France.

2. Althouse, R., Huff, J.E., Tomatis, L., and Wilbourn,
J.D. (1980). Report of an IARC Working Group. An
Evaluation of Chemicals and Industrial Processes Associated
with Cancer in Humans Based on Human and Animal Data: IARC
Monographs Volumes 1 to 20. Cancer Res. 40:1-12.

3. Armitage, P. (1971). Statistical Methods in Medical
Research. John Wiley & Sons, Inc, New York, pages 362-365.

4. Boyland, E. (1980). The History and Future of Chemical
Carcinogenesis. British Med. Bull. 36(1):5-10.

5. Carver, R.P. (1978). The Case Against Statistical
Significance Testing. Harvard Educat. Rev. 48(3):378-399.

6. Chu, K.C., Cueto, C., Jr., and Ward, J.M. (1981). Factors in the Evaluation of 200 National Cancer Institute Carcinogen Bioassays. J. Toxicol. Environ. Health 8:251-280.

7. Cox, D.R. (1972). Regression Models and Life Tables. J.R. Stat. Soc. B34:187-220.

8. Farber, E. (1982). Chemical Carcinogenesis: A Biological Perspective. Amer. J. Pathol. 106(2):271-296.

9. Fears, T.R., Tarone, R.E. and Chu, K.C. (1977). False Positive and False Negative Rates for Carcinogenicity Screens. Cancer Res. 37:1941-1945.

10. Food Safety Council (1980). Proposed System for Food Safety Assessment. Food Safety Council, Washington, DC, 160 pages.

11. Freireich, E.J., Geham, E.A., Rall, D.P., Schmidt, L.H., and Skipper, H.E. (1966). Quantitative Compairson of Toxicity of Anticancer Agents in Mouse, Rat, Hamster, Dog, Monkey, and Man. Cancer Chemother. 50:219-244.

12. Gart, J., Chu, K., Tarone, R. (1979). Statistical Issues in Interpretation of Chronic Bioassay Tests for Carcinogenicity. J. Nat. Cancer Inst. 62(4):957-974.

13. Haseman, J.K. (1977). Response to "Use of Statistics When Examining Lifetime Studies in Rodents to Detect Carcinogenicity". J. Toxicol. Environ. Health, 3:633-636.

14. Haseman, J.K. (1983a). Statistical Issues in the Design, Analysis, and Interpretation of Animal Carcinogenicity Studies (Draft NTP Document).

15. Haseman, J.K. (1983b). A Re-examination of False-Positive Rates for Carcinogenicity Bioassays. Fund. Appl. Toxicol. (in press).

16. Haseman, J.K., Huff, J.E., and Moore, J.A. (1983). Response to "The Lifetime Feeding Study in Mice and Rats--An Examination of its Validity as a Bioassay for Human Carcinogens". Fund. Appl. Toxicol. 3(3):3a-5a.

17. Huff, J.E. (1982). Carcinogenesis Bioassay Results from the National Toxicology Program. Environ. Health Perspect. 45:185-198.

18. Huff, J.E. and Moore, J.A. (1982). Toxicology Data Evaluation Techniques and the National Toxicology Program, 81-98. In: Cosmides, G. (ed.), Information Transfer in Toxicology, National Technical Information Service, Springfield, VA. Proceedings of a Symposium held at the National Library of Medicine in Bethesda, Maryland on 16-17 September 1981.

19. Huff, J.E., Haseman, J.K., McConnell, E.E., and Moore, J.A. (1983): The National Toxicology Program, Toxicology Data Evaluation Techniques, and the Long-Term Carcinogenesis Bioassay. In: Safety Evaluation of Drugs and Chemicals, Proceedings of a Symposium held at the Iowa State University on 1-3 June 1981, in press.

20. IARC (1972-1983). IARC Monographs on the Evaluation of the Carcinogenic Risk of Chemicals to Humans. Volumes 1-30, International Agency for Research on Cancer, Lyon, France.

21. IARC (1980). Long-Term and Short-Term Screening Assays for Carcinogens: A Critical Appraisal. IARC Monographs, Supplement 2, International Agency for Research on Cancer, Lyon, France, 426 pages.

22. IARC (1982). Working Group Report. Chemicals, Industrial Processes and Industries Associated with Cancer in Humans, IARC Monographs, Volumes 1-29, Supplement 4, 292 pages. Proceedings of February 1982 Meeting, International Agency for Research on Cancer, Lyon, France.

23. IARC (1983). Preamble, IARC Monographs on the Evaluation of the Carcinogenic Risk of Chemicals to Humans 30:11-31, International Agency for Research on Cancer, Lyon, France, 424 pages.

24. Kaplan, E.L. and Meier, P. (1958). Nonparametric Estimation of Incomplete Observations. J. Amer. Stat. Assoc. 53:457-481.

25. Kluwe, W.K., Haseman, J.K., and Huff, J.E. (1983). The Carcinogenicity of Di(2-ethylhexyl)Phthalate (DEHP) in Perspective. J. Toxicol. Environ. Health 12:(in press).

26. Mantel, N. and Haenszel, W. (1959). Statistical Aspects of the Analysis of Data from Retrospective Studies of Disease. J. Nat. Cancer Inst. 22:719-748.

27. Maronpot, R.R. and Boorman, G.A. (1982). Interpretation of Rodent Hepatocellular Proliferative Alterations and Hepatocellular Tumors in Chemical Safety Assessment. Toxicol. Pathol. 10(2):71-80.

28. McConnell, E.E. (1983a). Pathology Requirements for Two-Year Toxicology and Carcinogenesis Studies in Rodents. I. A Review of Current Practices. (Draft NTP Document).

29. McConnell, E.E. (1983b). Pathology Requirements for Two-Year Toxicology and Carcinogenesis Studies in Rodents II. Alternative Approach. (Draft NTP Document).

30. Miller, E.C. (1978). Some Current Perspectives on Chemical Carcinogenesis in Humans and Experimental Animals: Presidential Address. Cancer Res. 38:1479-1496.

31. Miller, E.C. and Miller, J.A. (1981). Mechanisms of Chemical Carcinogenesis. Cancer 47(5):1055-1064.

32. NAS (1977). Drinking Water and Health, 939 pages, National Academy of Sciences, Washington, DC.

33. NCI (1976-1980). Bioassay of 'Chemical' for Possible Carcinogenicity. Carcinogenesis Technical Report Series, Numbers 1-200, NCI, Bethesda, MD.

34. NTP (1981-1983). NTP Technical Report on the Carcino-genesis Studies of 'Chemical' (CAS NO.) in F344/N Rats and B6C3F$_1$ Mice (Dose Route), Technical Report Series, Numbers 201-300. National Toxicology Program, Research Triangle Park, NC 27709.

35. NTP (1982). National Toxicology Program Board of Scientific Counselor's Meeting, 10-12 March 1982, Minutes available from the NTP Public Information Office, P. O. Box 12233, Research Triangle Park, NC 27709.

36. NTP (1983a). NTP Third Annual Report on Carcinogens, 423 pages, National Toxicology Program, Research Triangle Park, NC (in press).

37. NTP (1983b). NTP Technical Report on the Carcinogenesis Studies of Benzyl Acetate (CAS NO. 140-11-4) in F344/N Rats and B6C3F$_1$ Mice (Gavage Study). Technical Report Number 250, National Toxicology Program, Research Triangle Park, NC 27709, USA (in press).

38. OSHA (1980). Identification, Clarification and Regulation of Potential Occupational Carcinogens, Occupational Safety and Health Administration. Federal Register 45(15):5001-5296.

39. OTA (1981). Assessment of Technologies for Determining Cancer Risks from the Environment, 239 pages, Office of Technology Assessment, Government Printing Office, Washington, DC.

40. Peraino, C., Staffeldt, E.F., and Ludeman, V.A. (1981). Early Appearance of Histochemically Altered Hepatocyte Foci and Liver Tumors in Female Rats Treated with Carcinogens One Day After Birth. Carcinogenesis 2:463-465.

41. Peto, R., Pike, M., Day, N., Gray, R., Lee, P., Parish, S., Peto, J., Richard, S., Wahrendorf, J. (1980). Guidelines for Simple, Sensitive, Significant Tests for Carcinogenic Effects in Long-Term Animal Experiments, In: Long-Term and Short-Term Screening Assays for Carcinogens: A Critical Appraisal, Supplement 2:311-426, IARC Monographs on the Evaluation of the Carcinogenic Risk of Chemicals to Humans, International Agency for Research on Cancer, Lyon, France. 426 pages.

42. Pitot, H.D. and Sirica, A.E. (1980). The Stages of Initiation and Promotion in Hepatocarcinogens. Biochem. Biophys. Acta 605:191-215.

43. Portier, C. and Hoel, D.G. (1983a). Optimal Design of the Chronic Animal Bioassay. J. Toxicol. Environ. Health. (in press).

44. Portier, C. and Hoel, D.G. (1983b). Design of the Chronic Animal Bioassay for Goodness-of-fit to Multistage Models. (Draft Document).

45. Rall, D.P. (1979a). Validity of Extrapolation of Results of Animal Studies to Man. Ann. N.Y. Acad. Sci. 329:85-91.

46. Rall, D.P. (1979b). The Role of Laboratory Animal Studies in Estimating Carcinogenic Risks for Man, 179-189. In: Davis, W. and Rosenfeld, C. (editors). Carcinogenic Risks/Strategies for Intervention, 283 pages, IARC Scientific Pub. No. 25, INSERM Vol. 74, International Agency for Research on Cancer, Lyon, France.

47. Salsburg, D.S., 1977. Use of Statistics When Examining Lifetme Studies in Rodents to Detect Carcinogenicity. J. Toxicol. Environ. Health, 3:611-628.

48. Salsburg, D.S., 1983. The Lifetime Feeding Study in Mice and Rats--An Examination of Its Validity as a Bioassay for Human Carcinogens. Fund. Appl. Toxicol., 3:63-67.

49. Sontag, J.M., Page, N.P., and Saffiotti, U. (1976). Guidelines for Carcinogen Bioassay in Small Rodents. NCI Carcinogenesis Technical Report (TR 1), DHEW, Washington, DC, 65 pages.

50. Tarone, R.E. (1975). Tests for Trend in Life Table Analysis. Biometrika 62:679-682.

51. Tomatis, L. (1979). The Predictive Value of Rodent Carcinogenicity Tests in the Evaluation of Human Risks. Ann. Rev. Pharmacol. Toxicol. 19:511-530.

52. Weinstein, I.B. (1981). Current Concepts and Controversies in Chemical Carcinogenesis. J. Supramol. Struct. Cell. Biochem. 17:99-120.

Industrial Hazards of Plastics and Synthetic Elastomers, pages 65–78
© 1984 Alan R. Liss, Inc., 150 Fifth Ave., New York, NY 10011

TRENDS IN CANCER MORTALITY AMONG WORKERS IN THE SYNTHETIC
POLYMERS INDUSTRY

William J. Nicholson, Paul K. Henneberger and
Diane Tarr

Environmental Sciences Laboratory, Mount Sinai
School of Medicine of City University of New York
New York, New York 10029, U.S.A.

INTRODUCTION

The reactive double bonded structure of ethylene-like
molecules allows a wide variety of chemicals to undergo
polymerization. Unfortunately, this same structure has been
found capable of transformation to an epoxide by the mammalian mixed function oxidase system (Bonse and Henschler,
1976). These epoxides or their reactive metabolites can
bind to cellular macromolecules and may be responsible for
the carcinogenicity of the parent molecule. Epoxide formation has been suggested as an intermediate in the carcinogenic action of vinyl chloride (Van Duuren, 1975) and vinylidene chloride (Maltoni, 1977), and in the mutagenic action
of styrene (Milvy and Garro, 1976). The epoxides of ethylene, styrene and vinyl chloride have been shown to be carcinogenic, as well as directly mutagenic in bacterial test
systems without the need for activation. The potential for
conversion of ethylene-like molecules to the epoxides is
greater for unsymmetrical structures such as vinyl chloride
and vinylidene chloride than for symmetrical structures,
such as ethylene, 1,2-dichloroethylene or tetrachloroethy-
lene. It is beyond the scope of this review to discuss the
structure-activity relationships of the monomers used in the
plastics industry. Nevertheless, available data suggest
that carcinogenicity depends on the metabolism of these
monomers to reactive intermediates and that these reactions
may be non-linear. However, when the metabolism of a compound is understood, a coherent picture of the dose and time
dependence of cancer should emerge.

At this time, data are available on both experimental and human carcinogenesis from exposure to vinyl chloride (VC) and on its metabolism that provide information important for the understanding of observed dose-response relationships. This paper will consider these data on VC in detail as they provide estimates of the trends in future disease potential from past exposures and information on the efficacy of current occupational standards. As human and animal data accumulate on the effects of exposure to other monomers, the approach suggested by VC can be applied to their evaluation.

DOSE-RESPONSE RELATIONSHIPS

VC is one of the best studied chemicals in animal systems. The magnificent research by Maltoni and associates (1981) on nearly 7,000 animals over a ten year period is virtually unmatched in experimental carcinogenesis. A principal feature of their results is summarized in Figure 1 which shows the dose-response relationship for the percentage of animals that developed hemangiosarcoma (HSA) of the liver from 4 hr/day, 5 day/wk, 52 wk exposures to different concentrations of VC. As can be seen, the relationship is a non-linear one with evidence of saturation at high con-

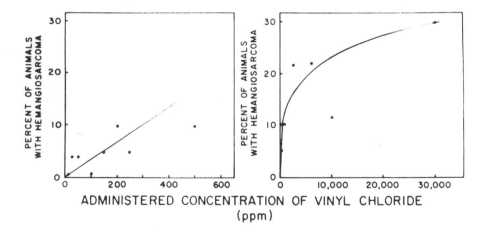

Figure 1. The percentage of rats developing liver hemangiosarcoma from 52 wk exposures to VC for 4 hr/day, 5 day/wk.

centrations. However, at concentrations of VC less than 500 ppm, a reasonably linear dose-response relationship obtains.

Gehring et al (1978) have explained the non-linearity in terms of Michaelis-Menten kinetics, in which the transformation of VC to a reactive intermediate follows the equation,

$$V = V_m S/(K_m + S) \tag{1}$$

V and V_m are the rate and maximum rate, respectively, for the biotransformation of VC, S is the concentration of VC in inspired air, and K_m, the Michaelis constant. K_m was determined experimentally to be 860 μg/l and V_m to be 5,706 μg/4 hr. Figure 2A displays the dose-response relationship between the percentage of animals with liver HSA and the quantity of VC metabolized according to Eq. 1. As can be seen, a direct liner relationship exists with no evidence of a threshold or altered slope at low doses. The possibility of a non-linear dose-response relationship from detoxification kinetic steps has been postulated (Gehring and Blau, 1977); and discussed in detail (Hoel et al, 1983), but no evidence exists for such non-linearity in the data yet available. The unweighted least squares regression equation for the dose-response relationship is

$$\% \text{ HSA} = -0.066 + 0.0039 \text{ V} \tag{2}$$

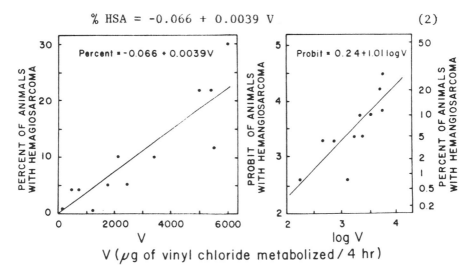

V (μg of vinyl chloride metabolized / 4 hr)

Figure 2. Linear and probit dose-response relationships for the quantity of VC metabolized/4 hr exposure (5 day/wk, 52 wk).

Gehring et al (1978) fitted the early data of Maltoni and Lefemine (1975) to a log-probit model. Figure 2B shows the log-probit plot using all available data from the studies from Maltoni et al (1981). The unweighted least squares regression line is

$$\text{Probit} = 0.24 + 1.01 \log V \tag{3}$$

While such a plot fits the observable data ($r^2 = 0.68$), the linear dose-response relationship fits the data somewhat better ($r^2 = 0.77$). Further, there is very limited biological rationale for the use of a log-probit relationship in carcinogenesis and its use as a means of extrapolation to predict effects at very low exposures would appear to be more an act of faith than of science. On the other hand, a linear dose-response relationship between the incidence of HSA and the quantity of VC metabolized is biologically plausible and fits all available data. Its use is strongly suggested.

TIME COURSE OF CANCER

Much of human cancer has been found to follow a power law relationship with age (Armitage and Doll, 1961; Cook et al, 1969),

$$R = bt^k \tag{4}$$

where R is the incidence rate of cancer at a specific site, t is age, and b and k are constants specific to site. In general, k is between 4 and 6 for most epithelial malignancies. While data for exposures to specific carcinogens are limited, bronchogenic carcinoma from cigarette smoking and mesothelioma from asbestos exposure also follow a power law of time from onset of exposure with an exponent between 3 and 5 (Doll and Peto, 1978; Newhouse and Berry, 1976; Peto et al, 1982). These findings have been interpreted in terms of a multistage model of carcinogenesis, the implications of which have been discussed by Peto (1977), Whittemore and Keller (1978), and Day and Brown (1980), among others. Deviations from the above time course occur with exposures to carcinogens that interact synergistically, such as asbestos and cigarette smoking in the production of lung cancer. This interaction can be incorporated in the multistage model, but a more complicated relationship obtains. However, for a rare tumor, such as HSA, interactive effects may not be important and a power law relationship should adequately describe the time course of risk following exposure.

Some data are available from the use of Thorotrast in Japan and Denmark that indicate the incidence rate of HSA does follow Eq. 4 (Mori et al, 1979a; Mori et al, 1979b; Faber, 1978). The material was used in these countries over a limited period of time, so the incidence per calendar year and estimates of the population at risk can be used to estimate incidence rates by time from onset of exposure. While the data are very limited, they are consistent with a power law dependence of risk and suggest an exponent of approximately 3. Three is also compatible with the incidence of liver HSA in the mortality study of polymerization workers described elsewhere in this volume (Nicholson et al, 1983). However, only nine cases are available for analysis.

PROJECTIONS OF FUTURE MORTALITY FROM PAST VC EXPOSURE

Sufficient data have accumulated on the pattern of mortality from past VC exposures to allow an estimate future mortality from these exposures of using a linear dose-response relationship and a time course for risk of death from liver HSA given by Eq. 4. Figure 3 shows the number of cases of HSA according to various measures of time that have been identified in the United States, Western Europe and the world (NIOSH, 1982). The distributions shown in Figure 3 are the result of the exposure to VC of various groups of individuals in different periods of time since 1935. Equation 4 indicates that the incidences (not incidence rates) according to calendar year, year of exposure, and year from onset of exposure, respectively, are:

$$I_j = {}_i\Sigma_j \; C_i \; t^k_{j-i} \; F_j(\text{Mort}) \tag{5a}$$

$$I_i = C_i \; {}_j\Sigma_i \; t^k_{j-i} \; F_j(\text{Mort}) \tag{5b}$$

$$I_{j-i} = t^k_{j-i} \; \Sigma \; C_i \; F_j(\text{Mort}) \tag{5c}$$

where i represents the quinquenium of exposure and j, the quinquenium of observation. i runs from 1 to 8, representing the years 1935-1974 and j from 1 to 9, extending the observations through 1979. The F_j(Mort) are the appropriate age and calendar year adjustments to the population in

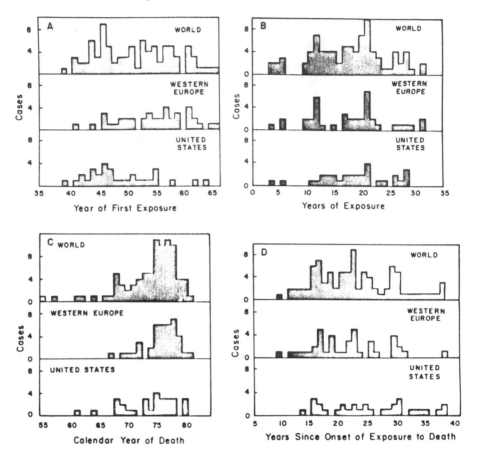

Figure 3. The number of cases of hemangiosarcoma of the liver in the U.S., Western Europe and the world according to several time criteria.

quinquenium j from normal mortality. The C_i's are proportional to the total population exposure, i.e., the average number of workers exposed in a given time period times the average VC concentration. Since the dose-response relationship for both inspired and metabolized VC is linear in the range of most worker exposures, the risk of HSA is proportional to the total population exposure; one need not know the number of workers exposed and their vinyl chloride exposure separately.

Relative values for the C_i's can be determined from two sets of data. The first is the incidence of HSA according to calendar period of first exposure (I_i). Here the C_i's are directly proportional to the incidence in a given calendar period and available data are sufficient to establish reasonable values of C_i for the time period 1935-1955. Additional data on C_i can be developed from published data on the production of VC monomer. Figure 4 displays the available information on production in the United States (S.P.I., 1975-1978; U.S. Tariff Commission, 1948-1968) and Western Europe (O.E.C.D., 1971). A first approximation to the population exposure in different years would be to consider the C_i's to be proportional to VC production. However, average VC concentrations changed over the years of concern (Table 1) and an adjustment for the different relative exposures in different times must be made. This adjustment is indicated in Table 1 and on Figure 4. Further, an adjustment must be made to take into account the different number of workers required to produce a metric ton of VC in different time periods. As it would be expected that more workers were employed per tonne of VC produced during earlier years, an adjustment is required to account for productivity. Initial estimates of this factor are also indicated in Figure 4. The relative population exposure, taken to be the product of production, the workforce productivity adjustment, and the exposure adjustment is shown by the solid

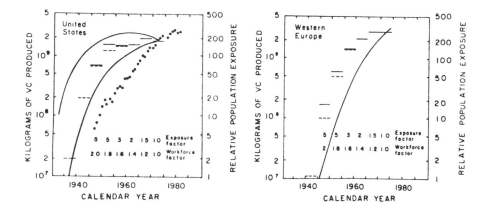

Figure 4. The production of VC in the U.S. and Western Europe along with estimates of the population exposures to VC polymerization workers in different calendar periods.

Table 1

Measured and estimated exposures to vinyl chloride
to polymerization workers in various time periods

Calendar Period	Vinyl chloride exposures (ppm)			Approximate relative exposure
	Barnes (1976)	Ott et al (1975)	Suciu et al (1975)	
Before 1950	1000			5
1950-54	1000	100 - 400		5
1955-59	400 - 500	100 - 400		3
1960-64	300 - 400	20 - 80	200 - 900	2
1965-69	300 - 400	20 - 80	40 - 50	1.5
1970-74	150 - 300		40 - 60	1

lines across each quinquenium. The dashed lines across each
quinquenium during earlier years are those determined by
fitting the observed HSA incidence to Eq. 5a and matched to
the value estimated from production data in the quinquenium
1955-1959. As can be seen, the comparison of the two sets
of data suggested that the population exposure prior to 1960
was slightly less in some quinquenia than that estimated by
the use of the adjustment factors indicated in Figure 4.

The procedure of estimating the relative values for C_i,
particularly in the years after 1960, is clearly an approxi-
mate one. To consider how sensitive any projections of
future mortality are to the choices of C_i's, alternate
choices are shown by the light solid lines in Figure 4a.
Any realistic estimates of the C_i's must lie between the two
lines.

Relative values of I_j, and I_{j-i} were calculated using
the relative values of C_i shown in Figure 4, values of k
between 2 and 4, and absolute values determined by matching
to the incidence data of HSA found in Figure 3. In this
calculation, the age distribution used for time of first
exposure was: 15-19, 8.5%; 20-24, 26%; 25-29, 26%; 30-34,
15%; 35-39, 11%; 40-44, 7%; 45-49, 4%; 50-54, 2.5%. This
distribution was that of 740 VC workers examined by Mount
Sinai School of Medicine personnel during 1974. The pattern
of duration of employment was assumed to be a decreasing
exponential with an average employment time of 12 years.
This corresponds to typical patterns of employment for
long-term workers in the chemical industry (Nicholson et al,
1982; Wong, 1982). Separate calculations were made for the

United States and Western Europe. The results of this procedure, combining the data for the United States and Western Europe, are shown in Figure 5. As can be seen I_{j-i}, the incidence according to years from onset of exposure is best fit by a value of k = 2. A value of 3 is compatible with the data, but values greater than 4 can be ruled out. I_j is relatively insensitive to the choice of k, but a value of 4 fits the data best.

An interesting feature of this calculation is that the separate determination of the C_i's for Western Europe and the United States indicates that the population exposures per tonne of VC produced were approximately four times greater in Western Europe than the United States. This would suggest that more intense exposures occurred in some European plants or that more workers were exposed per tonne of VC produced.

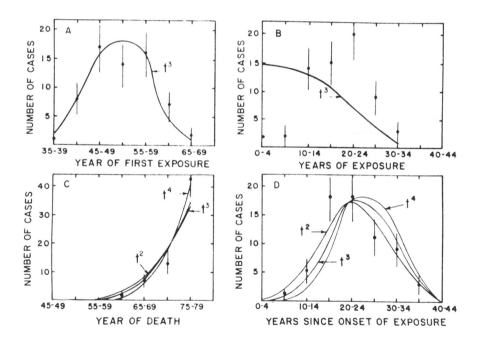

Figure 5. A comparison of the calculated incidence of hemangiosarcoma of the liver with that observed in the U.S. and Western Europe according to several time criteria and models for calculation.

One set of data that differs significantly from that calculated is the distribution of cases according to years of VC exposure. As mentioned previously, we assumed the distribution of employment times in the VC industry would be a decreasing exponential with a mean employment time of 12 years. The significant deficit of cases with employment times less than 10 years suggest that the available information on duration of VC exposure may not be correct, that our assumed employment distribution may be in error, that there may be an underascertainment of cases with shorter exposures, or that there may be proportionately less risk for shorter exposures than would be predicted on a linear dose-response relationship. It should be mentioned that duration of employment is not an important variable in assessment of population risk. Shorter employment times would have required more men to be exposed, but their average exposure would be proportionately lower.

Using the values of C_i's determined by the preceding analysis and values of k from 2 to 4, the mortality from liver HSA is calculated to the year 2040, using Eq. 4a. These data are listed in Table 2, separately for Western Europe and the United States. Also shown in the data for the United States are projections using values of C_i's indicated by the solid curves of Figure 4 and projections assuming that the risk of HSA will increase quadratically with age of exposure. This age dependence was suggested by experimental results of Groth et al (1981). We also considered a time course for HSA that increased as t^3 for only 45 years and remained constant thereafter. As can be seen, the pro-

Table 2

Projections of mortality in the United States
and Western Europe to the year 2040 from
exposures to vinyl chloride prior to 1975

Model	Total Projected Mortality	
	United States[a]	Western Europe[b]
t^2	190	540
t^3	340	1190
t^4	630	2780
t^3, to 45 years from onset of exposure	310	1120
t^3, upper exposure curve, Fig. 4	240	
t^3, lower exposure curve, Fig. 4	450	
t^3, + Age^2	260	

[a] 26 deaths have occurred through 1979
[b] 39 deaths have occurred through 1979

jected numbers of HSA for the United States range from 200 to 600 and, for Western Europe, from 550 to 2,800. (The greater range for Europe is the result of the more recent usage of pattern.) The most probable projection for future disease is felt to be that represented by a power of 3, a choice suggested by Thoratrast data and the very limited mortality data on HSA in the study by Nicholson et al (1983). Lower values are also reasonable, but the fit to the data would suggest that the use of a power of 4 may be inappropriate.

Obviously, many caveats exist in the consideration of these projections. The estimates strongly depend upon a reasonable ascertainment of cases through 1979. The concerns for VC-induced HSA in recent years would suggest that ascertainment was fairly good, at least for long term employees and pensioners. However, some cases in short term workers may have been missed. The projections also depend on the choices of the C_i and the k. We have projected mortality based on reasonable choices for these parameters. However, other choices cannot be absolutely excluded. While these uncertainties exist, the data indicate that, within a factor of 2 or 3, future HSA mortality from exposures prior to 1975 will be about 350 deaths in the United States and 1,200 in Western Europe. Further, these deaths will occur in a relatively small population. In the United States, the group at highest risk would be comprised of fewer than 5,000 individuals. Among this heavily exposed group, HSA may account for 10% of all deaths (Nicholson et al, 1983). Clearly, any intervention techniques that might be developed to reduce this projected risk could be efficiently applied.

OCCUPATIONAL STANDARDS FOR VC

Nicholson et al (1983) have shown that liver HSA accounts for at least 50% of all VC-induced malignancies. Thus, it would appear that average exposures of 200-500 ppm in previous years will lead to 1,000-4,000 excess cancer deaths in all workers exposed to VC in Western Europe and the United States prior to 1975. If a standard of 1 ppm is met, the average exposure of all the workers would be between 0.2-0.5 ppm, 1,000 times less than that which existed previously. One would expect the VC-induced malignant risk to be reduced by a corresponding amount. This implies that, <u>if the VC industry complies with a 1 ppm standard</u>, cancer from employment therein would be virtually eliminated. However, there will still remain a risk of developing HSA of the order of 10^{-4} per individual for a working lifetime, based on a United States or European workforce of about 10,000 workers.

SUMMARY

A high risk of death from liver HSA has been documented from past exposures to VC. Similar to other carcinogens, the risk of VC-induced liver HSA appears to increase as the second or third power of time from onset of exposure. It is possible to project future mortality using this power relationship, estimates of VC exposure, and observed mortality to 1980. These projections suggest that 200-600 deaths may occur in the United States and 550-2,800 in Western Europe from liver HSA. These projections also suggest that a 1 ppm standard in the VC industry will go far to protecting workers from future malignant disease.

REFERENCES

Armitage P, Doll R (1961). Stochastic models for carcinogenesis. In: Proceedings of the Fourth Berkeley Symposium on Mathematical Statistics and Probability. (Ed. Neyman J) Univ Calif Press, Berkeley pp. 19-38.

Barnes AW (1976). Vinyl chloride and the production of PVC. Proc Roy Soc Med 69:277-280.

Bonse G, Henschler D (1976). Chemical reactivity, biotransformation and toxicity of polychlorianted aliphatic compounds. CRC Crit Rev Toxicol 5:395.

Cook PJ, Doll R, Fellingham SA (1969). A mathematical model for the age distribution of cancer in man. Int J Cancer 4:93-112.

Day NE, Brown CC (1980). Multistage models and primary prevention of cancer. J Natl Cancer Inst 64:977-989.

Doll R, Peto R (1978). Cigarette smoking and bronchial carcinoma: dose and time relationships among regular smokers and lifelong non-smokers. J Epidem Comm Health 32:303-313.

Faber M (1978). Malignancies in Danish Thorotrast patients. Health Physics 35:153-158.

Gehring PJ, Blau GE (1977). Mechanisms of carcinogenesis: dose response. J Environ Path Toxicol 1:163-179.

Gehring PJ, Watanabe PG, Park CN (1978). Resolution of dose-response toxicity data for chemicals requiring metabolic activation: example - vinyl chloride. J Toxicol Appl Pharmacol 44:581-591.

Groth DH, Coate WB, Ulland BM, Hornung, RW (1981). Effects of aging on the induction of angiosarcoma. Environ Health Persp 41:53-57.

Hoel DG, Kaplan NL, Anderson MW (1983). Implication of nonlinear kinetics on risk estimation in carcinogenesis. Sci 219:1032-1037.

Maltoni C, Lefemine G (1975). Carcinogenicity assays of vinyl chloride: current results. Ann NY Acad Sci 246: 195-224.

Maltoni, C (1977). Recent findings on the carcinogenicity of chlorinated olefins. Environ Health Persp 21:1-5.

Maltoni C, Lefemine G, Ciliberti A, Cotti G, Carretti D (1981). Carcinogenicity bioassays of vinyl chloride monomer: a model of risk assessment on an experimental basis. Environ Health Persp 41:3-29.

Milvy P., Garro AJ (1976). Mutagenic activity of styrene oxide (1,2-epoxyethylbenzene), a presumed styrene metabolite. Mutat Res 40:15-18.

Mori T, Kato Y, Shimamine T, Watanabe S (1979a). Statistical analysis of Japanese Thorotrast-administered autopsy cases. Environ Res 18:231-244.

Mori T, Maruyame T, Kato Y, Tahahashi S (1979a). Epidemiological follow-up study of Japanese Thorotrast cases. Environ Res 18:44-54.

National Institute of Occupational Safety and Health (U.S.) (October,1982). Reported cases of angiosarcoma of the liver among vinyl chloride polymerization workers.

Newhouse ML, Berry G (1976). Prediction of mortality from mesothelial tumors in asbestos factory workers. Brit J Indus Med 33:147-151.

Nicholson WJ, Perkel G, Selikoff IJ (1982). Occupational exposure to asbestos: population at risk and projected mortality - 1980-2030. Am J Indust Med 3:259-311.

Nicholson WJ, Henneberger P, Seidman H. Occupational hazards in the VC-PVC industry. This volume.

Organization for Economic Cooperation and Development, Chemical Industry (1971). Quoted in: Levinson C. Work hazard: vinyl chloride. ICF Geneva.

Ott MG, Langner RR, Holder BB (1975). Vinyl chloride exposure in a controlled industrial environment. Arch Environ Health 30:333-339.

Peto R (1977). Epidemiology, multistage models and short-term mutagenicity tests. In: Origins of Human Cancer (Eds. Hiatt HH, Watson JD, Winsten JA). Cold Spring Harbor Laboratory pp. 1403-1430.

The Society of the Plastics Industry, Inc (1975-1982). Facts and Figures of the U.S. Plastics Industry, New York.

Suciu I, Prodan EI, Paduraru A, Pascu L (1975). Clinical manifestations in vinyl chloride poisoning. Ann NY Acad Sci 246:53-69.

U.S. Tariff Commission (1948-1968). Polyvinyl chloride and copolymer production data.

Van Duuren B (1975). On the possible mechanism of carcino-genic action of vinyl chloride. Ann NY Acad Sci 246:258-267.

Whittemore AS, Keller JB (1978). Quantitative theories of carcinogenesis. Society for Industrial and Applied Mathematics Review 20:1-30.

Wong O (1981). An epidemiologic study of workers potentially exposed to brominated chemicals: with a discussion of a multifactor adjustment. In: Quantification of Occupational Cancer (Eds. Peto R, Schneiderman M). Banbury Report 9 Cold Spring Harbor Laboratory pp. 359-378.

Industrial Hazards of Plastics and Synthetic Elastomers, pages 79–87
© 1984 Alan R. Liss, Inc., 150 Fifth Ave., New York, NY 10011

REPRODUCTVE HAZARDS AND PLASTICS INDUSTRY

K. Hemminki, M.-L. Lindbohm, T. Hemminki and
H. Vainio
Institute of Occupational Health

Haartmaninkatu 1, SF-00290 Helsinki 29, Finland

Introduction

The output of plastics has increased markedly over the
last three decades and the production of plastics goods has
become one of the main branches of employment in industrial-
ized countries. The health effects of the ingredients in
plastics, and of process and pyrolysis emissions are poorly
known, and pose a standing challenge to occupational health
personnel.

The present article surveys the available knowledge,
published and unpublished, on the reproductive effects of
employment in plastics industry. Some environmental studies
are also discussed as well as ethylene oxide, a metabolite of
ethene, which is used in chemical sterilization rather than
in plastics industry. The reproductive outcomes surveyed
include spontaneous abortions and malformations in the
offspring.

Published Studies

Vinyl chloride. Infante et al. (1976) studied the
effects of paternal exposure to vinyl chloride on the
frequency of spontaneous abortions in the wives. Information
on spontaneous abortions was obtained by interview of the
husbands. The frequency of spontaneous abortions, adjusted to
paternal age, was 15.8 % when the husbands were employed in
vinyl chloride polymerization as compared to 6.1 % prior to
the employment. In a control population, the wives of

polyvinyl chloride fabrication and rubber workers the
respective rates were 8.8 and 6.9 %.

Environmental exposure to vinyl chloride and the
prevalence of congenital malformations has been a subject of
three studies. Infante (1976) studied Ohio communities, where
vinyl chloride polymerization plants were in operation. The
prevalence rates of malformations per 1000 births were:
10.14 in the state of Ohio, 17.37 in Asthabula, 18.10 in
Painesville and 30.33 in Avon Lake, the three communities
with vinyl chloride polymerization plants. The excess in the
three communities was significant statistically (p < 0.01).
Yet even higher prevalence rates of malformations were
observed in two communities without vinyl chloride plants.

A case-control study was conducted by Edmonds et al.
(1975) in Painesville. 15 babies born with central nervous
system malformations in the Painesville hospital in 1970 - 74
were compared with 30 control babies. The parents of the
cases were interviewed and information on the controls was
collected from clinical records. None of the case parents and
2 of the control parents worked at a vinyl chloride polymerization plant; more control mothers than case mothers
worked within 10 miles of the Painesville plant.

Another case-control study was conducted by Edmonds et
al. (1978) in the Kanawha County (U.S.A.): incident cases of
central nervous system malformations occurring in 1970 - 74
were compared to a group of control babies matched by
paternal education, maternal age, month of birth, race and
social status. Parents were interviewed by telephone on their
work in vinyl chloride polymerization plants and the place of
residence. Two out of 41 case fathers and 2/41 control
fathers worked in vinyl chloride plants during the conception
time. More cases than controls lived within 3 miles from the
vinyl chloride facilities (p < 0.02).

Styrene. In an ongoing study based on the Finnish
Register of Congenital Malformations, 63 case parents and
a similar number of control parents were interviewed about
chemical exposures. Two mothers with children having central
nervous system defects had been employed in the reinforced
plastics industry during pregnancy. A third case mother was
also found, who had been exposed to styrene at home
(Holmberg, 1977). No control mothers reported exposure to
styrene.

Women employed in plastics industry in Finland were
found to have more spontaneous abortions, treated in
a hospital, as compared to all Finnish women or all members
of the Union of Chemical Workers (Hemminki et al. 1980a). The
analysis was based on the hospital discharge register from
years 1973 to 1976. Women employed in styrene workplaces,
mainly reinforced plastics workshops, had a higher rate of
spontaneous abortions than all plastics workers. However,
only a few spontaneous abortions were obtained (6 for styrene
workplaces and 52 for the whole Union) and no information on
individual exposure was available. In this article, which was
the first one in our ongoing study based on hospitalized
spontaneous abortions, the diagnoses numbers taken for
spontaneous abortions and for births were fewer than the ones
used by us subsequently (Hemminki et al. 1980b, c).

Härkönen and Holmberg (1982) interviewed 67 female
lamination workers at a fertile age and a similar number of
textile and food production workers. The number of deliveries
was found to be decreased among the styrene workers,
partially explained by the higher number of induced abortions
in the styrene-exposed group. The groups were reported not to
differ in their menstrual behaviour nor in the number of
spontaneous abortions. The lamination workers had 4 spontaneous abortions per 16 pregnancies (25 %) and per 4 births;
the referent population had 4 spontaneous abortions per
22 pregnancies (18 %) and per 14 births.

Ethylene oxide. A study was carried out in Finland
about the reproductive effects of chemical sterilizing agents
among female hospital personnel (Hemminki et al. 1982).
Questionnaires were sent to the identified sterilizing staff
and to a referent population of nursing auxiliaries from the
same hospitals. The participation rate was about 90 % both
among the sterilizing staff and the nursing auxiliaries. The
rate of spontaneous abortions among the sterilizing staff was
11.3 % as compared to 10.6 % for the nursing auxiliaries. The
respective numbers of pregnancies were 1440 and 1180. When
the pregnancies of the sterilizing staff were classified
according to the exposure to sterilizing agents, significantly more spontaneous abortions (16.7 %) were observed in the
exposed than in the non-exposed pregnancies (6.0 %). The
excess number of spontaneous abortions was contributed mainly
by the pregnancies when ethylene oxide was used as the
sterilizing agent. The effect of ethylene oxide exposure
remained when the data were controlled for age and parity of

women, consumption of coffee and alcohol, tobacco smoking and
the decade of the reported pregnancy. An excess of spontaneous abortions in pregnancies taking place during exposure
to ethylene oxide was confirmed when the reproductive events
were retrieved from the hospital discharge register.

Unpublished Studies

      The Nordic working group on occupational reproductive
hazards initiated a study on the maternal effects of work in
plastics industry on malformations in the offspring,
stillbirth, perinatal mortality and spontaneous abortions. In
Sweden and Norway lists of female employees, including office
and production personnel were obtained from the industry. In
Sweden about 1400 and in Norway about 300 pregnancies were
scored from the birth register, while the women were employed
in the plastics industry. In both countries the initial
analysis of malformations in the offspring, stillbirths and
perinatal mortality revealed no deviations from the expected
numbers (T. Bjerkedal and B. Källen, personal communications). Further studies are being carried out to ascertain
possible exposures of the employees.

      In Finland information on the employment was collected
from the membership register of the Union of Chemical
Workers. Information on malformations was obtained from the
Finnish Register on Congenital Malformations from years
1973 - 1979. However, as the Register has a limited coverage,
probably about 50 %, we could not use the prevalence rates
directly. Instead, to each chemical worker with a malformed
child two control women with a delivery were obtained from
the Hospital Discharge Register. The data of delivery
($\pm$ 1 year) and the Union chapter were matched. Two cases
could not be analysed because controls were not found.
According to the case-control design it was analysed, whether
the mothers of malformed or healthy children had been working
(e.g. whether they were Union members) at the time of early
pregnancy (Table 1). All the relative risks were close to
unity; for women employed in pharmaceutical industry the
relative risk was 1.7 (95 % confidence limits 0.1 - 21.2).

      The types of malformations found in the children of
chemical workers are shown in Table 2. Hip luxation was the
main type of malformation found, also being the most common
malformation in the Finnish Register of Congenital

Table 1. Relative risks for malformations in children of the
          members of Chemical Workers' Union 1973 - 1979

Branch of employment	Number of triplets	Number of exposed[1] mothers		Relative risk	95 % confidence limits
		Cases	Controls		
Plastics industry	11	9	18	1.0	0.2 - 5.5
Pharma-ceutical industry	4	3	5	1.7	0.1 - 21.2
Laundries	4	3	6	1.0	0.1 - 18.9
Other industries	3	2	4	..	..

[1]Exposed = member of the Union during pregnancy.

Malformations. Among other types of malformation 4 cases of
oral clefts, and 2 cases of each of central nervous system
defects, gastro-intestinal atresias and club feet were
detected in the children of the exposed chemical workers.

Spontaneous abortions of the chemical workers were
analysed from the Hospital Discharge Register that was
supplemented with spontaneous abortions treated policlinical-
ly. The diagnoses numbers used were 643 and 645 (with no
preceeding induced abortion) for spontaneous abortion,
640 - 642 for induced abortion, and 650 - 662 for birth. Two
frequencies were calculated: rate = number of spontaneous
abortions x 100/number of pregnancies, and ratio = number of
spontaneous abortions x 100/number of births (see Hemminki et
al. 1980b, c).

The age dependence of the spontaneous abortion rate is
shown in Fig. 1A separately for the women who were and who
were not Union members during the first two months of their
pregnancy. The Union members had more spontaneous abortions
in the younger age groups (< 30 years) than the

Table 2. Types of malformations in the offspring of chemical
workers

During Union membership		Before/after membership	
Plastics industry			
Hip luxation	(3)	Hip luxation	(1)
Club foot	(2)	Undefined malformation	(1)
Hydrocephalus + other	(1)		
Oral cleft	(1)		
Duodenal atresia	(1)		
Multiple malformations	(1)		
Pharmaceutical industry			
Oral cleft	(2)	Multiple malformations	(1)
Polydactylia	(1)		
Laundries			
Hip luxation	(1)	Hip luxation	(1)
Meningomyelocele	(1)		
Malformations of circulatory system	(1)		
Anal atresia	(1)		
Other			
Hip luxation	(2)	Hip luxation	(1)
Oral cleft	(1)		

The figure in the parenthesis refers to the number of cases.

non-members. The rate for all members was 9.0 % and for the
non-members 8.2 %. The lack of increase in the rate of
spontaneous abortions with increasing age of the women is
abnormal among the members, and is likely to suggest the
presence of some selection.

Further information on the operation of some selection
mechanisms can be seen in Fig. 1B, when the rate is plotted
according to the year of observation. Among members, the rate

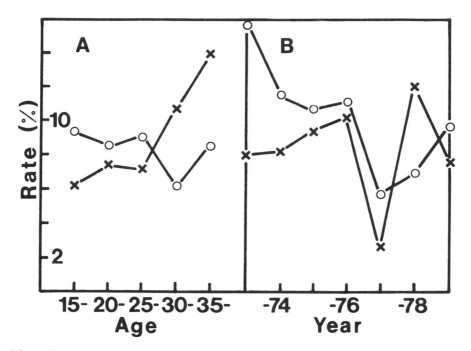

Fig. 1. Age-dependence (Fig. 1A) and annual rate (Fig. 1B) of spontaneous abortions during (-o-) and before/after (-x-) Union membership.

declined from 11 to 16 % in 1973 - 1976 to 6 % in 1977. The reasons for such changes are unclear.

The rate and ratio of spontaneous abortions among the members of the Union of Chemical Workers is shown in Table 3. 149 spontaneous abortions were scored during Union membership, and 110 before or after the membership. The rate and ratio in viscose rayon, pharmaceutical and other branches were higher than the Union total, and also higher in pregnancies taking place during Union membership as compared to those before or after membership. In plastics and styrene workplaces the rate and ratio tended to be smaller for members than for non-member's pregnancies. However, we have to analyse the data carefully in relation to the suspected selection mechanisms and actual exposure before the possible hazards can be evaluated.

Table 3. Spontaneous abortions among the members of the
Union of Chemical Workers according to Union
membership during pregnancy 1973 - 1979

Branch of employment	Pregnancy during membership		
	No. of spontaneous abortions	Rate of spontaneous abortions	Ratio of spontaneous abortions
Plastic industry	59	8.6	13.3
Styrene production and use	9	6.8	10.2
Viscose rayon industry	18	12.5	22.0
Laundries	21	7.5	11.4
Pharmaceutical industry	14 (12)[1]	11.4* (9.3)[1]	17.1° (14.7)[1]
Other industries	37	8.8*	13.6°
All Union members	149	9.0	14.0
	Pregnancy before/after membership		
Plastic industry	65	10.1	17.3
Styrene production and use	19	12.2	22.1°
Viscose rayon industry	6	10.3	18.2
Laundries	21	8.6	15.7
Pharmaceutical industry	5	4.1	6.9
Other industries	13	4.7	7.4
All Union members	110	8.2	13.9

[1] Figures in the parenthesis only include a single
spontaneous abortion per woman.
°p < 0.10, *p < 0.05, pregnancies during membership
compared with pregnancies before/after membership, x^2-test.

Conclusions

    Plastics industry is likely to expand also in years to
come and a special research effort is needed in this new
field to evaluate its health effects. Reproductive epidemiology has the potential of detecting occupational hazards

with reasonable power relatively shortly after exposure.
However, reproductive epidemiology is a new dicipline, and it
may be compounded by unknown selection mechanisms relating to
pregnancy and employment that need to be resolved before
causal relationships can be established.

References

Edmonds LD, Anderson CE, Flynt JW, James LM (1978).
Congenital central nervous system malformations and vinyl
chloride monomer exposure: a community study. Teratology
17:137.
Edmonds LD, Falk H, Nissim JF (1975). Congenital malformations and vinyl chloride. Lancet ii:1098.
Hemminki K, Franssila E, Vainio H (1980a). Spontaneous
abortions among female chemical workers in Finland. Int Arch
Occup Environ Health 45:123.
Hemminki K, Mutanen P, Saloniemi I, Niemi M-L, Vainio H
(1982). Spontaneous abortions in hospital staff engaged in
sterilising instruments with chemical agents. Br Med J
285:1461.
Hemminki K, Niemi M-L, Koskinen K, Vainio H (1980c).
Spontaneous abortions among women employed in the metal
industry in Finland. Int Arch Occup Environ Health 47:53.
Hemminki K, Niemi M-L, Saloniemi I, Vainio H, Hemminki E
(1980b). Spontaneous abortions by occupation and social
class in Finland. Int J Epidemiol 9:149.
Holmberg PC (1977). Central nervous defects in two children
of mothers exposed to chemicals in the reinforced plastics
industry: chance or causal relation? Scand J Work Environ
Health 3:212.
Härkönen H, Holmberg PC (1982). Obstetric histories of women
occupationally exposed to styrene. Scand J Work Environ
Health 8:74.
Infante PF (1976). Oncogenic and mutagenic risks in
communities with polyvinyl chloride production facilities.
Ann NY Acad Sci 271:49.
Infante PF, Wagoner JK, McMichael AJ, Waxweiler RJ, Falk H
(1976). Genetic risks of vinyl chloride. Lancet i:734.

CHAPTER II. POLYVINYL AND RELATED POLYMERS

Industrial Hazards of Plastics and Synthetic Elastomers, pages 91–98
© 1984 Alan R. Liss, Inc., 150 Fifth Ave., New York, NY 10011

PRODUCTION AND PROCESSING OF PVC

Ariel Hoff

Department of Polymer Technology,
The Royal Institute of Technology
S-100 44 Stockholm, Sweden

In commercial practice vinyl chloride is polymerised by
free radical mechanisms in bulk in suspension and emulsion.

Bulk Polymerisation

The only commercial process of bulk polymerisation was
one operated by St. Gobain in France. This was one-stage
process. Vinyl chloride was polymerised with 0.8% of its
own weight of bensoyl peroxide in a rotating cylinder containing steel balls for 17 hours at 58°C. A new technology
involving a two-stage process has been developed by the same
company. The first stage is carried out as a liquid with up
to about 15% conversion whilst the second stage carried out
as a powder takes the conversion to 80-85%. The presence of
two stages allows considerable flexibility for the process,
particle characteristics usually being determined by the
operation of the first stage and the average molecular
weight by the second stage. By the early 1970's at least 20
companies throughout the world had taken licences to operate
this process.

Suspension Polymerisation

Suspension polymerisation is generally easier to control than bulk polymerisation. In this process, which has
been the leading industrial method of vinyl chloride polymerisation, the monomer and the oil soluble initiator are
suspended in water as small droplets in which the polymeri-

sation takes place. For this purpose it is necessary to have a suspending agent, or a protective colloid, or some type of surfactant that will allow the beads to be formed and to remain stable.

A typical polymerisation vessel is a stainless steel container with a heating jacket, offset stirrer, pressure gauge, temperature controller and inlets for the monomer, water and vacuum.

A typical charge would be:

Monomer	Vinyl chloride	30-50 parts
Dispersing agent	Gelatine or polyvinyl alcohol	0.001 part
Modifier	Trichlorethylene	0.1 part
Initiator	Caproyl peroxide	0.001 part
	Demineralised water	90 parts

In addition, buffer salts such as disodium hydrogen phosphate may be used to prevent the pH of the aqueous phase falling during polymerisation. Small amounts of an anti-foam agent may be employed to reduce frothing when discharging from the vessel at the end of the polymerisation process. The modifier is a solvent transfer agent used to control molecular weight.

Using the above recipe the dispersing agent is first dissolved in a known weight of water, and added to the kettle. The rest of water the peroxide and the modifier are then added to the kettle which is sealed down and evacuated to 28 in.Hg. The monomer is then drawn in from the weighting vessel. The vessel is heated to about $50^{\circ}C$. Through the rise in temperature a pressure of about 6-7 atm will be developed in the reactor. When the pressure has dropped to about 1 atm. excess monomer is vented off and the batch cooled down, discharged and dried. The product is checked for particle size, for colour, contamination and for viscosity in a dilute solution.

Emulsion Polymerisation

In this process the monomer is polymerised in an emulsion. As emulsifiers secondary alkyl sulphonates or alkali salts of alkyl sulphates are used. The use of redox initiating systems has made possible rapid reaction at temperatures as low as 20°C. Sodium persulphate, potassium persulphate and hydrogen peroxide are typical initiators, whilst bisulphites and ferrous salts are useful reducing agents. Modifiers are often employed to control the molecular weight. Reaction times are commonly of the order of 1-2 hours. After the polymerisation the particles are normally spray dried. There will thus be residual emulsifier which will adversely affect clarity and electrical insulation properties.

Production

About 90% of the PVC produced is used in the form of a homopolymer. The other 10% are copolymers, mainly with vinyl acetate. These are used for phonograph records and for vinyl floor tiles.

Approximately 50% of PVC produced is used in rigid pipe extrusion applications, most commonly as suspension homopolymers of high bulk density compounded as powder blends. Flexible extrusion is the second largest application. The properties of plastized PVC depends largely on the amounts and chemical types of plasticizers added. Wire and cable insulation and clear film for meat packaging and stretch film are the major applications for flexible PVC.

Compounding Ingredients

In the massive form PVC is colourless rigid material with limited heat stability and with a tendency to adhere to metallic surfaces when heated. For these and other reasons it is necessary to compound the polymer with other ingredients to make useful plastic materials. By such means it is possible to produce a wide range of products including rigid piping and soft elastic cellular materials. A PVC compound may contain the following ingredients:

Polymer	Fillers
Stabilizers	Pigments
Plasticizers	Polymeric processing aids
Extenders	Impact modifiers
Lubricants	

It is obvious that the range of possible formulations based on PVC is very wide indeed. As an example two formulations are given below. The first is for a transparent calandering compound, the second gives a typical rigid opaque formulation suitable for pipes.

	1		2
Suspension polymer	100	Suspension polymer	100
D.I.O.P.	40	Tribasic lead sulphate	6
Ba-Cd phenate	3	Lead stearate	1
Trisnonyl phenal		Glyceryl monostearate	0.4
phosphite	1	Acrylic process aid	2
Epoxidised oil	5		
Stearic acid	0.4		

Unplasticised PVC (UPVC)

The processing of UPVC is more critical than that of the plasticised material since UPVC only becomes processable in the temperature range at which decomposition occurs at a measurable rate. Most UPVC compounds are prepared by blending of powders. High-speed mixers are preferred, frictional heating causing a temperature rise bringing the PVC above its T_g and thus facilitating rapid absorption of liquid and semi-solid additives. Too early addition of lubricants may retard the heat build up and the point of their addition may be critical. The blends continue to increase in temperature as mixing proceeds causing agglomeration and an increase in bulk density. This leads to increasing output in, for example, extruders but also reduces the thermal stability of the compound. An accurate temperature control is clearly important. Although it is possible to extrude rigid PVC sheet, it is commonly made by compression moulding techniques. Such sheet may be welded using hot gas welding guns to produce industrial equipment.

Plasticised PVC

The melt processing of plasticised PVC normally involves the following stages:

1. Pre-mixing polymer and other ingredients.
2. Fluxing the ingredients.

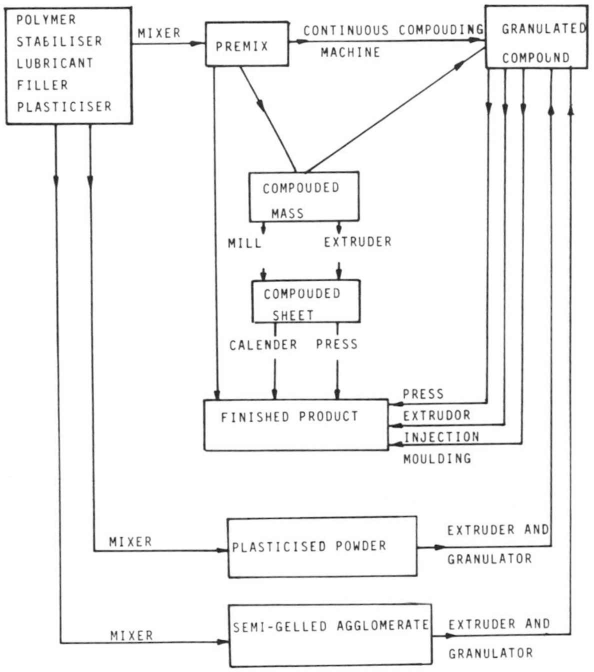
POLYMER
STABILISER
LUBRICANT
FILLER
PLASTICISER
MIXER
PREMIX
CONTINUOUS COMPOUDING
MACHINE
GRANULATED
COMPOUND
COMPOUDED
MASS
MILL
EXTRUDER
COMPOUDED
SHEET
CALENDER
PRESS
FINISHED PRODUCT
PRESS
EXTRUDOR
INJECTION
MOULDING
MIXER
PLASTICISED POWDER
EXTRUDER AND
GRANULATOR
MIXER
SEMI-GELLED AGGLOMERATE
EXTRUDER AND
GRANULATOR

3. Converting the fluxed product into a suitable shape for
   further processing, e.g. granulating for injection moulding or extrusion.
4. Heating the product to such an extent that it can be formed
   by such processes as calendering, extruding, etc. and
   cooling the formed product before removal from the shaping
   zone.

Routes from raw materials to finished products illustra-
ting different compounding techniques with PVC compounds
(modified from G.A.R. Matthews, Advances in PVC compounding
and processing, (ed. M. Kaufman, Maclaren, London, p. 53
(1962)

DEGRADATION CF PVC

PVC is the least stable of commercially available polymers. In air atmosphere degradation of PVC commences at temperatures above $150^{\circ}C$. At processing temperatures used in
practice $(150-200^{\circ}C)$ sufficient degradation may take place
during standard processing operations to render the product
useless. There is a great deal of uncertainity as to the
mechanism of PVC degradation, but certain aspects have emer-
ged. Firstly dehydrochlorination occurs at an early stage in
the degradation process. The comparison of PVC with model
compounds, e.g. 2-chlorpropane would seem to suggest that
the normal structure of PVC is very stable. Anomalous weak
links are, thus, probably responsible for the degradation
of the polymer. Some of the probable weak links are:

vinyl chlorine at $-CH = CH-Cl$
chain ends
allylic chlorine $-CH_2-CH-CH = CH-CH_2-$
in the chain $|$
                     Cl

As it was mentioned above the first step in the PVC degradation is dehydrochlorination

$$\left(-\overset{\overset{\displaystyle H}{|}}{\underset{\underset{\displaystyle H}{|}}{C}} - \overset{\overset{\displaystyle H}{|}}{\underset{\underset{\displaystyle Cl}{|}}{C}} - \overset{\overset{\displaystyle H}{|}}{\underset{\underset{\displaystyle H}{|}}{C}} - \overset{\overset{\displaystyle H}{|}}{\underset{\underset{\displaystyle Cl}{|}}{C}} - \overset{\overset{\displaystyle H}{|}}{\underset{\underset{\displaystyle H}{|}}{C}} - \overset{\overset{\displaystyle H}{|}}{\underset{\underset{\displaystyle Cl}{|}}{C}}-\right)_n \xrightarrow{-HCl} \left(-\overset{\overset{\displaystyle H}{|}}{C} = \overset{\overset{\displaystyle H}{|}}{C} - \overset{\overset{\displaystyle H}{|}}{C} = \overset{\overset{\displaystyle H}{|}}{C} - \overset{\overset{\displaystyle H}{|}}{C} = \overset{\overset{\displaystyle H}{|}}{C}-\right)_n$$

PVC colourless   PVC (dark coloured)

The first physical manifestation of degradation, used in
the widely sence, is a change in colour. Initially water-white,
on heating it will turn in sequence pale, yellow, orange,
brown and black. The main volatile degradation product of PVC
is hydrogen chloride followed by benzene.

The hydrochloric acid that is released catalyzes additional decomposition in what is called a zipper reaction.
Heavy metal salts or soaps which are also used as lubricants,
may react with HCl and hence are effective scavengers. However,
many of these additives are incompatible and exude. Therefore, more compatible phenolic complexes of barium or zink
are often used. Besides, organotin compounds such as dibutyl
tin dilaurate are used. Epoxidised unsaturated oil, such as
soy bean oil, which acts as HCl scavenger is used.

In case of a rigid polymer which contains very few
additives all the other volatile products formed during thermal degradation of PVC consists only a fraction of the amount
of benzene. (Y. Isuchio and K. Sumi, 1967).

During the oxidation of PVC at elevated temperatures
apart from HCl and benzen, considerable amounts of carbon
oxide and carbon dioxide are formed. Among the other volatile products aldehydes, alcohols and acids are discovered.

Vinyl chloride has only been found during thermal oxidation of PVC at temperatures above $350^{\circ}C$ (7–33 ppm) rather
than thermal degradation of the polymer (I. Michal, 1976).
However, there is some other data showing that vinyl chloride
is found on thermal degradation of PVC at temperatures
above $225^{\circ}C$ (B. Wakeman and J. Johnson, 1978). It was found
that the amount of vinyl chloride formed at temperatures
up to $400^{\circ}C$ is not exceeding 70 ppm (B. Wakeman and H. Johnson,
1978).

The degradation of flexible PVC which contains plasticizers also results in a formation of phtalic acid anhydride
and C_8-olefins.

REFERENCES

1. Isuchio Y, Sumi K (1967) J Appl Polym Sci 17:304
2. Matthews GAR (1962) Advances in PVC Compounding and Processing
   Ed Kaufman M, London: Maclaren, p 53.
3. Michal I (1976) Fire and Materials, 1:57.
4. Wakeman B, Johnson MR (1978) Polym Eng and Sci, 18:5.

Industrial Hazards of Plastics and Synthetic Elastomers, pages 99–112
© 1984 Alan R. Liss, Inc., 150 Fifth Ave., New York, NY 10011

THE TOXICOLOGY OF MONOMERS OF THE POLYVINYL PLASTIC
SERIES

Professor Bo Holmberg

National Board of Occupational Safety and Health
Unit of Occupational Toxicology. Research Dept.

S-17184 Solna, Sweden.

Among the thermoplastics, the polyvinyl series is of
great commercial and technical importance. The series
can be divided into three main groups (Miles, Briston
1979):

- Polyvinyl chloride (homopolymer)

- Vinyl chloride copolymers (using e g comonomers
vinyl acetate, vinylidene chloride, ethylene or
propylene)

- Polyvinylidene chloride, polyvinyl acetate,
polyvinyl alcohol, polyvinyl acetals, polyvinyl
ethers.

Of these, the homopolymer and the copolymers vinyl
chloride/vinylidene chloride are commercially the most
important. The monomers used in these two groups are
also the most studied in terms of toxicity compared to
the other monomers belonging to the series.

Vinylchloride (VC) is a gas at room temperature, boiling
at -13,8°C (Torkelson, Rowe 1981). Its production is
estimated (IARC, 1979) to amount around 4000 million kg
in Western Europe and in USA around 2600 million kg. The
main use of VC has been for plastic production; a few
percent of the yearly production have been used as an
aerosol propellant and in drug and cosmetic products.

The acute toxicity of VC is low in terms of lethality.
10 mins inhalation to 240000-290000 ppm is the LC_{low} for
mice (Peoples, Leake, 1933) and the primary acute
physiological effect (Torkelsson, Rowe 1981) is CNS
depression. The explosion limit, 4-22 % in air, made VC
a low priority candidate for use in surgical anesthesia.
Besides CNS effects, liver injury could occur in experimental animals even after single high dose exposures
(Prodan et al 1975). This effect is, however, of particular interest to chronic exposures at low doses both in
animals (for review see IARC 1979) and man (Holmberg,
Molina, 1974). In acute experiments, cytochrome P450
inducers enhance the hepatoxic effect of VC (IARC
1979).

VC is rapidly absorbed by the lung and readily metabolized in rats (Hefner et al 1975). 68 % of the radioactivity is excreted in urine (Watanabe et al 1976) within
72 h of rats exposed to 10 ppm for 6 h and only a small
fraction is exhaled unchanged. The cytochrome P450
system is predominantly involved in the metabolism of VC
(IARC 1979). N-acetyl-S-(2-hydroxyethyl)cysteine and
thiodiglycolic acid are the main urinary metabolites
obtained in rats (review by Bolt et al 1980). Metabolic
elimination in the rat is determined by first-order
kinetics (Filser, Bolt 1979) below the saturation point
250 ppm in air. VC is metabolized by epoxidization
(Hefner et al, 1975; Bonse, Henschler 1976). Alkylations
of N^4-cytidine, N^6-adenosine (Green, Hathway 1978), and
N^7-guanosine (Osterman-Golkar et al 1977) in liver DNA
have been identified. Alkylation of macromolecules
occurs in mouse liver, kidney, spleen, and pancreas
(Bergman 1982) after i p administration of VC.

VC is mutagenic (reviews Bartsch et al 1976; Fabricant,
Legator 1981) in Salmonella (Rannug et al 1974), Escherichia (Greim et al 1975), Sacharomyces (Loprieno 1977),
Drosophila (Magnusson, Ramel 1978), and V79 Chinese
hamster cells (Drevon, Kuroki 1979). The VC metabolites
2-chloroethylene oxide, and 2-chloroacetaldehyde, and
2-chloroethanol show all mutagenic properties (Malavielle
et al 1975, Rannug et al 1976) towards Salmonella.
Chromosome breaks (Funes-Cravioto et al 1975), fragmen-
tations and rearrangements (Ducatman et al 1975) have
also been observed in peripheral lymphocytes of exposed
PVC workers.

Many experimental studies have been made on VC carcinogenicity with many mammalian species, administration
routes and doses (Viola et al 1971, Maltoni, Lefemine
1974 - recently updated by Maltoni et al 1981 - Keplinger
et al 1975, Holmberg et al 1976, Lee et al 1977, Suzuki
1978). A number of sites appear with tumors after VC
inhalation of animals (Maltoni et al 1981). VC is certainly a multipotential carcinogen. Tumor frequencies are
dose related (Maltoni et al 1981); however, hemangiosarcomas in different organs are more common at low doses
and liver hemangiosarcomas appear to be in minority
(Holmberg et al 1976). Significant excesses of tumors
appear for Zymbal gland carcinoma, liver angiosarcoma,
nephroblastoma, mammary gland carcinoma, and forestomach
papilloma (Maltoni et al 1981). 50 ppm induces a significant frequency of liver hemangiosarcomas and one animal
out of 120 have the same type of tumor at 10 ppm (Maltoni
et al 1981). Telangiectasis of the liver and hemocoelia
due to rupture of blood vessels was a common feature in
VC exposed mice, suggesting that blood vessels might be
the target for VC carcinogenicity (Holmberg et al 1976).
Hepatocytes may, however, metabolically activate VC to
the ultimate carcinogen, which acts on sinusoidal cells
(Ottenwälder, Bolt 1980). The suggestion that blood
vessels are target tissue is corroborated from human
data showing that skin capillary changes (Maricq et al
1976), Raynaud's phenomenon and acroosteolysis (Torkelson, Rowe 1981) occur among PVC workmen. VC induced
liver hemangiosarcomas become more frequent after
ethanol pretreatment of rats (Radike et al 1977) reflec-
ting a metabolic interaction of VC by ethanol (Hefner et
al 1975).

Early human toxicity data showed that the main recognized biological effect effect of VC was on the CNS
(reviews Holmberg, Molina 1974; Torkelson, Rowe 1981).
In the middle of the 1960's acroosteolysis was described
among PVC workmen, particularly autoclave cleaners.
Raynaud's phenomenon and scleroderma was common among
victims. Liver involvement in the long term toxicity
picture was found later (Gabor et al 1964, Kramer,
Mutchler 1972) and was the reason for lowering the
occupational standards in many countries.

Epidemiological studies and case reports from many
countries have overwhelmingly shown that VC vorkers are

under carcinogenic risk (review Infante 1981). Liver hemangiosarcomas, brain and lung tumours, and possibly tumors of the lymphatic and hematopoietic system and of skin (malignant melanomas; Heldaas et al 1983) among workers occupied with VC and PVC synthesis fit into the statement of VC being a multipotential carcinogen. Studies on processing industry workers indicate that tumors of the gastrointestinal system (Chiazze et al 1977, Chiazze, Ference 1981, Molina et al 1981) may also be associated with low VC exposures. In two studies on VC exposed workmen (Byrén et al 1976, Molina et al 1981), death risks in myocardial infarctions/cardiovascular diseases were elevated.

One experimental VC inhalation study (John et al,1981) with mice, rats and rabbits with and without simultaneous p o administration of ethanol did not reveal a teratogenic activity of VC at exposure levels between 50 and 2500 ppm. Epidemiological data, however, suggest a teratogenic effect (Infante et al 1976) through exposed male workers.

The occupational standards for VC have been considerably reduced during the last 20 years, the greatest step taken in 1974/1975 in USA and Sweden (Holm et al, 1982) as in many other Western countries. The occupational standard has been based mainly on technological feasibility criteria.

Vinylidene chloride (VDC; 1,1-dichloroethylene) is a colourless liquid at room temperature, boiling at $31,7^{\circ}$C (Torkelson, Rowe 1981). Around two million kg are produced annually (IARC, 1979) in USA. Japan produces 28 million kg.

Liquid VDC is irritating to skin in rabbits (Torkelson, Rowe 1981). Vapors cause nasal irritation in rats (Gage, 1970) at 200 ppm for 6 h and 20 d exposure. 500 ppm also caused hepatic cell degeneration. An almost continuous 7 d exposure (Short et al 1977) to 60 ppm was lethal for male mice but not for male rats. ALAT and ASAT were increased in mice indicating liver injury. Several studies illuminate the liver toxicity of VDC (for review see Haley 1975). Hepatic (focal necrosis, hemosiderin deposition) and renal (nuclear hypertrophy of tubular epithelium) lesions are induced in animals (Prendergast et al 1967; Jenkins, Andersen 1978). The fasted rat

(Jaeger et al, 1974) has an enhanced sensitivity and
ultrastructural studies (Reynolds, Moslen 1977) of rat
liver show swollen mitochondria, cytoplasmic vacuolization,
plasma membrane detachment and chromatin segregation.
Organic sulphur compounds, such as diethyldithiocarbamate,
thiram, methionin, and particularly disulfiram protect
against the toxic effects of VDC (Short et al 1977).

Rats inhaling VDC exhibit a dose dependent retention of
^{14}C at 72 h post exposure. Fasted rats showed a lower
retention than fed rats inhaling 200 ppm VDC for 6 h
(McKenna et al 1977). Fasted rats did not biotransform
VDC to the same extent as fed rats, which may be a part
of the explanation for the higher hepatotoxicity of VDC
to fasted animals. 70-80 % of inhaled VDC radioactivity
is eliminated via urine within 72 h. GSH plays probably
a major role in the detoxification also of VDC (Jaeger
et al, 1974; McKenna 1977). N-acetyl-S-(2- hydroxyethyl)
cysteine and thiodiglycolic acid are identified urinary
metabolites in rats (McKenna et al 1977). VDC metabolites
are covalently bound to liver macromolecules (McKenna et
al 1977) in fasted 200 ppm rats, suggesting the role of
reactive metabolites in producing hepatotoxic and carcinogenic effects. The metabolism of VDC has been postulated
(Leibman, Ortiz 1977) to be similar to that of trichloroethylene, i e VDC may be epoxidized and transformed to
monochloroacetic acid. VDC seems however not to cause
DNA alkylation or DNA repair synthesis (Reitz et al
1980) but does induce DNA synthesis in mouse kidney and
liver.

Mutagenicity studies (see also review by Fishbein 1976)
show that VDC is mutagenic to Salmonella (Bartsch et al
1975) after microsomal activation, Escherichia (Greim et
et al 1975), Sacharomyces (Bronzetti et al 1981), and
V79 Chinese hamster cells (Drevon, Kuroki 1979).

VDC was not teratogenic by inhalation of 20, 80 and
160 ppm in rats and rabbits or by peroral administration
of 200 ppm in water to rats (Murray et al 1979).

As can be expected from data obtained in mutagenicity
studies, VDC should be carcinogenic. This is also the
case for mice, rats, and hamsters (IARC 1979). In a
peroral study (Ponomarkov, Tomatis 1980) with rats where
pregnant rats were given a single dose of 100 mg/ kg bw

and the progeny weekly administrations of 50mg VDC/kg
bw for life, liver and meningeal tumors were more
frequently observed. Inhalation studies (Lee et al
1978; Hong et al 1981; Maltoni 1977) have been made with
mice, rats and hamsters. In mice, renal adenocarcinomas,
liver hemangiosarcoma, and mammary gland tumors appear
(Maltoni 1977; Lee et al 1978; Hong et al 1981). In one
study VDC induces lymph node and subcutaneous tissue
hemangiosarcomas in rats (Lee et al 1978) after inhalation.

Data on human populations are limited. No significant
mortality was found in a small (138 individuals) cohort
study (Ott et al 1975) on VDC production workers. No
conclusion on the possible human cancer risk due to VDC
exposure can be made on this study.

Vinyl acetate (VA) is a colourless liquid at room
temperature with a boiling point around 72^{0}C (NIOSH,1978).
The annual production in Western Europe is about 290
million kg and in USA 673 million kg. It is used for
the production of homo- and copolymers and for the
synthesis of polyvinyl alcohol and polyvinyl butyral.

The acute toxicity of VA is moderate (Documentation of
of TLVs, 1981) for rats (LC_{50}:3987 ppm), mice (1546
ppm), guinea pigs (6215 ppm), and rabbits (2511 ppm)
during 4 h inhalation (NIOSH, 1978). Irritation of the
eyes and respiratory system appears in rats at 2000 ppm
(Gage, 1970), but 100 ppm did not affect the animals.
Russian studies (see NIOSH, 1978) on inhalation and p o
administration of VA show effects on CNS of exposed
rabbits, mice and rats. When used as a control substance
in the test series (Maltoni, Lefemine, 1974) 2500 ppm
was used as maximum tolerated dose for rats. No carcinogenic effect was reported (Maltoni, Lefemine, 1975). A
Salmonella microsomal activation test (Bartsch et al
1976) was negative.

Inhaled VA is rapidly equilibrated in blood of rabbits
(NIOSH 1978). 70 % of inhaled VA is retained, but may be
rapidly hydrolyzed by nonspecific blood esterases. GSH
levels of rat livers were initially depressed by 70 %
and then elevated to 149 % (Boyland, Chasseaud 1970).

VA has in other metabolism studies (Chasseaud 1973) been
shown to be enzymatically conjugated with GSH. Human

exposure to VA vapor (Deese, Joyner 1969) gives irritation in the eye and nose. The irritation threshold
seems to be between 10 and 22 ppm. Exposure to liquid
may result in skin irritation (Deese, Joyner 1969). The
odor is clearly detectable by most individuals at levels
lower than 3 ppm but olfactory fatigue may occur at
about 20 ppm (NIOSH, 1978).

Vinylcyclohexene dioxide (VCHD) is a colourless liquid
at 20°C with a boiling point at 228°C. It has been used as a
copolymer for PVC.

It is of low acute toxicity for rabbits by peroral administration in the only study done (Documentation of TLVs
1981). VCHD can penetrate rabbit skin and is more toxic by
this route (0,62ml/kg bw). It also causes skin injury.
Besides being a skin irritant, VCHD is also irritating to
eyes and the respiratory system. The monomer causes skin
tumors by topical application in mice. Although limited
studies are available, the toxicological information on
this diepoxide warrants low exposure limits. ACGIH has a
recommended value of 10 ppm and lists VCHD as a suspected
carcinogen.

References

Bartsch H, Malaveille C, Montesano R (1976). The predictive value of tissue-mediated mutagenicity assays to
assess carcinogenic risk of chemicals. In: Montesano R,
Bartsch H, Tomatis L (Eds): Screening Tests in Chemical
Carcinogenesis. Lyon: IARC Sci. Publ. No 12, pp 467-491

Bartsch H, Malaveille C, Montesano R, Tomatis L (1975).
Tissue-mediated mutagenicity of vinylidene chloride and
2-chlorobutadiene in Salmonella typhimurium. Nature
255:641-643

Bergman K (1982). Reactions of vinyl chloride with RNA
and DNA of various mouse tissues in vivo. Arch Toxicol
49:117-129

Bolt HM, Filser JG, Laib RJ, Ottenwälder H (1980).
Binding kinetics of vinyl chloride and vinyl bromide at
very low doses. Arch Toxicol Suppl 3:129-142

Bonse G, Henschler D (1976). Chemical reactivity, biotransformation, and toxicity of polychlorinated aliphatic compounds. CRC Crit Rev Toxicol 4:395-409

Boyland E, Chasseaud LF (1970). The effect of some carbonyl compounds on rat liver glutathione levels. Biochem Pharmacol 19:1526-1528

Bronzetti G, Bauer C, Corsi C, Leporini C, Nieri R, del Carratore R (1981). Genetic activity of vinylidene chloride in yeast. Mut Res 89:179-185

Byrén D, Engholm G, Englund A, Westerholm P (1976). Mortality and cancer morbidity in a group of Swedish VCM and PVC production workers. Env Health Persp 17: 167-170

Chasseaud LF (1973). Nature and distribution of enzymes catalyzing the conjugation of glutathione with foreign compounds. Drug Metabol Rev 2:185-219

Chiazze Jr L, Ference LD (1981). Mortality among PVC-fabricating employees. Env Health Persp 41:137-143

Chiazze Jr L, Nichols WE, Wong O (1977). Mortality among employees of PVC fabricators. JOM 19:623-628

Deese DE, Joyner RE (1969). Vinyl acetate: A study of chronic human exposure. Am Ind Hyg Assoc J 30:449-457

Documentation of TLVs (1981). Supplemental documentation Cincinnati: American Conference of Governmental Industrial Hygienists, Inc

Drevon C, Kuroki T (1979). Mutagenicity of vinyl chloride, vinylidene chloride and chloroprene in V79 Chinese hamster cells. Mut Res 67:173-182

Ducatman A, Hirschhorn K, Selikoff IJ (1975). Vinyl chloride exposure and human chromosome aberrations. Mut Res 31:163-168

Fabricant JD, Legator MS (1981). Mutagenicity studies of vinyl chloride. Env Health Persp 41:189-193

Filser JG, Bolt HM (1979). Pharmacokinetics of halogenated ethylenes in rats. Arch Toxicol 42:123-136

Fishbein L (1976). Industrial mutagens and potential mutagens. I. Halogenated aliphatic derivatives. Mut Res 32:267-308

Funes-Cravioto F, Lambert B, Lindsten J, Ehrenberg L, Natarajan AT, Osterman-Golkar S (1975). Chromosome aberrations in workers exposed to vinyl chloride. Lancet i: 459

Gabor S, Radu M, Preda N, Abrudean S, Ivanof L, Anea Z, Valaczky C (1964). Comments on the biochemical alterations in workers of the vinyl chloride synthesis and polymerization industry. Igiena (Bucaresti) 13:409-418

Gage JC (1970). The subacute inhalation toxicity of 109 industrial chemicals. Br J Ind Med 27:1-18

Green T, Hathway DE (1978). Interactions of vinyl chloride with rat-liver DNA in vivo. Chem-Biol Interact 22:211-224

Greim H, Bonse G, Radwan Z, Reichert D, Henschler D (1975). Mutagenicity in vitro and potential carcinogenicity of chlorinated ethylenes as a function of metabolic oxirane formation. Biochem Pharmacol 24:2013-2017

Haley TJ (1975). Vinylidene chloride: A review of the literature. Clin Toxicol 8:633-643

Hefner Jr RE, Watanabe PG, Gehring PJ (1975). Preliminary studies of the fate of inhaled vinyl chloride monomer in rats. Ann NY Acad Sci 246:135-148

Heldaas SS, Langård S, Anderssen A (1983). Incidence of cancer among vinyl chloride and polyvinyl chloride workers. Brit J Ind Med, in press.

Holm L, Westlin A, Holmberg B (1982). Technical control measures in the prevention of occupational cancer: An example from the PVC industry. In: Prevention of Occupational Cancer -International symposium. Geneva: Occupational Safety and Health Series No 46. International Labour Office.

Holmberg B, Molina G (1974). The industrial toxicology of vinyl chloride. A review. Work-environm-hlth 11: 138-144

Holmberg B, Kronevi T, Winell M (1976). The pathology of vinyl chloride exposed mice. Acta Vet Scand 17:328-342

Hong CB, Winston JM, Thornburg LP, Lee CC (1981). Follow-up study on the carcinogenicity of vinyl chloride and vinylidene chloride in rats and mice. Tumor incidence and mortality subsequent to exposure. J Toxicol Environ Health 7:909-924

IARC (1979). Monographs on the Evaluation of the Carcinogenic Risk of Chemicals to Humans. Vol 19. Some monomers, Plastics and Synthetic Elastomers and Acrolein. Lyon: International Agency for Research on Cancer.

Infante PF (1981). Observations of the site-specific carcinogenicity of vinyl chloride to humans. Env Health Persp 41:89-94

Infante PF, Wagoner JK, Waxweiler RJ (1976). Carcinogenic, mutagenic and teratogenic risks associated with vinyl chloride. Mut Res 41:131-142

Jaeger RJ, Conolly RB, Murphy SD (1974). Effect of 18 HR fast and glutathione depletion on 1,1-dichloroethylene-induced hepatotoxicity and lethality in rats. Exp Mol Pathol 20:187-198

Jenkins LJ, Andersen ME (1978). 1,1-dichloroethylene nephrotoxicity in the rat. Toxicol Appl Pharmacol 46:131-141

John JA, Smith FA, Schwetz BA (1981). Vinyl chloride: Inhalation teratology study in mice, rats and rabbits. Env Health Persp 41:171-177

Keplinger ML, Goode JW, Gordon DE, Calandra JC (1975). Interim results of exposure of rats, hamsters, and mice to vinyl chloride. Ann NY Acad Sci 246:219-224

Kramer CG, Mutchler JE (1972). The correlation of clinical and environmental measurements for workers exposed to vinyl chloride. Am Ind Hyg Ass J 33:19-30

Lee CC, Bhandari JC, Winston JM, House WB, Peters PJ, Dixon RL, Woods JB (1977). Inhalation toxicity of vinyl chloride and vinylidene chloride. Env Health Persp 21:25-32

Lee CC, Bhandari JC, Winston JM, House WB, Dixon RL, Woods JS (1978). Carcinogenicity of vinyl chloride and vinylidene chloride. J Toxicol Environ Health 4:15-30

Leibman KC, Ortiz E (1977). Metabolism of halogenated ethylenes. Env Health Persp 21:91-97

Loprieno N (1977). The use of yeast cells in the mutagenic analysis of chemical carcinogens. Colloq Int CNRS 256:315-331

Magnusson J, Ramel C (1978). Mutagenic effects of vinyl chloride on Drosophila melanogaster with and without pretreatment with sodium phenobarbiturate. Mut Res 57:307-312

Malaveille C, Bartsch H, Barbin A, Camus AM, Montesano R, Croisy A, Jacquignon P (1975). Mutagenicity of vinyl chloride, chloroethylene oxide, chloroacetaldehyde and chloroethanol. Biochem Biophys Res Commun 63:363-370

Maltoni C (1977). Recent findings on the carcinogenicity of chlorinated olefins. Env Health Persp 21:1-5

Maltoni C, Lefemine G, (1974). Carcinogenicity bioassays of vinyl chloride - I. Research plan and early results. Env Res 7:387-405

Maltoni C, Lefemine G (1975). Carcinogenicity bioassays of vinyl chloride: Current results. Ann NY Acad Sci 246:195-218

Maltoni C, Lefemine G, Ciliberti A, Cotti G, Carretti D (1981). Carcinogenicity bioassays of vinyl chloride monomer: a model of risk assessment on an experimental basis. Env Health Persp 41:3-29

Maricq HR, Johnson MN, Whetstone CL, Le Roy EC (1976). Capillary abnormalities in polyvinyl chloride production workers. J Am Med Assoc 236:1368-1371

McKenna MJ, Watanabe PG, Gehring PJ (1977). Pharma-cokinetics of vinylidene chloride in the rat. Env Health Persp 21:99-105

Miles DC, Briston JH (1979). Polymer Technology. New York: Chemical Publ Corp

Molina G, Holmberg B, Elofsson S, Holmlund L, Maasing R, Westerholm P (1981). Mortality and cancer rates among workers in the Swedish PVC processing industry. Env Health Persp 41:145-151

Murray FJ, Nitschke KD, Rampy LW, Schwetz BA (1979). Embryotoxicity and fetotoxicity of inhaled or ingested vinylidene chloride in rats and rabbits. Toxicol Appl Pharmacol 49:189-202

NIOSH criteria for a recommended standard (1978). Occupational exposure to vinyl acetate. Washington US Dept of Health, Education, and Welfare, DHEW (NIOSH) Publ. No 78-205

Osterman-Golkar S, Hultmark D, Segerbäck D, Calleman CJ, Göthe R, Ehrenberg L, Wachtmeister CA (1977). Alkylation of DNA and proteins in mice exposed to vinyl chloride. Biochem Biophys Res Commun 76:259-266

Ott MG, Fishbeck WA, Townsend JC, Schneider EJ (1975). A health study of employees exposed to vinylidene chloride. JOM 18:735-738

Ottenwälder H, Bolt HM (1980). Metabolic activation of vinyl chloride and vinyl bromide by isolated hepatocytes and hepatic sinusoidal cells. J Env Pathol Toxicol 4:411-417

Peoples AS, Leake CD (1933). The anesthetic action of vinyl chloride. J Pharmacol Exp Ther 48:284

Ponomarkov V, Tomatis L (1980). Long-term testing of vinylidene chloride and chloroprene for carcinogenicity in rats. Oncology 37:136-141

Prendergast JA, Jones RA, Jenkins Jr LJ, Siegel J (1967). Effects on experimental animals of long-term inhalation of trichloroethylene, carbon tetrachloride,

1,1,1-trichloroethane, dichlorodifluoromethane, and 1,1-dichloroethylene. Toxicol Appl Pharmacol 10:270-289

Prodan L, Suciu I, Pislaru V, Ilea E, Pascu L (1975). Experimental acute toxicity of vinyl chloride (monochloroethene). Ann NY Acad Sci 246:154-158

Radike MJ, Stemmer KL, Brown PG, Larson E, Bingham E (1977). Effect of ethanol and vinyl chloride on the induction of liver tumors: preliminary report. Env Health Persp 21:153-155

Rannug U, Göthe R, Wachtmeister CA (1976). The mutagenicity of chloroethylene oxide, chloroacetaldehyde, 2-chloroethanol and chloroacetic acid, conceivable metabolites of vinyl chloride. Chem-Biol Interact 12:251-263

Rannug U, Johansson A, Ramel C, Wachtmeister CA (1974). The mutagenicity of vinyl chloride after metabolic activation. Ambio 3:194-197

Reitz RH, Watanabe PG, McKenna MJ, Quast JF, Gehring PJ (1980). Effects of vinylidene chloride on DNA synthesis and DNA repair in the rat and mouse: A comparative study with dimethylnitrosamine. Toxicol Appl Pharmacol 52:357-370

Reynolds ES, Moslen MT (1977). Damage to hepatic cellular membranes by chlorinated olefins with emphasis on synergism and antagonism. Env Health Persp 21:137-147

Short RD, Winston JM, Minor JL, Seifter J, Lee CC (1977). Effect of various treatments on toxicity of inhaled vinylidene chloride. Env Health Persp 21:125-129

Suzuki Y (1978). Pulmonary tumors induced in mice by vinyl chloride monomer. Environ Res 16:285-301

Torkelson TR, Rowe VK (1981). Halogenated aliphatic hydrocarbons containing chlorine bromine and iodine. In: Patty's Industrial Hygiene and Toxikology 3rd ed, vol 2B: Toxicology. Clayton GD, Clayton FE (eds). New York: John Wiley & Sons, 3433-3601.

Watanabe PG, McGowan GR, Madrid EO, Gehring PJ (1976). Fate of (^{14}C) vinyl chloride following inhalation exposure in rats. Toxicol Appl Pharmacol 37:49-59

Viola PL, Bigotti A, Caputo A (1971). Oncogenic response of rat skin, lungs and bones to vinyl chloride. Cancer Res 31: 516-522

Industrial Hazards of Plastics and Synthetic Elastomers, pages 113–135
© 1984 Alan R. Liss, Inc., 150 Fifth Ave., New York, NY 10011

TOXICITY OF THE COMPONENTS OF POLY(VINYLCHLORIDE) POLYMERS
ADDITIVES

Lawrence Fishbein

Department of Health and Human Services, Food
and Drug Administration, National Center for
Toxicological Research, Jefferson, AR (USA)

INTRODUCTION

Many of the general aspects of the occupational and
environmental health hazards in the plastics industry have
been reviewed (Eckardt, 1973, 1976; IARC, 1979; Hemminki et
al, 1979; Karstadt, 1976; Vainio et al, 1980). There is
increasing recognition that we are now at a crucial stage in
regard to the necessity of having more definitive studies of
the effects of plastics and particularly plastics components
on human health. It is generally acknowledged that human
exposure to massive doses of vinyl chloride (VCM) whether
due to uncontrolled venting of polyvinyl chloride (PVC)
polymerization vessels in the factory or the voluntary
venting in one's home of an aerosolized VCM product is
essentially over (Choan, 1975; Jones, 1981; Karstadt, 1976).
In general, the principal hazards of plastics are associated
with their monomers and with the wide variety of additives
that are used in them (Eckardt, 1976; Kardstadt, 1976;
Withey, 1977).

The complex plastic system called PVC is the best
example where a broad spectrum of additives of an even
larger array of chemical classes and individual components
and mixtures are employed. It is readily acknowledged by
many that there is a paucity of toxicological data both in
laboratory animals and humans for the vast majority of these
additives. It is particularly relevant to examine PVC
because of its huge production volume as well as those who
are at potential risk via exposure to PVC in its production
as well as in contact with its myriad products. For example,

"total world employment in the VCM and PVC producing industries is likely to be well over 70,000 workers.   Those
employed in industries which use PVC as a basic element are
likely to total in the millions.   Those who come in contact
with PVC every day in some form or other probably make up
about at least one-third of the human race" (Levinson,
1974).

In the case of PVC, end products result from three
distinct phases of manufacturing: synthesis of VCM, polymerization of vinyl chloride and processing of PVC resins
into finished products.   PVC resins are produced by four
basic processes: suspension, emulsion, bulk and solution
polymerization.   Suspension polymerization is the major
process used for the manufacture of PVC resins and is used
for about 80-85% of U.S. production (Cohan, 1975; IARC,
1979; Seymour, 1975; Wheeler, 1981).

The production of PVC in the U.S. in 1982 is believed to
range from 5.2 to 5.5 billion pounds with the following use
patterns of major fabricated forms: extrusions nearly two
thirds in pipe, 60%; calendered sheet and film, 10%; coatings, 5% and moldings, 5% (Anon, 1982a).

According to OSHA, the major categories of PVC additives
used in both production and processing of PVC are plasticizers, stabilizers, flame retardants, colorants and fibrous
reinforcements (U.S. Dept. of Labor, 1974).

The major objective of this overview is to highlight the
toxicological aspects of a number of the major additives
used in PVC production and processing, representative of a
spectrum of use categories, principally plastiziers, stabilizers, initiators and flame retardants.   This admittedly
represents but a small fraction of the chemical substances
employed within the 14 classes of additives used in synthetic polymers noted earlier in this issue (Fishbein, 1982).

TOXICITY OF PVC ADDITIVES

Plasticizers

Plasticizers (e.g., phthalates, adipates, trimellitates,
azelates, polyesters, glutarates and epoxidized esters) are
the most important category of additives employed with PVC
and vinyl chloride resins from a viewpoint of both the scope

of usage as well as the toxicity of several important members of this category (e.g., phthalates, adipates and trimellitates) (IARC, 1979, 1982). Addition to PVC and vinyl
chloride copolymer resins at typical elvels of 50%, is the
dominant market for all plasticizers, representing about 85%
of total U.S. use in 1979. For phthalate plasticizers this
represented about 90% of the total plasticizer use (IARC,
1982). Approximately 400,000 metric tons of phthalates were
produced in the U.S. in 1982 which represents about 64% of
the total production of the major types of plasticizers
(Anon, 1982a).

Di(2-ethylhexyl)phthalate (DEHP) (dioctyl phthalate)
(DOP) is by far the most important plasticizer employed.
The estimated 303 million pounds (138 million kg) of DEHP
used in the U.S. in 1979 represented nearly 26% of the total
U.S. phthalate plasticizer consumption (IARC, 1982). Because
of its properties (e.g., low extraction by oil and water,
good stability to heat and light and fast production rates,
DEHP is the plasticizer used preferentially in many flexible
PVC products (e.g., calendering film, sheeting, and coated
fabrics). It is also frequently used in blends with other
plasticizers such as adipates when special properties are
required (IARC, 1982).

There is a burgeoning literature concerning the environmental fate, biological and toxicological effects of DEHP as
well as its major metabolite, monoethyl hexyl phthalate
(MEHP) (IARC, 1982; Lee and Falk, 1973; Thomas and Northrup
1982; Thomas et al, 1978).

In a recent evaluation of the carcinogenicity of DEHP,
IARC (1982) concluded that there is sufficient evidence for
the carcinogenicity in mice and rats. DEHP significantly
increased the incidence of benign and malignant liver-cell
tumors in B6C3F1 mice and and Fischer 344 rats (National
Toxicology Program, 1982).

Several studies in rats have demonstrated that DEHP can
cause testicular damage in rats (Grey et al, 1977; National
Toxicology Program, 1982; Shaffer et al, 1945) with the
degree of damage being directly proportional to both the
size of the dose and exposure. For example, administration
of 20,000 mg/kg in the diet produced seminiferous tubular
degeneration and testicular atrophy within 7 days (Oishi and
Hiraga, 180a); 12,500 mg/kg produced similar effects within

90 days while 6000 mg/kg produced the same effects by the end of 2 years of exposure (National Toxicology Program, 1982).

DEHP as well as its principal metabolite mono(2-ethylhexyl)phthalate (MEHP) are teratogenic and embryolethal to mice (ICR, ICR-JCL and CBA strains) and Sprague-Dawley rats (Nakamura et al, 1979; Oishi and Hiraga, 1980; Peters and Cook, 1973; Singh et al, 1975; Yagi et al, 1980).

DEHP was not mutagenic to Salmonella typhimurium strains TA 1535, TA 1537, TA 1538, TA 98 or TA 100 with or without metabolic activation (Simmon et al, 1977) and did not induce chromosomal aberrations in Chinese hamster cells (Abe and Sasaki, 1977; Ishidate and Odashima, 1977), human leucocytes (Stenchever et al, 1976; Tsuchiya and Hattori, 1976) or human fetal lung cells (Stenchever et al, 1976). It induced a non-dose dependent but significant (greater than two-fold background) increase in sister chromatid exchanges in Chinese hamster cells (Abe and Sasaki, 1977). Although DEHP caused dominant lethal mutations in mice after systemic administration (Dillingham and Autian, 1973), it, as well as MEHP, showed no dominant lethal activity on oral administration to ICR mice (Hamano et al, 1979).

The metabolism of DEHP in different biological systems has been reviewed by Thomas (1978). The metabolites which are primarily excreted in the urine, differ from species to species (Albro et al, 1981). The metabolism of DEHP following its oral administration to rats is shown in Figure 1 (Albro et al, 1973). Albro et al (1973) isolated four major urinary metabolites of DEHP that were derived during $\omega-$ and $\omega-1$ oxidation of MEHP. The alcohol and the ketone metabolites did not undergo conjugation while free phthalic acid accounted for approximately 3% of the urinary metabolites (Figure 1). The metabolism and biodistribution of MEHP in the rat has been further studied by Chu et al (1978) and MEHP was shown to undergo $\omega-$ and $\omega-1$ oxidation to yield the same metabolites as DEHP. In plasma there is an equilibrium between MEHP absorbed to albumin and in free solution whereas DEHP is bound to liproproteins. MEHP orally administered to rats results in mild hepatic changes with no bioaccumulation of the monoester (Lake et al, 1977). MEHP is reportedly responsible for the increased lipotropic activity of the liver while 2-ethyl hexanol, another major metabolite of DEHP, is responsible for hepatic peroxisomal changes

Figure 1.   Metabolism of DEHP (Albro et al., 1973)

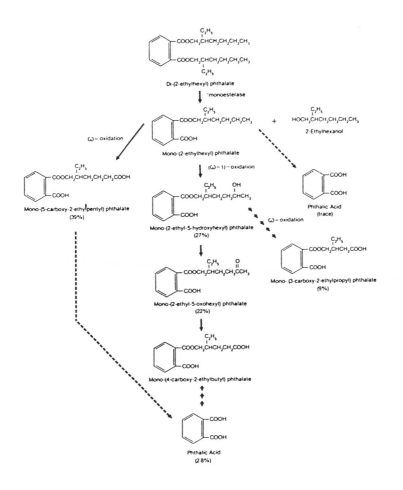

(Morton and Rubin, 1980). MEHP has shown no teratogenic effects when administered to rats or rabbits (Thomas et al, 1979; Villeneuve et al, 1978).

MEHP was not mutagenic to S. typhimurium strains TA 1535, TA 100, TA 1537, TA 1538 or TA 98 with or without metabolic activation (Ruddick et al, 1981). When DEHP, MEHP, 2-ethyl hexanol and phthalic acid were recently tested for clastogenic activity in cultured Chinese hamster ovary (CHO) cells, only MEHP was found to cause chromosome damage at relatively high concentrations and only within a narrow dose range including cytotoxic concentrations (0.8 to 1.75 mM). MEHP was without effect in the SCE and HGPRT mutation test in CHO cells (Phillips et al, 1982). Although MEHP was without effect in the sister chromatid (SCE) and point-mutation tests (HGPRT), the clastogenic potential of this compound might explain, according to Phillips et al (1982), the reported dominant lethal effect of DEHP (Singh et al, 1974) and possibly its carcinogenic effect (National Toxicology Program, 1982).

The lack of mutagenic response of DEHP in several test systems would for the present suggest limited value for such tests in predicting the carcinogenic potentials of other phthalates in rodent bioassays. One plausible theory of DEHP carcinogenicity suggests that the DEHP-induced proliferation of hepatic peroxisomes (subcellular organelles) results in the increased presence of hydrogen peroxide or hydroxyl radicals which could subsequently injure both genetic and non-genetic cellular materials (National Toxicology Program, 1982b). It should be noted that peroxisomes are not present in 9000 xg liver supernatant fractions (S-9) or other mammalian activating systems commonly used in mutagenicity assays.

In 1980, the number of U.S. workers potentially exposed to DEHP was estimated to be about 625,000 (NIOSH, 1980). Among the major industries in which the exposure was estimated to be the greatest were in plastics and rubber products manufacture. Occupational exposure to DEHP probably occurs primarily during its manufacture, its use as a plasticizer, its use in dielectric fluids for electrical capacitors and in the further processing or use of plasticized products containing it. EPA has estimated that there are approximately 300 to 400 production workers and up to 10,000 plastics goods workers involved with the use of six phthalate esters

(Dickson, 1980). However, it should be noted that the reported widespread occurrence of DEHP in ambient air, in drinking river and ocean waters, in industrial effluents and in blood stored in plasticized bags, indicates potential environmental exposure and exposure to the general human population (IARC, 1982).

Butyl benzyl phthalate (BBP) is used as a plasticizer generally at levels of 50-75 parts per 100 parts of resin. Over half of the BBP produced goes into PVC from which vinyl floor tiles are made. Other polymers that can be plasticized with BBP include: acrylic resins, ethyl cellulose, polyvinyl formal and polyvinyl butyral. It is currently produced commercially by only one U.S. company, the production of which was estimated to have been approximately 150 million pounds (IARC, 1982b). It has been estimated that over 66,000 workers in the U.S. are potentially exposed to BBP (National Institute for Occupational Safety and Health, 1980b).

BBP was not carcinogenic for B6C3F1 mice of either sex when administered in the diet at levels of 6,000 or 12,000 ppm for 28 to 103 weeks. Administration of the above levels of BBP to Fischer 344 rats 28 to 103 weeks resulted in an increased incidence of mononuclear cell leukemia in female rats (National Toxicology Program, 1981a). IARC (1982b) recently concluded that the available carcinogenicity studies reported with BBP (National Toxicology Program, 1981; Thiess et al, 1977) were inadequate to evaluate its carcinogenicity to mice and rats.

BBP has not been found mutagenic for E. coli, B. subtilis (Omori, 1976) or S. typhimurium strains TA 1535, TA 1538, TA 98 or TA 100 nor to Saccharomyces cerevisiae D4 in the presence or absence of metabolic activation (Anon, 1976) or in the mouse lymphoma assay (Anon, 1977). In a chick embryo system test for reproductive effects, BBP was shown to be lethal by Haberman et al (1968), but not by Bower et al (1970). Abnormal development was not observed (Bower et al, 1970; Haberman et al, 1968) and neonatal death rates at 21 days of age were not affected (Haberman et al, 1968).

Di(2-ethylhexyl)adipate (DEHA) (dioctyl adipate) (DOA), is used frequently in blends with DEHP in products intended for low-temperature use. It is used primarily with polyvinyl resins in films, sheeting, extrusions and plastisols.

The single most important market for DEHA is in stretchable PVC film used for wrapping meat (IARC, 1982c) (such film may contain 20-30% of the plasticizer) (Boettner and Ball, 1980). DEHA is also used in plasticizing polystyrene, polyvinyl butyral, and cellulose acetate butyrate (Monsanto Co., 1973). Twelve U.S. companies produced nearly 46 million pounds of DEHA in 1979.

It has been estimated that the number of U.S. workers potentially exposed to DEHA in 1980 was approximately 11,000 (National Institute for Occupational Safety and Health, 1980). Occupational exposure to DEHA occurs during its production, its use as a plasticizer, as well as its use as a lubricant and functional fluid. However, its presence as a plasticizer in polyvinyl film used for wrapping food may result in exposure to segments of the general public (IARC, 1982c).

DEHA, when administered in the diet for 103 weeks to male and female B6C3F1 mice at levels of 12,000 or 25,000 mg/kg, induced significant increases in the incidence of liver-cell tumors in both sexes. However, no increase in tumor incidence was found when DEHA was administered in the diet to male and female F344 rats at levels of 12,000 or 25,000 mg/kg for 103 weeks (National Toxicology Program, 1981b). IARC (1982c) concluded from the above studies that there is limited evidence that DEHA is carcinogenic in mice.

DEHA induced dominant lethal mutations and reduced fertility in mice (Singh et al, 1975b). DEHA did not induce reverse mutations in S. typhimurium strains TA 1535, TA 1537, TA 1538, TA 98 or TA 100 with or without metabolic activation (Singh et al, 1975). When Sprague-Dawley rats were given injections of 0.9, 4.5 or 9.2 g/kg of DEHA on days 5, 10 and 15 of pregnancy, no embryolethality occurred but reduced fetal weight was found with the two higher doses (Singh et al, 1973). DEHA caused significant growth inhibition in cultures of human diploid cells (Jones et al, 1975).

In elimination, distribution and metabolism studies in rats, orally administered DEHA was found to be rapidly hydrolyzed to form adipic acid, without any accumulation of mono-(2-ethylhexyl)adipate (MEHA) (Takahashi et al, 1981).

Trimellitic anhydride (1,2,4-benzenetricarboxylic acid, 1,2-anhydride) (TMA) is used as a curing agent for epoxy and other resins and in vinyl plasticizers. The largest market for TMA is as an intermediate in the production of trimellitate plasticizers, the estimated consumption of which in 1982 was 13,000 metric tons (Anon, 1982b).

The triisoctyl and triisodecyl esters of trimellitic acid are used as vinyl plasticizers where permanency is required, as in PVC wire insulation, upholstery refrigerator gasketing and thin fabric coatings. These specialty plasticizers are used primarily in PVC resins where performance over a wide temperature range is necessary (EPA, 1980). NIOSH (1978) estimated that 20,000 workers in the U.S. are potentially exposed to TMA in its various applications. Sensitization and a spectrum of respiratory symptoms including pulmonary edema, have been found in workers exposed to TMA by inhalation (NCI, 1978; Rice et al, 1977). Bladder stone induction resulted in weanling Fischer 344 rats when treated with terephthalic aid or dimethyl terephthalate (DMT) in their diet (Chin et al, 1981; Heck, 1981). Dimethyl terephthalate was reported not to be carcinogenic when administered to mice and rats in their diet (NCI, 1979).

Heat Stabilizers

The largest use of heat stabilizers is in PVC, e.g., barium-cadmium stabilizers for flexible PVC and organotin stabilizers for rigid PVC where the stabilizer content can be up to 10%. The consumption of the major heat stabilizers in the US in 1982 (in metric tons) was: barium-cadmium, 14,000; tin, 9,600; lead, 9,400; calcium-zinc, 1,970; and antimony, 410 (Anon, 1982). The major organotins that are employed are di-N-octyltinmercaptide and dilauryl and dibutyl tin maleates. Specific compounds which are the most widely used organotins for stabilizing PVC and chlorinated PVC (CPVC) potable water pipe include: methyl-, butyl- and octyltin esters particularly of lauric, maleic and thioglycolic acids. Organotin stabilizers are used in the range of 0.3 to 1.5 parts per 100 parts resin for PVC pipe and fittings and in the range of 1.5 to 3.5 parts resin for CPVC pipe and fittings (Boettner et al, 1982).

Boettner et al (1982) reported that alkyl tin species were extractable from PVC and CPVC pipes by water. The amount of dimethyl tin (as the dichloride) leached from PVC into pH 5 extractant water at 37°C was 35 ppb for day one, and decreased from approximately 3.0 to 0.25 ppb per 24 hours in a biphasic manner from days 2 through 22. The amount of dibutyltin (as the dichloride) leached from CPVC into pH 5 extractant water at 72°C was 2.6 ppb for day one and decreased from 1.0 to 0.03 ppb per 24 hours, in a biphasic manner from days 2 through 21. The exact organotin compounds from which these alkyltin species arose were indefinite, although there was evidence that they existed in ionic form in the extractant water. There was no evidence of the sulfur-containing part of the stabilizer (from the original dialkyltin-bis-isooctylthioglycolate compounds tested) in the extractant water. It should also be noted that in this study volatile organic solvents such as methyl ethyl ketone, tetrahydrofuran, and cyclohexaneone used in the sealing cements applied to PVC and CPVC pipe joints continued to leach into water supplies for more than 14 days, in quantities which ranged from 10 ppb to 10 ppb during the 15 days of sampling. Sufficient toxicological data are not currently available to assess the health significance to man of very low levels of organotin and cement solvent chemicals found in this study. However, there are considerable toxicological data on the toxicity of the dialkyltins (NIOSH, 1976; Kimbrough, 1976; Piver, 1973; Barnes and Stoner, 1959; WHO, 1980; Zuckerman et al, 1978).

Acute and chronic toxicity studies in animals with dialkyltin showed them to be less toxic than the trialkyl- and triaryl tin derivatives. A clear distinction can be drawn between the toxic effects of the trialkyl tin compounds, which predominantly affect the central nervous system and of the dialkyl tin compounds which produce bile duct lesions in vivo (Barnes and Stoner, 1959).

By contrast with the monoalkyl derivatives, dialkyl tin compounds from ethyl to butyl are considerably more toxic orally but show decreasing toxicity as the alkyl chain lengthens to eight carbons. Parenteral toxicity was found to be considerably greater than oral in acute studies with dialkyl tins. It should also be noted that diorganotins are potent immunosuppressants in vivo (Hill et al, 1980; Sernin et al, 1977; Sernin and Penninks, 1979). The degree of dialkyl-induced thymus atrophy was strongly related to the chain length of the alkyl group, the greatest effect occurring at four to eight carbon atoms. One of the more important actions of di-n-octyl tin chloride is to deplete lymphoid function (Sernin et al, 1977; Sernen and Penninks, 1979).

In recent in vitro and genotoxicity studies with dibutyl tin dichloride (DBTC) it was found that the compound greatly decreased the antibody production by lymphocytes at noncyto- toxic doses which may explain the immunosuppressive effect of the compound in vivo. DBTC also induced mutation in Chinese hamster ovary cells (Li et al, 1982).

Single oral doses of dioctyl tin-S,S'-bis(2-ethylhexyl- mercaptoacetate) and dioctyl tin-S,S'-bis(butyl mercaptoace- tate) administered to mice produced a slight fatty degenera- tion of the renal cortical tubular epithelium in mice (Peli- kin and Cerny, 1970). Feeding rats dioctyl tin-S,S'-bis- (isooctylmercaptoacetate) at a dietary level of 200 mg/kg for 12 months produced an increase in kidney weight in fe- male rats only (Nikonorow et al, 1973). Mazeav and Korolen (1969) also reported dystrophic changes in the kidneys of rats treated with dibutyl tin dichloride at 0.1 and 1.0 mg/kg body weight. Nikonorow et al (1973) reported toxic effects of dioctyl tin-S,S'-bis(isooctyl mercaptoacetate) on the embryo and fetus at daily oral doses of 20 and 40 mg/kg administered for about 3 months. A distinct increase in the incidence of fetal deaths was observed but no teratogenic effects were noted.

In vitro studies of disubstituted organotins from di- methyl- to dioctyltin have shown that most of them inhibit mitochondrial respiration by preventing the oxidation of keto acids, presumably via the inhibition of alpha-ketooxi- dase activity, leading to the accumulation of pyruvate (Aldridge, 1976; Piver, 1973).

In general, organotin compounds are more readily ab- sorbed from the gut than inorganic tin compounds. It should be noted in this regard that great variations exist between different compounds and different species. As a rule, tin compounds with a short alkyl chain are more readily absorbed from the intestinal tract (WHO, 1980).

Many organotin compounds are transformed to some extent in the tissues. The dealkylation and dearylation of tetra-, tri- and disubstituted organo tin compounds seem to occur in the liver. However, the dealkylation of diethyl tin compounds appears to occur both in the gut and in tissues of other organs (WHO, 1980).

Lead stabilizers account for approximately 60% of the total stabilizer consumption. Commonly used lead stabilizers include dibasic lead phthalate, lead chlorosilicate, basic lead carbonates, lead maleate and lead stearate. They are generally used at 3-8 phr in PVC wire and cable insulation and in flooring (Deanin, 1975; Fischbein et al, 1982). Lead poisoning is a common problem in the United States (Baker et al, 1979; Lilis et al, 1977). While improved engineering controls have rendered the occurrence of acute lead poisoning with severe and life-threatening symptoms relatively rare, it should also be stressed that lead exposures associated with more subtle health effects continue to concern the medical community (Fischbein et al, 1982; Needleman, 1980).

Blowing Agents

Foaming or blowing agents are widely employed in the preparation of foamed plastics which have excellent resilience and thermal insulating properties. Both physical and chemical agents are employed. A number of thermally labile compounds have been employed as blowing agents the most prominantly employed are azobisisobutyronitrile, azobisformamide, oxybisbenzenesulfonyl hydrazides. The largest use for these agents is in the preparation of foamed PVC plastics. Although thermally labile blowing agents are designed to decompose completely in the plastics processing, there is no assurance that they do so (Withey, 1977). Azobis isobutyronitrile has been found to produce lesions in the liver and kidney (Motoc et al, 1971).

Azodicarbonamide is used as a blowing agent for calendered and plastisol vinyl foams, cellular polyolefins, polystyrenes and other unicellular products requiring a uniform cell structure. It is also used as a thermoplastic blowing agent for styrene-butadiene rubbers, nitrile-butadiene and isobutylene-isoprene. The consumption of azodicarbonamide and modified azodicarbonamide in the U.S. in 1982 was approximately 4000 tons and 650 tons respectively (Anon, 1982). Azodicarbonamide was inactive in a 2-year rat feeding study which was judged to be inadequate for a conclusive evaluation of potential carcinogenicity. In this study, animals were fed bread prepared from flour treated with either a single low dose of azodicarbonamide (10 ppm) or brurea (a hydrazo product formed from the reduction of azodicarbon amide during the dough-making process) at doses of 750, 3750 or 7500 ppm. The exact levels of azodicarbonamide or biurea ingested by the rats was not reported nor was azodicarbonamide or biurea fed to the animals directly. Azodicarbonamide is currently or test for carcinogenicity by the National Toxicology Program.

Withey (1977) has noted the paucity of information on the toxicological and oncogenic properties of the blowing agents and their decomposition products.

Free-Radical Initiators

A broad spectrum of peroxides and peroxy compounds have been employed to initiate the copolymerization of vinyl and diene monomers, e.g., vinyl chloride, styrene, ethylene, acrylonitrile and butadiene as well as in the curing or cross-linking of rubbers such as styrene-butadiene rubber. The major organic peroxides which have been mainly employed as initiators for vinyl monomer polymerizations and copolymerizations for vinyl chloride include: benzoyl peroxide, didecanoyl peroxide and dilauroyl peroxide. The respective consumption of these peroxides in the US in 1982 (for all uses) was 3,800, 120 and 630 metric tons (Anon, 1982). Additionally, mixtures of organic peroxides are often used, e.g., the above organic peroxides and acetyl sulfonyl peroxide for low temperature polymerizations.

The major peroxide used in the suspension process for PVC production is dilauroyl peroxide. For the preparation of higher molecular weight PVC low-temperature catalysts are employed. Among those principally used are diisopropyl- and diethylperoxy dicarbonates as well as acetylcyclohexyl sulfonyl peroxide. An organic peroxide that is principally employed in the suspension process for the production of PVC is tert.butylperoxy pivalate.

There is a paucity of information regarding the carcinogenicity of the organic peroxides. A recent study cited the skin tumor-promoting activity of benzoyl peroxide (Slaga et al, 1981). Benzoyl peroxide promoted both papillomas and carcinomas when it was topically applied to mice after

7,12-dimethylbenz[a]anthracene initiation although it was inactive on the skin as a complete carcinogen or as a tumor initiator. Benzoyl peroxide, like other known tumor promoters, also inhibited metabolic cooperation (intercellular communication in Chinese hamster cells). Another free radical generating organic peroxide (e.g., lauroyl peroxide) has also been shown to be an effective skin tumor promoter (Slaga et al, 1981). In earlier skin-painting studies, benzoyl peroxide had shown negative results as a carcinogen (Van Duuren et al, 1963).

A number of organic peroxides have been found to be mutagenic. For example, tert.butyl hydroperoxide is mutagenic in Drosophila (Altenberg, 1940, 1958), E. coli (Chevallier and Luzatti, 1960) and Neurospora (Dickey et al, 1949). Cumene hydroperoxide is mutagenic in E. coli (Chevallier and Luzatti, 1960) and Neurospora (Jensen et al, 1951; Latarjet et al, 1958) while di-tert.butyl peroxide is mutagenic in Neurospora (Jensen et al, 1951).

Flame Retardants

Poly(vinyl chloride) and its copolymers contain an appreciable amount of halogen (56.8% in PVC). Thus, incorporation of antimony oxide gives an effective flame-retardant composition. While unplasticized PVC is considered a self-extinguishing plastic because of the high levels of chlorine contained in the polymer flexible, PVC is not self-extinguishing. Generally, plasticizers commonly used to flexibilize PVC significantly increase its flammability. Antimony oxide is used at levels of approximately 3.0 parts by weight as a flame retardant with plasticized PVC. Other flame retardants that are used with plasticized PVC include: zinc borate, antimony trioxide-zinc borate, molybdenum oxide, molybdenum oxide-antimony trioxide (Avento and Touval, 1980; Larsen, 1980; Pearce and Liepins, 1975). Other very effective retardants are phosphorus-based systems which are used to balance off phthalate plasticizer effects (Pierce and Liepins, 1975).

Polychlorinated paraffins are principally employed as secondary plasticizers in PVC for wire and cable coatings and calendered products. These flame retardants are mainly compounds containing high (50-85 weight%) levels of either chlorine or bromine. Among the polychlorinated paraffins that have been employed in this fashion are decabromodiphenyl oxide (DBDPO) (Larsen, 1980).

There is a paucity of data a relating to the toxicity of flame retardants currently in use (Ulsamer et al, 1980).

There is little information on the bioavailability of DBDPO. Although DBDPO has been reported to be a poor skin penetrant, data were not presented (Norris et al, 1975; Ulsamer et al, 1980). Incorporation of DBDPO at concentrations equivalent to 8, 80 or 800 mg/kg/day for 30 days, produced no pathological effects in rats at the 8 mg/kg/day level. However, liver enlargements and thyroid hyperplasia were noted in rats at 80 mg/kg/day. The liver enlargement consisted of centrolobular hepatocellular cytoplasmic enlargement and vacuolation. This was enhanced at 800 mg/kg/day for 30 days and accompanied by renal hyaline degenerative cytoplasmic changes (Norris et al, 1974). Chronic and carcinogenic studies with DBDPO have not been reported.

DBDPO has not been found mutagenic in the Ames test (Mischutin, 1977) nor has it been found to be teratogenic in studies involving rats treated with 3, 30 or 100 mg/kg for 90 days. (This regimen was continued through mating, gestation, and lactation.) (Norris et al, 1974, 1975).

SUMMARY

The salient features of the toxicity of a number of additives used in polyvinyl chloride polymers were reviewed with primary emphasis on the toxicity of plasticizers (e.g., diethylhexyl phthalate and its metabolites, butylbenzylphthalate and di(2-ethylhexyl)adipate), heat stabilizers (e.g., organotin and lead stabilizers), blowing agents (e.g., azodicarbonamide), free-radical initiators (e.g., benzoylperoxide, lauroyl peroxide, ter.butylhydroperoxide and di-tert.butylperoxide, and flame retardants (e.g., decabromodiphenyl oxide). The paucity of toxicity data on the vast majority of PVC additives should be stressed.

REFERENCES

Abe S, Sasaki M (1977). Chromosome aberrations and sister chromatid exchanges in Chinese hamster cells exposed to various chemicals. J Natl Cancer Inst 58:1635.

Albro PW, Corbett JT, Schroeder J, Jordan S, Matthews HB (1981). The fate of di(2-ethylhexyl)phthalate in rats. Toxicologist 1:55.

Albro PW, Thomas R, Fishbein L (1973). Metabolism of diethylhexylphthalate by rats. Isolation and characterization of the urinary metabolites. J Chromatog 76:321.

Aldridge WN (1976). The influence of organo tin compounds on mitochondrial functions. In: (ed) Zuckerman JJ "Organo Tin Compounds: New Chemistry and Applications", Washington, DC, American Chemical Society, pp. 186-196.

Altenberg LS (1940). The production of mutations in Drosophila by tert.butyl hydroperoxide. Proc Natl Acad Sci 40:1037.

Altenberg LS (1958). The effect of photoreacting light on the mutation rate induced in Drosophila by tert.butyl hydroperoxide. Genetics 43:662.

Anon (1976) Mutagenicity evaluation of Bio-76-17 Saniticizer 160 NB 259784. Final Report. Submitted to the US Environmental Protection Agecy under Section 8(d) of the Toxic Substances Control Act of 1976, 8dHQ-1078-0282.

Anon (1977). Mutagenicity Evaluation of Bio-76-243 CP 731 (Sanitizer 160) in the mouse lymphoma assay. Final Report. Submitted to the US Environmental Protection Agecy under Section 8(d) of the Toxic Substances Control Act of 1976, 8dHQ-1078-0282.

Anon (1982a). Polyvinyl chloride. Chem Eng News Sept 6, p. 13.

Anon (1982b). Chemicals and additives. Modern Plastics 59:55.

Avento JM, Touval I (1980). Flame retardants-Antimony and other Inorganic Compounds. In: (ed) Grayson M. "Kirk-Othmer Encyclopedia of Chemical Technology", Wiley, New York, pp. 355-372.

Baker EL, Landrigan PJ, Barbour AG (1979). Occupational lead poisoning in the United States: Clinical and biochemical findings related to blood lead levels. Brit J Ind Med 36:314.

Barnes JM, Stoner HB (1959). The toxicology of tin compounds. Pharmacol Rev 11:211.

Boettner EA, Ball GL (1980). Thermal degradation products from PVC film in food wrapping operations. Am Ind Hyg Assoc J 41:513.

Boettner EA, Ball GL (1982). Organic and organo tin compounds leached from PVC and CPVC pipe. Project Summary EPA-600/SI-81-062. Health Effects Research Laboratory, US Environmental Protection Agency, Cincinnati, OH, February, pp. 1-9.

Bower RK, Haberman S, Minton PD (1970). Teratogenic effects in the chick embryo caused by esters of phthalic acid. J Pharmacol Exper Therap 171:314.

Chevallier MR, Luzatti D (1960). The specific mutagenic action of 3 organic peroxides on reversae muations of 2 loci in E. coli. Compt Rend 250:1572.

Chin TY, Tyl RW, Heck Hd'A (1981). Chemical urolithiasis. 1. Characteristics of bladder stone induction by terephthalic acid and dimethyl terephthalate in weanling Fischer 344 rats. Toxicol Appl Pharmacol 58:307.

Chu I, Villeneuve DC, Secours V, Franklin C, Roch G, Viau A (1978). Metabolism and tissue distribution of mono-2-ethylhexyl phthalate in the rat. Drug Metab Dispos 6:146.

Cohan GF (1975). Industrial preparation of poly(vinyl chloride). Env Hlth Persp 11:53.

Curto KA, Thomas JA (1982). Comparative effects of diethylhexyl phthalate or monoethylhexyl phthalate on male mouse and rat reproductive organs. Toxicol Appl Pharmacol 62:121.

Deanin RD (1975). Additives in plastics. Environ Htlh Persp 11:35.

Dickey FH, Cleland GH, Lotz C (1949). The role of organic peroxides in the induction of mutations. Proc. Natl Acad Sci 35:581.

Dickson D (1980). More tests required on new chemicals. Nature 285:60.

Dillingham EO, Autian J (1973). Teratogenicity, mutagenicity and cellular toxicity of phthalae esters. Environ Hlth Persp 3:81.

Eckardt RE (1976). Occupational and environmental health hazards in the plastics industry. Env Hlth Persp 17:103.

Eckardt RE, Hindin R (1973). Health hazards of plastics. J Occup Med 15:808.

EPA (1980). Trimellitic anhydride. In: TOSCA Chemical Assessment Series. Chemical Hazard Information Profile (CHIPS), Washington, DC, pp. 273-276.

Fischbein A, Thornton JC, Berube L, Villa F, Selikoff IJ (1982) Lead exposure reduction in workers using stabilizers in PVC manufacture: Effects of a new encapsulated stabilizer. Am Ind Hyg Assoc J 43:652.

Fishbein L (1982). Additives in synthetic polymers: An overview, in press.

Gray TJB, Butterworth KR, Gaunt IF, Grasso P, Gangolli SD
(1977). Short-term toxicity study of di(2-ethylhexyl)-
phthalate in rats. Food Cosmet Toxicol 15:389.

Haberman S, Guess WL, Rowan DF, Bowman RO, Bower RK (1968).
Effects of plastics and their additives on human serum
proteins, antibodies and developing chick embryos. Soc
Plastics Eng J 24:62.

Hamano Y, Inoue K, Oda Y, Yamamoto H, Kunita N (1979).
Studies on the toxicity of phthalic acid esters. Part 2.
Dominant lethal tests for DEHP and MEHP in mice (Food
Hygiene Series No. 10) Osaka, Osaka Public Health and
Sanitation Research Centre, pp. 1-4.

Heck Hd'A (1981). Chemical urolithiasis. 2. Thermodynamic
aspects of bladder stone induction by terephthalic acid
and dimethyl terephthalate in weanling Fischer 344 rats.
Fundamental Appl Toxicol 1:299.

Hemminki K, Sorsa M, Vainio H (1979). Genetic risks caused
by occupational chemicals. Scand J Work Environ Hlth
5:307.

Hill JO, Giere MS, Pickrell JA, Hahn FF, Dahl RR (1980).
Toxic effects of organometallic compounds on lymphoid
tissue in vitro and in vivo. Fed Proc 39:623.

IARC (1979). Vinyl chloride, polyvinyl chloride and vinyl
chloride-vinyl acetate copolymers. In: IARC Monographs on
the Evaluation of the Carcinogenic Risk of Chemicals to
Humans. Vol. 19. Some Monomers, Plastics and Synthetic
Elastomers and Acrolein, Lyon, International Agency for
Research on Cancer, pp. 402-438.

IARC (1982a). Di(2-Ethylhexyl)phthalate. In: IARC Monographs on the Evaluation of the Carcinogenic Risk of
Chemicals to Humans. Vol. 29. Some Industrial Chemicals
and Dyestuffs, Lyon, International Agency for Research on
Cancer, pp. 270-294.

IARC (1982b). Butyl benzyl phthalate. In: IARC Monographs
on the Evaluation of the Carcinogenic Risk of Chemicals to
Humans. Vol. 29. Some Industrial Chemicals and Dyestuffs,
Lyon, International Agency for Research on Cancer, pp.
193-201.

IARC (1982c). Di(2-Ethylhexyl)adipate. In: IARC Monographs
on the Evaluation of the Carcinogenic Risk of Chemicals to
Humans. Vol. 29. Some Industrial Chemicals and Dyestuffs,
Lyon, International Agency for Research on Cancer, pp.
258-267.

IARC (1982d). The Rubber Industry. IARC Monographs on the
Evaluation of the Carcinogenic Risk of Chemicals to
Humans. Vol. 19. Lyon, International Agency for Research
on Cancer, p. 108.

Ishidate M Jr, Odashima S (1977). Chromosome tests with 134 compounds on Chinese hamster cells in vitro: A screening for chemical carcinogens. Mutat Res 48:337.

Jensen KA, Kirk I, Kolmark G, Westergaard M (1951). Chemically induced mutations in Neurospora. Cold Springs Harbor Symp Quant Biol 16:245.

Jones AH, Kahn RH, Groves JT, Napier EA Jr, (1975). Toxicol Appl Pharmacol 31:283.

Jones JH (1981). Workers exposure to vinyl chloride and poly(vinyl chloride). Env Hlth Persp 41:129.

Karstadt M (1976). PVC: Health implications and production trends. Env Hlth Persp 17:107.

Kay K (1977). Chemical agents from occupational sources. In: (eds) Lee DK and Falk H "Handbook of Physiology" Section 9. Reaction to Environmental Agents, Bethesda, American Physiology Society, pp. 181-191.

Kimbrough RD (1976). Toxicity and health effects of selected organo tin compounds: A review. Environ Hlth Persp 14:51.

Lake BG, Phillips JC, Linnell JC, Gangolli SD (1977). The in vitro hydrolysis of some phthalate diesters by hepatic and intestinal preparations from various species. Toxicol Appl Pharmacol 39:239.

Larsen ER (1980). Halogenated flame retardants. In: (ed) Grayson M, "Kirk-Othmer Encyclopedia of Chemical Technology", Vol. 10, 3rd ed., Wiley, New York, pp. 373-395.

Latarjet R, Rebeyrotte N, Demerseman P (1958). In: "Organic Peroxides in Radiobiology" (ed) Haissinsky, M, Oxford, Pergamon Press, p. 6.

Lawrence WH, Tuell SF (1979). Phthalate esters: The question of safety-an update. Clin Toxicol 15:447.

Levinson C (1974). Vinyl chloride: A case study of the new occupational health hazard. Geneva, International Chemical Federation, p. 15.

Li AP, Dahl AR, Hill JO (1982). In vitro cytotoxicity and genotoxicity of dibutyl tin dichloride and dibutyl germanium dichloride. Toxicol Appl Pharmacol 64:482.

Lilis RA, Fischbein A, Eisinger JE (1977). Prevalence of lead disease among secondary lead smelter workers and biological indicators of lead exposure. Environ Res 14:255.

Mazaev VT, Korolev AA (1969). Experimental standardization of the maximum permissible concentration of the dichlorobutyl tin in surface water. Prom Zagrjaznenija Vodoemov 9:15.

Mischutin V (1977). Safe flame retardant. Am Dyest Rep 66:51.

Monsanto Co. (1973). Technical Data Sheet: Dioctyl Adipate,
St. Louis, MO.
Motoc FS, Constantinescu G, Filipescu M, Dobre M, Richir E,
Pambuccian G (1971). Noxious effects of certain
substances used in the plastics industry (acetone
cyanohydrin, methyl methacrylate, azo(bisisobutyronitrile)
and anthracene oil. Arch Mal Prof Med Trav Secur Soc
32:653.
Morton SJ, Rubin RJ (1980). Hepatic effects of di(2-ethyl-
hexyl)phthalate (DEHP) and its primary metabolites:
Enhancement of ethanol oxidation and role of the
mitochondria in the phthalate-induced hypolysemia. Proc
19th Ann Meet Soc Toxicol March 9, p A72.
Nakamura Y, Yagi Y, Tomita I, Tsuchikawa K (1979). Teratogenicity of di(2-ethylhexyl)phthalate in mice. Toxicol
Lett 4:113.
National Toxicology Program (1981a). Bioassay of butyl
benzyl phthalate for possible carcinogenicity. NTP No.
81-25, DHHS Publication No. (NIH) 80-1769), Washington,
DC, US Department of Health and Human Services.
National Toxicology Program (1981b). Carcinogenesis
bioassay of di(2-ethyl hexyl)adipate (CAS No. 103-23-1).
Technical Report Series No. 212. DHHS Publication No.
(NIH)81-1768. US Department of Health and Human Services,
Research Triangle Park, NC.
National Toxicology Program (1981c). Carcinogenesis
bioassay of di(2-ethyl hexyl)phthalate (CAS No. 117-81-7)
in F334 rats and B6C3F1 mice (feed study). Technical
Report Series No. 217) (NIH Publ. No. 82-1173) Research
Triangle Park, NC.
National Toxicology Program (1982a). NTP Technical Bulletin
No. 8, July, p. 1.
National Toxicology Program (1982b). Carcinogenesis Testing
Program. Chemicals on Standard Protocol Management
Status. October 7, pp. 1-7.
NCI (National Cancer Institute) (1978). Summary of Data for
Chemical Selection, Clearinghouse on Environmental
Carcinogens, Chemical Selection Subgroup, Bethesda,
National Cancer Institute, December 7.
NCI (National Cancer Institute) (1979). Carcinogenesis
Technical Report Series, DHEW Publication (NCI) CG-TR-121.
Needleman HL (1980). Low Level Lead Exposure: The Clinical
Implications of Current Research, New York, Raven Press.
Nikonorow M, Mazur H, Piekacz H (1973). Effects of orally
administered plasticizers in the rat. Toxicol Appl
Pharmacol 26:253.

NIOSH (National Institute for Occupational Safety and Health) (1976). Criteria for a recommended standard: Occupational Exposure to Organo tin compounds, Washington, DC, US Department of Health, Education and Welfare.

NIOSH (National Institute for Occupational Safety and Health) (1978). Current Intelligence Bulletin 21: Trimellitic Anhydride (TMA), DHEW (NIOSH) Publication No. 78-121, Cincinnati, OH, National Institute for Occupational Safety and Health.

NIOSH (National Institute for Occupational Safety and Health) (1980a). Projected Number of Occupational Exposure to Chemical and Physical Hazards, Cincinnati, OH, p. 64.

NIOSH (National Institute for Occupational Safety and Health) (1980b). National Occupational Hazard Survey, Cincinnati, OH, Division of Surveillance Hazard Evaluations and Field Studies, p. 107.

Norris JM, Ehrmantraut JW, Gibbons CL, Kociba RJ, Schwetz BA, et al (1974). Toxicological and environmental factors involved in the selection of decabromodiphenyl oxide as a fire retardant chemical. J Fire Flammability/Combust Toxcicol 1:52.

Norris JM, Kociba RJ, Schwetz BA, Rose JQ, et al (1975). Toxicology of octabromobiphenyl and decabromodiphenyl oxide. Environ Hlth Persp 11:153.

Oishi S, Hiraga K (1980a). Testicular atrophy induced by phthalic acid esters: Effect on testosterone and zinc concentrations. Toxicol Appl Pharmacol 53:35.

Oishi S, Hiraga K (1980b). Effect of phthalic acid esters on mouse testes. Toxicol Lett 5:413.

Omori Y (1976). Recent progress in safety evaluation studies on plasticizers and plastics and their controlled use in Japan. Environ Hlth Persp 17:203.

Pearce EM, Liepins R (1975). Flame retardants. Environ Hlth Persp 11:59.

Peliktan Z, Cerny E (1970). The toxic effects of some di- and mono-n-octyl tin compounds on white mice. Arch Toxicol 26:196.

Peters JW, Cook RM (1973). Effect of phthalate esters on reproduction in rats. Env Hlth Persp 3:91.

Phillips BJ, James TEB, Gangolli SD (1982). Genotoxicity studies of di(2-ethylhexyl)phthalate and its metabolites in CHO cells. Mutation Res 102:297.

Pivers WT (1973). Organo tin compounds: Industrial applications and biological investigation. Environ Hlth Persp 4:61.

Rice RL, Jenkins DE, Gray TM, Greenberg, SD (1977). Chemical pneumonitis secondary to inhalation of epoxy pipe coating. Arch Environ Hlth 32:173.

Ruddick JA, Villeneuve DC, Chu I, Nestmann E, Miles D (1981). An assessment of the teratogenicity in the rat and mutagenicity in Salmonella of mono-2-ethylhexyl phthalate. Bull Environ Contam Toxicol 27:181.

Seinen W, Penninks A (1979). Immune suppression as a consequence of a selective cytotoxic activity of certain organometallic compounds on thymus and thymus-dependent lymphocytes. Ann NY Acad Sci 320:499.

Sernen W, Vos JG, Van Spanje I, Snock M, Brands R, Hoykaus H (1977). Comparative in vivo and in vitro studies with various organo tin and organo lead compounds in different animal species with special emphasis on lymphocyte toxicity. Toxicol Appl Pharmacol 42:197.

Seymour WE (1975). Poly(vinyl chloride) and related polymers. In: Modern Plastics Technology, Reston, VA Reston Publ. pp. 192–204.

Shaffer CB, Carpenter CP, Smyth, HF Jr (1945). Acute and subacute toxicity of di(2-ethylhexyl)phthalate with note upon its metabolism. J Ind Hyg Toxicol 27:130.

Simmon VF, Kauhanen K, Tardiff RG (1977). Mutagenic activity of chemicals identified in drinking water. Dev Toxicol Env Sci 2:429.

Singh AR, Laurence WH, Autian J (1973). Embryonic-fetal toxicity and teratogenic effects of adipic acid esters in rats. J Pharm Sci 62:1596.

Singh AR, Laurence WH, Autian J (1975a). Maternal-fetal transfer of ^{14}C-di-2-ethylhexyl phthalate and ^{14}C-diethyl phthalate in rats. J Pharm Sci 64:1347.

Singh AR, Laurence WH, Autian J (1975b). Dominant lethal mutations and antifertility effects of di-2-ethylhexyl adipate and diethyl adipate in male mice. Toxicol Appl Pharmacol 32:566.

Slaga TG, Klein-Szanto AJP, Triplett LL, Yotti LP, Trosko JE (1981). Skin tumor-promoting activity of benzoyl peroxide, a widely used free radical generating compound. Science 213:1023.

Stenchever MA, Allen MA, Jerominski L, Petersen RV (1976). Effects of bis(2-ethylhexyl)phthalate on Chromosome of human leukocytes and human fetal lung cells. J Pharm Sci 65:1648.

Takahashi T, Tanaka A, Yamaha T (1981). Elimination, distribution and metabolism of di(2-ethylhexyl)adipate (DEHA) in rats. Toxicology 22:223.

Theiss JC, Stoner GD, Shimkin MB, Weisburger EK (1977).
Test for carcinogenicity of organic contaminants of US
drinking waters by pulmonary tumor responses in strain A
mice. Cancer Res 37:2717.

Thomas JA, Darby TD, Wallin RF, Garvin PJ, Martis L (1978).
A review of the biological effects of di(2-ethylhexyl)-
phthalate. Toxicol Appl Pharm 45:1.

Thomas JA, Northup SJ (1982). Toxicity and metabolism of
monoethylhexyl phthalate and diethylhexyl phthalate: A
survey of recent literature. J Toxicol Env Hlth 9:141.

Thomas JA, Schern LG, Gupta PK, McCafferty RE, Felice PR,
Donovan MP (1979). Failure of monoethylhexyl phthalate to
cause teratogenic effects in offspring of rabbits.
Toxicol Appl Pharmacol 51:523.

Tomita I, Nakamura Y, Yagi Y (1977). Phthalic acid esters
in various foodstuffs and biological materials.
Ecotoxicol Environ Safety 1:275.

Tsuchiya K, Hattori K (1976). Chromosomal study on human
leukocytes cultures treated with phthalate acid esters.
Hoeka 26:114.

Ulsamer AG, Osterberg RE, McLaughlin J (1980). Flame-
retardant chemicals in textiles. Clin Toxicol 17:101.

US Department of Labor (1974). Standard for Exposure to
Vinyl Chloride. Occupaional Safety and Health Administration, Washington, DC, Federal Register 29:35896 (October
4).

Vainio H, Pfäffli P, Zitting A (1980). Chemical hazards in
the plastics industry. J Toxicol Env Hlth 6:1179.

VanDuuren BL, Nelson N, Orres L, Palmer ED, Schmitt FL
(1963). Carcinogenicity of epoxides, lactones and peroxy
compounds. J Natl Cancer Inst 31:41.

Villeneuve DC, Franklin CA, Chu I, Yagminas H, Marino IA,
Ritter L, Ruddick JA (1978). Toxicity studies on mono-2-
ethylhexyl phthalate. Toxicol Appl Pharmacol 45:250.

Wheeler RN (1981). Poly(vinyl chloride) processes and
products. Env Hlth Persp. 41:123.

WHO (1980). Environmental Health Criteria 15. Tin and
Organo tin Compounds. Geneva, World Health Organization,
pp. 109.

Withey JR (1977). Mutagenic, carcinogenic and teratogenic
hazards arising from human exposure to plastics additives.
In: (eds) Hiatt HH, Watson JD, Winsten JA "Origins of
Human Cancer" Book A. Cold Spring Harbor, NY, Cold Spring
Harbor Laboratory, pp. 219-240.

Zuckerman JJ, Reisdorf RP, Ellis HV III, Wilkinson RR
(1978). Organo tins in biology and the environment. In:

(ed) Brinckman FE, American Chemical Society Symposium Series Organometals and Organometalloids, Washington DC, American Chemical Society, pp. 388-434.

Industrial Hazards of Plastics and Synthetic Elastomers, pages 137–154
© 1984 Alan R. Liss, Inc., 150 Fifth Ave., New York, NY 10011

Phthalate Esters Carcinogenicity
in F344/N Rats and B6C3F$_1$ Mice

J. E. Huff and W. M. Kluwe

National Toxicology Program

Research Triangle Park, NC, USA

Plasticizer: a chemical added to plastics, rubbers, and resins to facilitate processing and to impart flexibility, workability, or stretchability.

Plastics and plastic-containing products are realities of modern life that have transformed both the home and work environments. In addition to the polymerized molecules that form the matrix of the plastic, however, other chemicals (adjuvants) are commonly incorporated into plastic products to impart specific use or performance characteristics such as combustion resistance, color enhancement, or resiliency. Ortho esters of phthalic acid, generally referred to as phthalate esters, have been used to impart flexibility to plastics and may comprise as much as 50% by weight of the final product. Such plasticizers are not polymerized into the plastic matrix and may, with time and use, migrate from the plastic into the external environment (although the plastic products are obviously designed to retain the plasticizer and thereby remain flexible).

Di(2-ethylhexyl)phthalate (DEHP) is the most commonly used plasticizer for polyvinyl chloride products, including syringes, dialysis tubing, and other medical devices. DEHP is also used as a dielectric fluid in transformers and as a hydraulic fluid in heavy machinery. About 126 million killograms of DEHP are produced annually in the United States (USITC, 1983). Since plastic products are ultimately disposed of in the environment, human exposure to DEHP and other components of plastics can occur via contact with water and soil (and air, in some instances), as well as through direct contact with plastic products.

In September 1972, the National Institute of Environmental Health Sciences held a Symposium in Pinehurst, NC on phthalic acid esters (EHP, 1973). The toxic potentials to humans of DEHP and other phthalate esters were generally perceived to be quite low until the results were made known in 1980 that chronic ingestion of DEHP via feed (dietary incorporation) caused neoplasms in both sexes of rats and mice (NTP, 1982c). The controversy and public interest that surrounded these findings, and their impact on regulations to protect public health, prompted the NTP in collaboration with the U.S. Interagency Regulatory Liaison Group (IRLG) and the Chemical Manufacturers Association (CMA) to convene a conference to discuss the toxicology of phthalate esters. The conference was held on 9-11 June 1982 in Washington, D.C. (EHP 1982).

This paper describes select chemical properties, pharmacokinetics data, and planned research and testing as well as summarizes the carcinogenesis studies[a] of six phthalate esters and related compounds (see Kluwe et al. 1982b):

   o butylbenzylphthalate (NTP, 1982a)

   o diallylphthalate (NTP, 1983; NTP, 1984 [DRAFT])

   o di(2-ethylhexyl)adipate (NTP, 1982b; Huff, 1982)

   o di(2-ethylhexyl)phthalate (NTP, 1982c; Kluwe, et al. 1982a; Kluwe, Haseman, and Huff, 1983)

   o phthalamide (NCI, 1979a)

   o phthalic anhydride (NCI, 1979b)

Chemical Properties -- Structurally, phthalate esters consist of paired ester groups on a cyclohexatriene ring (benzene-dicarboxylic acid). The meta and para configurations are known as isophthalates and terephthalates, respectively. The ortho configuration, however, is implied in the generic use of the term "phthalate esters".

---

[a] Single copies of the complete Technical Reports on each chemical are available without cost, while supplies last, from the National Toxicology Program, Public Information Office, P. O. Box 12233, Research Triangle Park, NC, 27709, USA.

Phthalate esters are synthesized commercially by condensation of appropriate alcohols with phthalic anhydride. In the related plasticizers, the adipate esters, a four-carbon chain separates the two ester groups.

Specific phthalate esters with short alkyl groups, such as dimethyl and di-n-butyl phthalates, are apprecialby soluble (e.g., 0.5 g/100 ml) in water. Most other dialkyl phthalates, including DEHP, are relatively insoluble in aqueous mediums because of their lipophilic structures. Volatilities at standard temperature and pressure are generally low, particularly for the long-chain and branched compounds, such as DEHP.

Pharmacokinetics (Kluwe, 1982) -- Phthalate esters are generally well absorbed from the gastrointestinal tract following oral adminstration. Hydrolysis to the corresponding monoester metabolite, with release of an alcoholic substitutent, largely occurs prior to intestinal absorption of the longer-chain alkyl derivatives such as DEHP. Phthalate esters are widely distributed in the body, with the liver being the major initial repository organ. Clearance from the body is rapid and there is only a slight cumulative potential. Short-chain dialkyl phthalates, such as dimethyl phthalate, can be excreted in an unchanged form or following complete hydrolysis to phthalic acid. Longer-chain compounds such as DEHP, however, are converted principally to polar derivatives of the monoesters by oxidative metabolismm prior to excretion. A marked species difference in DEHP metabolism exists: primates (humans, monkey) and some rodent species glucuronidate DEHP at the carboxylate moiety following hydrolysis of a single ester linkage, whereas rats appear unable to glucuronidate the monoester metabolite, oxidizing the residual alkyl chain instead to various ketone and carboxylate derivatives. The major route of phthalate ester elimination from the body is urinary excretion. Certain phthalate esters are excreted in the bile but undergo enterohepatic circulation. The relationships of phthalate ester pharmacokinetics to their toxicological actions are unknown at the present time, due largely to a lack of elucidated mechanisms of toxic action.

Carcinogenicity Testing Rationale -- The National Toxicology Program (NTP) and its predecessor, the National Cancer Institute Carcinogenesis Bioassay Program, tests chemicals for carcinogenic potential as part of a broad initiative to aid in protecting the public health. Among the criteria used in choosing chemicals for testing are volume of production, number of persons exposed, and frequency and level of anticipated human exposure. Phthalate esters and related compounds were tested because of their presence in high concentrations in plastics; the ubiquity of phthalate ester-plasticized products in the home, work, and out-of-doors environments; and because of a general lack of previous chronic toxicity and carcinogenicity testing with these agents.

Methods -- Male and female rats of the inbred Fischer 344 strain and male and female mice of the hybrid B6C3F$_1$ strain, obtained from Frederick Cancer Research Center (Frederick, MD), were used in all of the studies. Background tumor rates for untreated controls have been tabulated and are given elsewhere in These Proceedings (Huff and Haseman, 1983). Except for diallylphthalate, the test chemicals were incorporated into the feed and supplied to the animals for 102-106 consecutive weeks beginning at 4-6 weeks of age. Diallylphthalate was given by gavage in corn oil 5 days per week for 103 weeks. In general, each treatment group consisted of 50 animals of each sex and species at the estimated maximally tolerable dose (referred to as the high dose), one-half of the estimated maximally tolerable dose (the low dose) and untreated or vehicle (diallylphthalate) controls (control groups for phthalamide and phthalic anhydride consisted of 20 animals, control groups for other chemicals consisted of 50 animals). Estimated maximally tolerable doses were determined by preceeding 90-day studies.

Table 1 shows the chemicals tested, companies where the chemicals were procured, percent purity, the laboratories where the studies were done, concentrations used, and the Technical Report numbers.

All animals that died during the study (and were not excessively cannibalized or autolyzed) and all survivors at the end of the study were subjected to a gross necropsy and a complete micropathological examination. Statistical comparisons of the incidences of animals with tumors at a specific anatomical site and of survival and body weight gain were made using both pairwise comparisons and trend tests. The basic designs of the studies conformed to the recommendations of the National Cancer Institute (Sontage, Page, and Saffiotti, 1976). Complete technical details of each of the carcinogenesis studies are available in the form of

Table 1. Phthalate Esters Test Chemicals, Sources, Compositions, and Doses Used in the Carcinogenesis Studies

Test Chemical	Source of Chemical	Purity (%)	Testing Laboratory	Concentration in Feed (ppm) or Gavage (mg/kg)	Reference
butylbenzyl-phthalate	Missouri Solvents & Chemicals	>98	EG&G Mason Research Institute	M & F Rats & Mice 0;6000;12,000 (Feed)	NTP (1982a) TR 213
diallyl-phthalate	Hardwick Chemical Company	>99	Litton Bionetics Inc.	M & F Rats: 0;50;100 M & F Mice: 0;150;300 (Gavage)	NTP (1983 & 1984) Rats: TR 284 Mice: TR 242
di(2-ethyl-hexyl)adipate	W.R.Grace & Co.	>98	EG&G Mason Research Institute	M & F Rats and Mice 0;12,000;25,000 (Feed)	NTP (1982b) TR 212
di(2-ethyl-hexyl)phthalate	W.R.Grace & Co.	>99	EG&G Mason Research Institute	M & F Rats: 0;6,000;12,000 M & F Mice: 0;3,000;6,000 (Feed)	NTP (1982c) TR 217
phthalamide	Sherwin-Williams, Inc.	>99	Frederick Cancer Rsch. Center	M Rats: 0;15,000;30,000 F Rats: 0;5,000;10,000 M Mice: 0;25,000;50,000 F Mice: 0;6,200;12,500; 25,000 (Feed)	NCI (1979a) TR 161
phthalic anhydride	Koppers Co.	>99	Frederick Cancer Rsch. Center	M & F Rats: 0;7,500;15,000 M Mice: 0;16,000;33,000 F Mice: 0;12,000;24,000 (Feed)	NCI (1979b) TR 159

Technical Reports from the NTP (NCI, 1979a; NCI, 1979b; NTP 1982a; NTP, 1982b; NTP, 1982c; NTP, 1983).

RESULTS

Butylbenzylphthalate (BBP) -- Body weight gains in male and female mice and in female rats were reduced by the ingestion of BBP, but survivals were unaffected. In contrast, excessive numbers of BBP-treated male rats died from apparent internal hemorrhaging after approximately 3 months of exposure. The study in male rats was terminated and no evaluation of tumorigenic response could be made. The increase of tumors at specific anatomical sites in BBP-treated male and female mice did not differ significantly from those in controls. The incidence of animals with mononuclear cell leukemia was greater (P<0.05) for the high-dose female rats than for the controls (control, 7/49, 14%; low dose, 7/49, 14%; high dose 18/50, 36%), and the trend was significant (P<0.01) as well.

Diallylphthalate (DAP) -- Mean body weights and survival of male and female rats administered diallylphthalate were essentially the same as those of the vehicle controls throughout the 2-year studies, although hepatotoxicity was produced in both sexes by the high (100 mg/kg) dose. Male and female rats receiving the high dose of DAP developed chronic liver disease characterized by periportal fibrosis, periportal accumulation of pigment, and severe bile duct hyperplasia. DAP increased (P<0.05) the occurrence of mononuclear cell leukemia in high dose female rats (vehicle control, 15/50, 30%; low dose, 15/43, 35%; high dose, 25/49, 51%). An increased occurrence of mononuclear cell leukemia was not observed in male rats administered DAP (13/50, 26%; 12/49, 24%; 14/49, 29%).

Survivals and mean body weights of dosed mice were not different from those of controls, and pathological lesions unrelated to proliferative changes were not observed. The incidences of lymphoma in dosed male mice were not significantly greater than those in the controls (P>0.05), but the trend tests were significant (P<0.05). The incidence of lymphomas in the high-dose male mice was 12/50 (24%) in comparison with 6/50 (12%) in the controls and 5/50 (10%) in the low dose group. Since the incidence of high-dose male mice with lymphoma was not signficantly greater than that of concurrent or the laboratory historical controls, this marginal increase was considered to be equivocally related to DAP.

Increased incidences of squamous cell papillomas, hyperplasia, and inflammatory lesions of the forestomach were observed in DAP-dosed mice of both sexes in a dose-related manner: papillomas (vehicle controls, 0%; low dose 1/47, 2%; high dose 2/49, 4%) in mice of both sexes; hyperplasia (male: 0%; 7/47, 15%; 9/49, 18%; female: 4/48, 8%; 1/47, 2%; 14/49, 29%) chronic inflammation (male: 0%, 4/47, 9%; 8/49, 16%; female: 2/48, 4%; 1/47, 2%; 9/49, 18%). Because of the numerical elevation of forestomach papillomas in high-dose mice of both sexes, the concomitant observation of dose-related forestomach hyperplasia, and the rarity of this tumor in corn oil (gavage) control $B6C3F_1$ mice, the development of squamous cell papillomas of the forestomach may have been related to diallyl phthalate administration.

Di(2-ethylhexyl) Adipate (DEHA) -- Ingestion of DEHA decreased body weight gains in male and female rats and mice. Animal survivals were not compromised. The incidences of tumors at specific anatomical sites in DEHA-treated rats were not increased relative to the controls, but liver tumors (hepatocellular carcinomas in female mice, hepatocellular adenomas in male mice) occurred more frequently in DEHA-treated mice than in the controls (Table 2).

Table 2. Incidences of Mice with Liver Neoplasms in the Carcinogenesis Studies of Di(2-ethylhexyl)adipate

		Incidence		
Sex	Liver Neoplasm	Control	Low Dose	High Dose
Male	Carcinoma	7/50	12/49	12/49[a]
	Adenoma[b]	6/50	8/49	15/49[d]
	Combined[c]	13/50	20/49	27/49[e]
Female	Carcinoma[c]	1/50	14/50[e]	12/49[e]
	Adenoma	2/50	5/50	6/49
	Combined[c]	3/50	19/50[e]	18/49[e]

[a] One other male had a neoplasm NOS (not otherwise specified).
[b] Significant dose related trend (P<0.05).
[c] Significant dose related trend (P<0.01).
[d] Significantly greater than controls (P<0.05).
[e] Significantly greater than controls (P<0.01).

Di(2-ethylhexyl)phthalate (DEHP) -- DEHP treatment decreased body weight gains for female mice and for male and female rats. Animal survivals were not compromised by dietary ingestion of DEHP. Liver neoplasms occurred in both DEHP-treated rats and mice at incidences significantly greater than in the respective controls (Table 3). The incidences of animals with hepatocellular carcinomas was significantly greater for female rats and for male and female mice treated with DEHP than for the controls.

Table 3. Incidences of Animals with Liver Neoplasms
in the Carcinogenesis Studies of Di(2-ethylhexyl)Phthalate

Species/ Sex	Liver Neoplasm	Incidence		
		Control	Low Dose	High Dose
Rat/Male	Carcinoma[a]	1/50	1/49	5/49
	Neoplastic Nodule	2/50	5/49	7/49
	Combined[b]	3/50	6/49	12/49[c]
Rat/Female	Carcinoma[b]	0/50	2/49	8/50[d]
	Neoplastic Nodule[a]	0/50	4/49	5/50[c]
	Combined[b]	0/50	6/49[c]	13/50[d]
Mouse/Male	Carcinoma[a]	9/50	14/48	19/50[c]
	Adenoma	6/50	11/48	10/50
	Combined[b]	14/50	25/48[c]	29/50[d]
Mouse/Female	Carcinoma[b]	0/50	7/50[d]	17/50[d]
	Adenoma	1/50	5/50	1/50
	Combined[b]	1/50	12/50[d]	18/50[d]

[a] Significant dose related trend ($P<0.05$).
[b] Significant dose related trend ($P<0.01$).
[c] Significantly greater than controls ($P<0.05$).
[d] Significantly greater than controls ($P<0.01$).

Phthalamide -- Phthalamide administration did not decrease body weight gain nor lessen survival in either sex of rats or male mice. Survival for female mice at the end of two years was only 36% in the high dose group in comparison to 80% for the controls. The incidences of phthalamide-treated animals with tumors at a specific anatomical site did not differ signficantly from those of the controls.

Phthalic Anhydride -- Body weight gains in male rats and in both sexes of mice were reduced by phthalic anhydride, but survival in all of the treated groups of animals was un-affected. The incidences of phthalic anhydride-treated animals with tumors at a specific anatomical site did not differ significantly from those of the controls.

DISCUSSION

BBP was considered probably carcinogenic in female F344 rats due to the increased incidence of mononuclear cell leukemia. For male rats, the study was judged to be inadequate because of chemically induced early deaths. No evidence of BBP carcinogenicity was observed in B6C3F1 mice.

DAP produced chronic liver disease in male and female F344 rats characterized by periportal fibrosis and pigment accumulation and severe bile duct hyperplasia. DAP caused an increased incidence in mononuclear cell leukemias in female F344 rats. The development of chronic inflammation and hyperplasia of the forestomach in both male and female B6C3F1 mice was considered to be a consequence of the administration of DAP. Squamous cell papillomas of the forestomach may also have been related to chemical administration, but the available data were insufficient to indicate a clear cause and effect relationship. An increase in the incidence of male mice with lymphomas was observed, but this increase was considered equivocally related to DAP. In total, these results do not clearly indicate that diallyl phthalate is carcinogenic in B6C3F1 mice. Increases in hematopoietic lesions in phthalate treated rodents (DAP: female rats and male mice; BBP: female rats) provide some evidence of an effect of these esters on the hematopoietic system.

DEHP was carcinogenic in both Fischer 344 rats and B6C3F1 mice because of the significantly increased incidences of hepatocellular carcinomas in male and female mice and in female rats. The incidence of male rats with either hepatocellular carcinomas or neoplastic nodules (total liver tumors) was also significantly greater in the high-dose group than in the controls. The absence of nonneoplastic lesions in the liver and the failure of DEHP to reduce survival suggests that the liver tumors were not caused by an indirect hepatotoxic effect.

Groups involved in the manufacture, marketing, and use of DEHP or DEHP-containing materials are understandably concerned over the report of carcinogenic effects of DEHP, lest the public be deprived unnecessarily of useful products, or private interests of the opportunity to market such products. Northrup and colleagues (1982) published a review of the metabolism and biological effects of DEHP and "comments" on its carcinogenic potential, including a critique of a preliminary draft of the rodent carcinogenicity study, subsequently reported by Kluwe, et al. (1982a). Some of the questions and criticisms raised by Northrup et al. (1982) on the draft Technical Report were dispatched in the final published report (Kluwe et al., 1982a; NTP, 1982c). Others have been discussed, and in a more recent communication the significance of the carcinogenic effects of DEHP in rodents is evaluated in light of current knowledge about DEHP, chemical carcinogens in general, and experimental detection of induced neoplasia (Kluwe, Haseman, and Huff, 1983).

DEHA was carcinogenic in female B6C3F1 mice because of the increased incidence of hepatocellular carcinomas. In male B6C3F1 mice, the incidence of hepatocellular adenomas was significantly increased and the data were judged to be evidence of probable carcinogenicity. Carcinogenic effects of DEHA in F344 rats were not demonstrated.

Neither phthalamide nor phthalic anhydride was demonstrated to be carcinogenic for Fischer 344 rats or B6C3F1 hybrid mice, under the conditions of the standard carcinogenesis studies. The incidences of chemically treated animals with tumors at a specific anatomical site did not differ significantly from those in controls.

In summary, four of the six phthalate esters and related compounds tested by the National Toxicology Program and the National Cancer Institute for carcinogenic potential in rodent feeding studies --BBP, DAP, DEHP, and DEHA-- were judged to induce a carcinogenic response (Table 4). DEHP caused liver tumors in both rats and mice, while DEHA caused liver tumors in mice. Because of the similarities in chemical structure and the site of induced tumor formation, it can be speculated that DEHP and DEHA act through similar mechanisms. BBP and DAP both increased the occurrence of mononuclear cell leukemia in female Fischer 344 rats. DAP also was associated with increases in lymphomas in male B6C3F1 mice, and with stomach lesions in both sexes. The final two chemicals, phthalamide and phthalic anhydride, exhibited no evidence of carcinogenicity in these studies (Table 4).

The International Agency for Research on Cancer (IARC, 1982) rendered these evaluations on certain phthalate esters: for BBP, IARC (pages 193-201) indicated "a somewhat higher incidence of monocytic leukemias was observed in female rats", yet concluded that "the available studies were inadequate to evaluate the carcinogenicity of butylbenzylphthalate to mice

Table 4. Phthalate Esters Test Chemicals, Neoplasms, and Conclusions

Test Chemical	Sex/Species	Neoplasms	Interpretation[a]
butylbenzylphthalate	M Rats	--	Inadequate Study
	F Rats	Leukemia	Some Evidence
	M & F Mice	--	No Evidence
diallylpthalate	M Rats	--	No Evidence
	F Rats	Leukemia	Some Evidence
	M & F Mice	Lymphoma(M); Fore-stomach Lesions (M&F)	Equivocal Evidence
di(2-ethylhexyl)adipate	M & F Rats	--	No Evidence
	M Mice	Liver Adenomas	Some Evidence
	F Mice	Liver Carcinomas	Clear Evidence
di(2-ethylhexyl)phthalate	M Rats	Liver Neoplastic Nodules/Carcinomas	Some Evidence
	F Rats	Liver Carcinomas	Clear Evidence
	M & F Mice	Liver Carcinomas	Clear Evidence
pthalamide/phthalic anhydride	M & F Rats and Mice	--	No Evidence

a Evidence of Carcinogenicity -- Five categories of interpretative conclusions have been adopted for use in the NTP Technical Reports series to specifically emphasize consistency and the concept of actual evidence of carcinogenicity. For each definitive study result (male rats, female rats, male mice, female mice) one category is selected to describe the findings. This category refers to the strength of the experimental evidence and not to either potency or mechanism (Huff and Moore, 1983; These Proceedings).

and rats". The NTP interpreted these experimental results for female F344 rats as being "probably carcinogenic". For DEHA, IARC (pages 257-267) stated that "significant increases [were observed] in the incidence of liver-cell tumours in male and female mice" giving the conclusion that "there is limited evidence that di(2-ethylhexyl)adipate is carcinogenic in mice". DEHP was considered by IARC (pages 269-294) to have "significantly increased the incidence of benign and malignant liver-cell tumours in animals of both species, and a dose-response relationship was observed". And their evaluation further reads "there is sufficient evidence of the carcinogenicity of di(2-ethylhexyl)phthalate in mice and rats".

Earlier Studies -- Wilbourn and Montesano (1982) have reviewed in brief the reports on carcinogenicity tests carried out on various phthalate esters prior to the availability of the NCI/NTP carcinogenesis studies (Anon, 1950; Anon, 1968; Carpenter, Weil, and Smyth, 1953; Harris, et al. 1956; Hodge et al. 1953; Omori, 1976; Theiss, et al. 1977). All of these studies, according to Wilbourn and Montesano (1982), suffer from diverse limitations in design or reporting when compared to more recent experimental protocols and reports, and "do not allow any evaluation of the carcinogenicity of these chemicals in mice or rats".

Continued Evaluation of Ortho Phthalic Acid Esters (NTP 1982d) -- Recognizing the technical and economic importances of phthalates and the need for more appropriate extrapolation of rodent carcinogenesis data to human health effects, the NTP will continue to study the deleterious effects of DEHP and to probe for mechanisms of action, as well as to evaluate the toxic potentials of several other phthalate esters both individually and comparatively.

The research described below is complementary both to previous NTP studies and to the endeavors of other groups. Of particular interest to the NTP is an assessment of the mechanisms, dose-relationships, and species dependencies of the toxic responses to DEHP, parameters relevant to predicting human response to various levels of DEHP exposure. Also of interest are structure-activity correlates that would indicate the relative propensities of similar chemicals to produce "DEHP-like" toxic effects.

Genotoxic Potential -- The lack of mutagenic response of DEHP in several systems indicates limited value for such tests in predicting the carcinogenic potentials of other phthalates in rodent bioassays. One plausible theory of DEHP carcinogenicity is that the DEHP-induced proliferation of hepatic peroxisomes (subcellular organelles) results in the increased presence of hydrogen peroxide or hydroxyl radicals in hepatocytes. Such reactive molecules could injure both genetic and non-genetic cellular materials. Peroxisomes are not present in 9,000 x g liver supernatant fractions (S-9) or other mammalian activating systems commonly used in mutagenicity assays. Moreover, the DEHP-induced peroxisomes may differ functionally from normal peroxisomes.

Mechanisms of Carcinogenicity -- In the absence of a probable mechanism of carcinogenic action it is prudent to assume a linear relationship between chemical dose and the occurrence of tumors. The development of neoplasia, however, can be influenced by chemical effects other than direct malignant transformation, and such effects may occur at exposure levels that greatly alter normal anatomical, biochemical, or physiological interrelationships. Recognized biological effects of DEHP in addition to hepato-carcinogenicity include hypolipidemia, enzyme inhibition, peroxisome proliferation, and pituitary hypertrophy (male rats only). The relationships of these effects to the toxic manifestations of DEHP are unknown. To study the dependencies of toxic response on biochemical or physiological abnormalities, the NTP will conduct a subchronic (60-70 days) feeding study in rats and mice at various doses of DEHP, including those which ultimately produced liver neoplasms. Among the parameters measured will be circulating concentrations of several pituitary hormones, serum lipids, peroxisomes (visual enumeration and enzyme activities), and oxidative injury to the liver.

Carcinogenesis Studies -- 1) Butylbenzylphthalate. A six months subchronic feeding study in F344 male rats will help determine the cause or causes of early deaths in the previous study and assess reproductive function. The dose-dependent disposition and metabolism of BBP in rats will be studied to correlate toxicity with pharmacokinetics and to ascertain the amounts of n-butanol and benzyl alcohol released (the latter is currently in chronic bioassay) as metabolites of BBP. A two-year carcinogenicity study of BBP in male rats will be designed and initiated subsequent to a complete analysis of the subchronic study results. 2) Diallylphthalate. Pharmacokinetic studies in rats and mice

will assess the possible implications of metabolism of
diallylphthalate to known carcinogens or mutagens (allyl
alcohol, acrolein, various epoxides) and provide a mecha-
nistic basis for species comparisons in toxic response. 3)
Diethylphthalate. Two-year carcinogenicity studies in rats
and mice by dermal exposure will be initiated. Tests for
initiation and promotion of skin tumorigenesis for both
diethylphthalate and dimethylphthalate will be included in
this assessment.

Reproductive Toxicity -- The teratogenic actions of dietary
DEHP in rats and mice will be defined, as well as the potential for neonatally-administered (via maternal milk) DEHP to
produce morphological or behavioral abnormalities. Further
studies in a single species will define structure activity
relationships and evaluate possible mechanisms of action.

The chemosterilizing effects of DEHP and the potential for
dominant lethal mutations in germ cells will be studied in
male rats and mice administered multiple doses of DEHP by
the dietary route. The relationships of zinc and possible
endocrinolgoical abnormalities to the development of gonadal
lesions will be evaluated.

Dermal Absorption -- The NTP will conduct a cursory comparative study of chemical absorption by monitoring urinary
and fecal excretion of radioactivity following dermal application of radiolabelled phthalate esters. The excretions of
intravenously administered compounds will be measured for
comparative purposes. Structure-absorption characteristics
will be evaluated for:

butylbenzylphthalate	dimethylphthalate
diethylphthalate	di-n-butylphthalate
di(2-ethylhexyl)phthalate	di-n-hexylphthalate
diisobutyphthalate	di-n-octylphthalate
diisodecylphthalate	phthalic acid

NTP/EPA Clearinghouse on Phthalates -- A key recommendation
made at the Conference on Phthalates held in Washington,
D.C. in June 1981 was that the NTP should consider creating
a clearinghouse of toxicological data on this class of compounds. NTP consulted with scientists from federal health
regulatory and research agencies as well as with the
Chemical Manufacturers Association, all of whom supported
the clearinghouse and contributed to its development. The
Environmental Protection Agency (EPA), particularly com-

mitted to this project because of its involvement in the testing of alkyl phthalates under the Toxic Substances Control Act, together with the NTP sponsored the formation of the Clearinghouse. Persons or organizations planning, conducting, or having completed research on the toxicological effects of these compounds are requested to contact NTP/EPA Clearinghouse on Phthalates, Westwood Towers, Room 835B, National Institutes of Health, 9000 Rockville Pike, Bethesda, MD 20205.

Acknowledgements -- The authors acknowledge the contributions of personnel at the National Toxicology Program, Toxicology Research and Testing Program, Research Triangle Park, N.C.; the National Cancer Institute, Division of Cancer Cause and Prevention, Carcinogenesis Program, Bethesda, MD; Frederick Cancer Research Center, Frederick, MD.; E. G. and G. Mason Research Institute, Worcester, MA.; and Tracor Jitco Inc., Research Triangle Park, N.C., in the planning, conduct, and evaluation of these studies. The authors thank Ms. Pamela Chadwick for editorial, format, and typing assistance.

REFERENCES

1. Anonymous (1950). A Study of the Toxicity of Butyl Phthalyl Butyl Glycolate (Santicizer B-16). Submitted to the Environmental Protection Agency under section 8(d) of the Toxic Substances Control Act of 1976, 8D HQ-1078-0250.

2. Anonymous (1968). Effects After Prolonged Oral Administration of Santicizer 160 C (sic.) [Butylbenzyl-phthalate]. Submitted to the Environmental Protection Agency under section 8(d) of the Toxic Substances Control Act of 1976, 8D HQ-1078-0280.

3. Carpenter, C.P., Weil, C.S., and Smyth, H.F. (1953). Chronic Oral Toxicity of Di(2-ethylhexyl)phthalate for Rats, Guinea Pigs and Dogs. Arch. Ind. Hyg. Occup. Med. 8: 219-226.

4. EHP (1973). Perspective on PAEs. Proceedings of the Conference on Phthalic Acid Esters. Held on 6-7 September 1972 in Pinehurst, N.C. Environ. Health Perspect. 3: 1-182.

5. EHP (1982). Phthalate Esters: Proceedings of the Conference on Phthalates. Held on 9-11 June 1981 in Washington, D.C. Environ. Health Perspect. 45: 1-153.

6. Harris, R.S., Hodge, H.C., Maynard, E.A., and Blanchet, H.J., Jr. (1956). Chronic Oral Toxicity of 2-Ethylhexyl Phthalate in Rats and Dogs. Arch. Ind. Health 13: 259-264.

7. Hodge, H.C., Maynard, E.A., Blanchet, H.J., Hyatt, R.E., Rowe, V.K., and Spencer, H.C. (1953). Chronic Oral Toxicity of Ethyl Phthalyl Ethyl Glycolate in Rats and dogs. Arch. Ind. Hyg. Occup. Med. 8: 289-295.

8. Huff, J.E. (1982). Di(2-ethylhexyl)adipate: Condensation of the Carcinogenesis Bioassay Technical Report. Environ. Health Perspect. 45: 205-207.

9. Huff, J.E., and Haseman, J.K. (1983). Background Neoplasms in "Untreated" Fischer 344/N Rats and B6C3F$_1$ Mice (These Proceedings).

10. Huff, J.E. and Moore, J.A. (1983). Carcinogenesis Studies Design and Experimental Data Interpretation/Evaluation at the National Toxicology Program (These Proceedings).

11. IARC (1982). IARC Monographs on the Evaluation of the Carcinogenic Risk of Chemicals to Humans. Some Industrial Chemicals and Dyestuffs. 29: 1-416, International Agency for Research on Cancer, Lyon.

12. Kluwe, W.M. (1982). Overview of Phthalate Ester Pharmacokinetics in Mammalian Species. Environ. Health Perspect. 45: 3-10.

13. Kluwe, W.M., Haseman, J.K., Douglas, J.F., and Huff, J.E. (1982a). The Carcinogenicity of Dietary Di(2-ethylhexyl)phthalate in Fischer 344 Rats and B6C3F$_1$ Mice. J. Toxicol. Environ. Health 10: 797-815.

14. Kluwe, W.M., McConnell, E.E., Huff, J.E., Haseman, J.K., Douglas, J.F., and Hartwell, W.V. (1982b). Carcinogenicity Testing of Phthalate Esters and Related Compounds by the National Toxicology Program and the National Cancer Institute. Environ. Health Perspect. 45: 129-133.

15. Kluwe, W.M., Haseman, J.K., and Huff, J.E. (1983). The Carcinogenicity of Di(2-ethylhexyl)phthalate (DEHP) in Perspective. J. Toxicol. Environ. Health 12: in press.

16. NCI (1979a). Bioassay of Phthalamide for Possible Carcinogenicity (CAS No. 88-96-0). NCI Technical Report Series TR No. 161, National Cancer Institue, Bethesda, MD (USA) 114 pages.

17. NCI (1979b). Bioassay of Phthalic Anhydride for Possible Carcinogenicity (CAS No. 85-44-9). NCI Technical Report Series TR No. 159, National Cancer Institute, Bethesda, MD (USA). 107 pages.

18. Northrup, S., Martis, L., Ubricht, R., Gaber, J., Maripol, J., and Schmitz, T. (1982). Comment on the Carcinogenic Potential of Di(2-ethylhexyl)phthalate. J. Toxicol. Environ. Health 10: 493-518.

19. NTP (1982a) Carcinogenesis Bioassay of Butylbenzyl-phthalate (CAS No. 85-68-7) in F344 Rats and $B6C3F_1$ Mice (Feed Study). NTP Technical Report Series TR No. 213, National Toxicology Program, Research Triangle Park, N.C. (USA). 98 pages.

20. NTP (1982b) Carcinogenesis Bioassay of Di(2-ethylhexyl) adipate (CAS No. 103-23-1) in F344 and $B6C3F_1$ Mice (Feed Study). NTP Technical Report Series TR No. 212, National Toxicology Program, Research Triangle Park, N.C. (USA). 121 pages.

21. NTP (1982c). Carcinogenesis Bioassay of Di(2-ethyl-hexyl)phthalate (CAS No. 117-81-7) in F344 Rats and $B6C3F_1$ Mice (Feed Study). NTP Technical Report Series TR No. 217, National Toxicology Program, Research Triangle Park, N.C. (USA). 127 pages.

22. NTP (1982d). Ortho Phthalic Acid Esters. NTP Tech. Bull. 8: 1-3 (July).

23. NTP (1983). Carcinogenesis Bioassay of Diallylphthalate (CAS No. 131-17-9) in $B6C3F_1$ Mice (Gavage Study). NTP Technical Report Series TR No. 242, National Toxicology Program, Research Triangle Park, N.C. (USA). 96 pages.

24. NTP (1984 DRAFT). Toxicology and Carcinogenesis Studies of Diallylphthalate (CAS No. 131-17-9) in F344/N Rats (Gavage Study). NTP Technical Report Series TR No. 284, National Toxicology Program, Research Triangle Park, N.C. (USA).

25. Omori, Y. (1976). Recent Progress in Safety Evaluation Studies on Plasticizers and Plstics and their Controlled use in Japan. Environ. Health Perspect. 17: 203-209.

26. Sontag, J.M., Page, N.P., and Saffiotti, U. (1976). Guidelines for Carcinogen Bioassay in Small Rodents. NCI Carcinogenesis Technical Report Series TR No. 1, National Cancer Institute, Bethesda, MD (USA). 65 pages.

27. Theiss, J.C., Stoner, G.D., Shimkin, M.B., and Weisburger, E.K. (1977). Tests for Carcinogenicity of Organic Contaminants of United States Drinking Waters by Pulmonary Tumor Response in Strain A Mice. Cancer Res. 37: 2717-2720.

28. USTIC (1983). Preliminary Report on U.S. Production of Selected Synthetic Organic Chemicals (including Synthetic Plastics and Resins) March, April, and Cumulative Totals, 1983. United States Trade Commission, Washington, D.C. Series C/P-83-4.

29. Wilbourn, J. and Montesano, R. (1982). An Overview of Phthalate Ester Carcinogenicity Testing Results: The Past. Environ. Health Perspect. 45: 127-128.

Industrial Hazards of Plastics and Synthetic Elastomers, pages 155-175
© 1984 Alan R. Liss, Inc., 150 Fifth Ave., New York, NY 10011

OCCUPATIONAL HAZARDS IN THE VC-PVC INDUSTRY

William J. Nicholson, Paul K. Henneberger and
Herbert Seidman.

Environmental Sciences Laboratory, Mount Sinai
School of Medicine of CUNY, New York, New York
10029 (WJN, PH) and American Cancer Society, 4
W. 35th Street, New York, New York 10001 (HS).

INTRODUCTION

On January 24, 1974, The Wall Street Journal published an article describing the occurrence of three deaths
from hemangiosarcoma of the liver among polyvinyl chloride
(PVC) production workers at the B.F. Goodrich Tire and
Rubber Company plant in Louisville, Kentucky. This announ-
cement shattered the relatively complacent view toward
health effects associated with plastic production in
general and PVC production in particular. At the time,
U.S. and Western European production of vinyl chloride
(VC) exceeded 6×10^6 metric tons. Numerous mortality and
clinical studies were undertaken in the major producing
countries in an attempt to establish the extent of the
carcinogenic risk and to identify clinical parameters
useful for surveillance of exposed groups. Because of the
immediate concern in 1974, most of these studies were completed between 1974 and 1977. Several reviews and sympo-
sia on human health effects from VC exposure have been
published recently. A superb one is by Lelbach and Marsteller (1981).

The exposures were high that led to the disease
observed in these various studies. Typical concentrations
in the industry were estimated to be about 1,000 ppm prior
to 1955, from 300-500 during 1955-1970, and from 100-200
during 1970-1974 (Barnes, 1976). However, variations from
such exposures would have occurred in specific plants
(Rowe, 1975). While historical average exposures were
generally less than 1,000 ppm, peak exposures often ex-

ceeded 5-10,000 ppm (where workers lost consciousness)
and, on occasion, 40,000 ppm (where plants exploded).
During 1974, exposures were reduced to about 10-20 ppm in
the U.S. industry (Jones, 1981) and even further, following the promulgation of a 1 ppm standard by the Occupational Safety and Health Administration in 1974.

MORTALITY STUDIES OF VC-EXPOSED WORKERS

Table 1 shows the populations observed and the follow-
up characteristics of twelve cohort studies of vinyl chloride exposed workers. The studies were independent with
the exception that the portions of the population reported
in the Equitable Environmental Health Study (1978) were
included in some other U.S. studies. The proportionate
mortality study of Monson et al (1974) is not included as
the VC-exposed individuals studied therein were included
in the cohort mortality study of Waxweiler et al (1976).
The size of the cohorts varied greatly, from 255 in the
study of Nicholson et al (1975) to 9,677 in the Equitable
Environmental Health study. A notable feature of all of
the studies is that the populations followed were relatively young or recently employed, even though many plants
in the studies started production in the 1940s. Most
workers were hired after 1950, when U.S. and Western
European production increased sixfold in ten years
(Nicholson and Henneberger, 1983). Thus, few deaths
occurred among most of the groups observed and data on
effects 25 or more years from onset of exposure are limited. The total mortality exceeded 10% of the observation cohort in only three studies. Further, the inclusion
of recently employed individuals or those with short
employment diluted the effects from VC exposure. Only
five studies limited consideration to individuals with
more than one year of exposure. In all cases, however,
some individuals with more than 20 years from onset of
employment were available for observation. The follow-up
terminated in the mid-1970s for all studies.

Table 2 compares the results for cancer of all sites
and chronic liver disease in all 12 studies. Cancer is
elevated in most of the studies, although it does not
achieve a 0.05 level of significance except in the studies
by Waxweiler et al (1976) and Nicholson et al (1975). In
the study by Ott et al (1975), a highly exposed subgroup
with 15 years latency had 8 cancer deaths compared to 3.2

Table 1

Population and follow-up characteristics of twelve
studies of vinyl chloride exposed workers

Study	Country	Analysis cohort size	Percent additional untraced	Number of deaths analyzed	Percent of total	Minimum exposure (years)	Minimum latency (years)	Earliest possible exposure	Maximum follow-up (years)	Last year of follow-up
Bertazzi et al. 1979	ITAL	4777	13.8	62	1.3	0.5	0.5	1952	22	1977
Buffler et al. 1979	USA	464	0.0	28	6.0	0.2	0.2*	1948	27	1975
Byren et al. 1976	SWED	750	low	58	7.7	>0	>0*	1945	28	1974
Duck et al. 1975	UK	2113	0.3	136	6.4	>0	>0*	1948	20	1975
Equitable Env. Health, 1978	USA	9677	4.9	707	7.3	1	1*	(1935)	25+	1972
Fox and Collier 1976, 1977	UK	7409	1.1	393	5.1	>0	>0*	1940	35	1974
Masuda 1979	JAP	304	0.3	26	8.6	1	1	1949	20	1975
Nicholson et al. 1975	USA	255	0.8	24	9.4	5	10*	1947	25	1974
Ott et al. 1975	USA	522	0.0	79	15.1	>0	>0	1942	31	1973
Reinl et al. 1979	GER	6544	7.3	414	6.3	>0	>0	NA	NA	1974
Theriault and Allard, 1981	CAN	451	2.8	59	13.1	5	5*	1943	30	1977
Waxweiler et al. 1976	USA	1287	0.5	136	10.6	5	10*	1940	22	1973

* Longer latencies considered for some causes of death.

Table 2

Observed and expected deaths among vinyl
chloride exposed workers in twelve studies

Study	Cancer of all sites Deaths			Chronic liver cancer Deaths		
	Obser.	Expec.	SMR	Obser.	Expec.	SMR
Bertazzi et al	30	30.9	97	5	–	–
Buffler et al	8	5.19	154	0	–	–
5 yr. latency	6	4.34	138			
Byren et al	–	–	–	0	–	–
Duck et al	35	36.44	96	–	–	–
Equitable	139	141.39	104*	14	26.45	56††
Fox & Collier	115	126.77	91	1	2.68	37
Masuda	8	5.8	138	5	1.00	500[a]
Nicholson	9	3.9	230	1	(0.6)	167
Ott et al	13	16.0	81	3	2.7	111
15 yr. latency	9	9.2	98			
Reinl et al	94	90.6	112*	14	18.4	82
Theriault & Allard	20	16.37	122	–	–	–
Waxweiler et al	35	23.5	149†	2	4.0	50
15 yr. latency	31	16.9	184††			

------

* Adjusted for unknown causes of death
( ) = Estimated as a percentage of U.S. rates
 † $p < 0.05$
†† $p < 0.01$
 a SMR of control population equally high

expected ($p < 0.05$). The absence of significant findings
in other studies may be attributed to their low power.
The study of Bertazzi et al (1979) may be biased because
of low follow-up in the group. Fourteen percent of the
population were untraced and person-years at risk were
calculated for these individuals as if they were alive.
The low SMR of 44 for all causes of death suggests that
proportionately more deaths occurred in untraced groups
than in the traced. The studies by Buffler et al (1979),
Byren et al (1976), Masuda (1979), and Theriault and
Allard (1981) had very few deaths available for analysis.
That of Ott et al (1975) also was limited by the number of
deaths and further by virtue of a study group with relatively lower exposure (through better industrial hygiene
control). While having more deaths available for analysis
(136), the study by Duck et al (1975) was significantly

diluted by the inclusion of many individuals with very short and recent periods of exposure.

Turning to chronic liver disease, one remarkable finding is the absence of significantly elevated mortality from this cause in most of the populations under observation. The only study with a significant elevation is that of Masuda (1979) in which five deaths from chronic liver disease occurred where only one was expected. However, this must be considered in the light of an equally high mortality from liver disease (6 observed vs. 1.4 expected) in a comparison population followed for control purposes. Five of 62 deaths from chronic liver disease seen in the study by Bertazzi et al (1979) are unusual, but the limitations of this study and lack of details make evaluation difficult. The generally benign results in other studies contrast sharply with the severe liver disease from VC exposure documented in clinical studies (Marsteller et al, 1975). Hepatomegaly, hepatic fibrosis, portal hypertension, and bleeding esophageal varices have commonly been found in individuals heavily exposed to VC, even without concomitant exposure to alcohol.

Table 3 lists the mortality data for primary cancer of the liver and biliary passages and for cancer of the lung, trachea and bronchus. In the case of liver cancer, the overall data are consistent and dramatic. Hemangiosarcomas of the liver were found in eight of the twelve studies. In each of the eight, a very large and highly significant SMR for liver cancer was seen. Methodological limitations can account for negative data in the other four studies. The large SMR's observed, however, are largely the result of low values for the expected number of cases rather than a high incidence of observed cases. Only 28 separate liver hemangiosarcomas were identified in all twelve studies. As the overall excess number of deaths from liver and biliary cancer in all studies was 47, some hemangiosarcomas may not have been identified. The low numbers must also be considered in light of the limited follow-up times in most studies.

The evidence for lung cancer is less clear. There is an elevation in some studies, but at a level that does not achieve statistical significance, except in the 15 year latency population of Waxweiler et al (1976). This, in part, may be the result of the low power of many of the

Table 3

Observed and expected deaths from selected causes
among vinyl chloride-exposed workers

	Cancer of the liver and biliary passages				Cancer of the lung, trachea and bronchus		
	Obs.[a]	Exp.	SMR	Hemangio-sarcomas	Obs.	Exp.	SMR
Bertazzi	8	$(1.0)^b$	(800)+++	3	7	$(7.7)^c$	(91)
Buffler 5 yr. latency	0	(0.17)	--	0	5 4	1.73 1.49	289† 268
Byren 10 yr. latency	4 4	0.97 0.68	413† 589††	2 2	3	1.78	168
Duck 19 yr. latency	--	--	--	0	16 14	15.53 10.69	103 131
Equitable 15 yr. latency	10	(4.5)	(224)†	5	45 41	44.29 37.0	107 111
Fox and Collier 15 yr. latency	4	0.71·	563††	2	46 28	51.23 26.0	90 108
Masuda	1	0.6	167	0	1	(0.8)	(125)
Nicholson	3	(0.12)	(2500)+++	3	0	(1.1)	--
Ott	0	(0.5)	--	0	4(5?)	5.2	77(96?)
Reinl	12	0.9	1523+++	4	22	24.6	95^b
Theriault 15 yr. latency	8	(0.5)	(1600)+++	8	2 2	5.78 4.25	35 47
Waxweiler 15 yr. latency	7 7	0.6 0.4	1155+++ 1606+++	6 6	12 11	7.7 5.7	156 194†
Total of nonduplicated hemangiosarcomas							29

----

† < 0.05
†† < 0.01
††† < 0.001

a All verified liver cancer deaths, including those established by
  review of all available information.

b ( ) = Expected deaths estimated on the basis of 1950-1969 U.S.
        adjusted rates, ICD 155/ICD 140-205.

c ( ) = Expected deaths estimated on the basis of national age adjusted
        rates, ICD 162-163/ICD 140-205.

d One hemangiosarcoma occurred in a PVC fabricator.

e Includes cancer of the pancreas.

studies. Only two have an 80% power to detect an overall risk of 1.5 (Beaumont and Breslow, 1981). Of significance, however, are the very low SMR's in the groups studied by Theriault and Allard (1981), Reinl, et al (1979), and Nicholson et al (1975), cohorts that would be expected to manifest a high risk on the basis of the many hemangiosarcomas that were found. The four largest studies, although in some cases limited by inclusion of short-term and recently employed workers, also are noteworthy for the SMR's close to 100. Where available, data on subcohorts with longer latency (> 15 yr) suggest some increased risk.

Waxweiler et al (1981) undertook a detailed analysis of the exposure of those with lung cancer in their previously published study (Waxweiler et al, 1976) in an attempt to identify particular etiological agents. The analysis used a serially additive expected dose model (Smith et al, 1980) in which a dose measure during each year of exposure was accumulated for each study individual for a variety of potentially carcinogenic agents. The cumulative doses for those with lung cancer were compared with those of other individuals in the plant under study. The results showed that the greatest correlation of lung cancer was with exposure to PVC dust. Secondarily, exposure to vinylidene chloride appeared to be important, but only for large cell and adenocarcinoma. The serially additive dose for VC monomer differed little in those with lung cancer compared to others in the plant, except, possibly, for large cell cancers.

Thus, evidence to date does not establish that VC monomer is an important lung carcinogen in exposed worker populations, although it is recognized that limited long-term observation has so far been available. In all studies considered here, a slight deficit of cases was seen compared to the number expected. In the subcohorts with more than 15 years from onset of exposure, an overall excess of 10% was observed. If, in addition, one considers a "healthy worker effect," any excess lung cancer would still be considerably less than the excess of liver cancer. A qualification to this conclusion is that no study specifically considered cigarette usage. If cigarette smoking was much less common among VC workers than the general population, higher SMR's would have been seen if smoking specific data were available. However, this possibility is unlikely, considering the many different populations studied. The uncertainty in human data is also reflected in animal studies. Increased lung cancers have been seen in mice but not in rats or hamsters (Maltoni et al, 1981).

Table 4 shows the results for brain and central nervous system cancers and for cancers of the lymphatic and hematopoietic systems. Cancers of the brain and central nervous system were significantly elevated in a number of studies, although the results differed considerably across studies. Again, negative data may be simply the result of limited long-term follow-up or the low power of the study. In such cases the information is only sufficient to set an upper limit on relative risk of brain cancer. In contrast to lung cancer, however, the largest study group has a significantly elevated risk of brain and central nervous system malignancy. As with lung cancer, the data on brain and CNS cancer in animals are equivocal. Neuroblastomas and brain malignancies are observed in rats exposed to VC, but not among mice or hamsters (Maltoni et al, 1981). The human data are also mitigated by the recent finding of brain and central nervous system tumors in a variety of chemical plant exposure circumstances (Alexander et al, 1980; Selikoff et al, 1982). Excess brain malignancies, but not the etiological agents, have been identified in several Texas and Louisiana chemical/ petrochemical plants. VC exposure was documented for some cases, but it could not explain the overall findings. As individuals in many of the VC studies considered here were exposed to other chemicals and petrochemicals, the possible role of these agents cannot be excluded. Further, it has been suggested that some working groups, with employer-paid medical plans, may have better case ascertainment than is generally available (Greenwald et al, 1981) and, thus, more brain malignancies identified. In any case, the number of excess malignancies of the brain and central nervous system (approximately 10) in all studies is considerably less than the number of hemangiosarcomas identified in the same populations.

Similar results are obtained for malignancies of the lymphatic and hematopoietic system. Here again, the analysis is limited by the few deaths and disparate results which occurred in different studies. Overall, there would appear to be an elevated risk, but the influence of

Table 4

Observed and expected deaths from selected causes
among vinyl chloride exposed workers

	Cancer of the brain & central nervous system			Cancer of the lymphatic and hematopoietic system		
	Obser.	Expect.	SMR	Obser.	Expect.	SMR
Bertazzi	1	$(0.8)^a$	125	4	$(3.0)^b$	(133)
Buffler	0	(0.1)	–	0	(0.5)	–
Byren	2	0.33	612[+]	0	–	–
Duck	–	–	–	–	–	–
Equitable	12	5.90	203[+]	20	17.01	124
Fox & Collier	2	3.66	55	9	9.01	100
Masuda	0	(0.15)	–	0	(0.5)	–
Nicholson	1	(0.1)	(1000)	2	(0.4)	(500)
Ott	1	0.4	(250)	1	(1.6)	(63)
Reinl	2	1.3	162	15	7.7	214[++]
Theriault	0	0.6	–	1	1.67	60
Waxweiler	3	0.9	329	4	2.5	159
15 yr. latency	3	0.6	498[+]			

[+] < 0.05

[++] < 0.01

[a] ( ) = Expected estimated from the ratio of age standardized
U.S. rates ICD 193/ICD 140-205.

[b] ( ) = Expected estimated from the ratio of 1950-1969 U.S.
rates ICD 200-205/ICD 140-250.

confounding exposures precludes definitive statements.
The overall excess of such malignancies (about 10) is also
much less than those from primary hemangiosarcomas of the
liver.

EFFECT OF REDUCTION OF EXPOSURE TO VC

As mentioned previously, most mortality studies followed
populations only to the 1972-1975 period. No data exist
on the risk to previously exposed populations after cessation of exposure in 1974, although hemangiosarcomas have
been noted among retirees. We have recently completed a
follow-up through 1981 of the population reported in 1975
(Nicholson et al, 1975) to determine whether a high risk
of liver cancer continues, following significant reduction
in exposure. The original group employed at a VC polymer-

ization plant in Niagara Falls, New York, has been expanded by 40 additional workers, all exposed for five
years, who achieved ten years from onset of exposure
subsequent to April 1974. Additionally, 195 individuals
employed at a VC polymerization plant in South Charleston,
West Virginia, with five years of exposure and ten years
from onset in December, 1966, were identified and traced
through 1980.

Table 5 lists the observed and expected deaths by
cause for both groups with the deaths occurring after 1974
separately identified. (These are preliminary data; full

Table 5

Observed and expected deaths among
vinyl chloride polymerization workers

Niagara Falls, NY (N = 296)
(January 1, 1956 - December 31, 1981)

Cause of death	Observed 56-74	74-81	Total	Expected	SMR
All causes	23	21	44	40.87	108
All cancer	8	8	16	9.01	177[a]
Lung	0	2	2	3.25	62
Colon/rectum	1	2	3	1.39	216
Brain	1	0	1	0.33	303
Liver	3	3	6	0.19	3158[b]
Lymphoma	2	1	3	0.55	545[a]
Pancreas	1	0	1	0.50	200
Cirrhosis of liver	1	1	2	1.41	142
Cardiovascular disease	13	8	21	19.88	106

South Charleston, WV (N = 195)
(December 1, 1966 - December 31, 1980)

Cause of death	Observed 66-73	74-80	Total	Expected	SMR
All causes	12	24	36	44.74	80
All cancer	2	10	12	10.65	113
Lung	1	1	2	4.07	49
Colon/rectum	0	0	0	1.89	--
Brain	0	0	0	0.43	--
Liver	0	4	4	0.23	1739[c]
Lymphoma	0	0	0	0.59	--
Pancreas	0	0	0	0.67	--
Cirrhosis of liver	0	1	1	1.81	55
Cardiovascular disease	8	12	20	27.24	73

----
a $p < 0.05$
b $p < 0.001$
c $p < 0.0005$

pathological review of all available specimens has not been completed.) Among the 44 deaths that occurred in the Niagara Falls cohort, 6 were from primary cancer of the liver, including 5 hemangiosarcomas. Three of the hemangiosarcomas occurred in the period prior to 1974 and 2 subsequently. Similar findings occurred among the smaller group in West Virginia. Here, of 36 deaths, 4 were from hemangiosarcoma, all of which occurred subsequent to 1974. Thus, the risk of neoplastic VC disease continues undiminished, even though exposures to the monomer have been significantly reduced. The combined data from both groups are shown in Table 6 and demonstrate an excess risk of cancer, which is totally accounted for by the enormously increased risk of liver malignancy observed in each time from onset of exposure category. The excess lymphomas which achieved significance at the $p < 0.05$ level in the Niagara Falls group lose significance when combined with the data from South Charleston. A deficit of lung cancer was observed in both study groups and brain malignancies were about equal to the number expected.

It is not certain whether the results of these two plants will be reflected in the results of other plants in future years. The South Charleston plant was the first facility to commercially produce VC. The New York plant opened immediately following the cessation of World War II. Thus, we are observing effects in populations that include many individuals with long times from onset of exposure. There is no information on whether the exposures in these two plants were significantly different from those of the majority of other VC polymerization facilities. It is known that pre-1974 exposures in the New York plant were sufficiently high to cause loss of consciousness to some individuals (4.5% of those examined in the clinical survey of 1974) (Lilis et al, 1975).

MORBIDITY AND CLINICAL FINDINGS AMONG VC-EXPOSED WORKERS

Clinical abnormalities from VC exposure predated by 25 years the documentation of its carcinogenicity. Various VC-related abnormalities were reported in Eastern European literature, including hepatomegaly (Tribukh et al, 1949), angioneurosis (Filatova and Gronsberg, 1957), osteolytic lesions of distal phalanges (Smirnova, 1961), Raynaud's phenomenon and sclerodermalike skin lesions (Suciu et al, 1963). However, VC disease was not seri-

Table 6

Observed and expected deaths among vinyl chloride
exposed workers in two polymerization facilities
by time from onset of exposure

Years since onset of exposure

Cause of death	10 - 19		20 - 29		30+		Total		
	Obs.	Exp.	Obs.	Exp.	Obs.	Exp.	Obs.	Exp.	SMR
All causes	24	19.13	30	34.83	26	31.64	80	85.61	93
All cancer	7	3.72	9	7.98	12	8.03	28	19.66	142
Lung	1	1.26	1	2.96	2	3.08	4	7.31	55
Liver	2[a]	0.08	3[a]	0.18	5[b]	0.17	10	0.42	2381
Brain	1		0		0		1	0.76	132
Lymphoma	2	0.27	1	0.46	0	0.41	3	1.14	263
Cirrhosis of liver	0	0.76	3	1.24	0	0.87	3	2.85	105
Cardiovascular disease	14	8.74	15	17.58	12	16.40	41	42.73	96
Person years	2924		2734		1404				

a. hemangiosarcoma
b. 4 hemangiosarcomas and 1 hepatoma

ously considered in the West until the published description of Raynaud's syndrome, acroosteolysis, and pseudoscleroderma in two Belgium VC reactor cleaners (Cordier et al, 1966). Additional cases were soon noted (Wilson et al, 1967) and a comprehensive epidemiological study of 5,011 U.S. workers employed in production and polymerization was undertaken. It showed that 11.9% had possible X-ray signs of acroosteolysis, compared with 3.2% in a Michigan general population control group, with 2% definitely having Raynaud's phenomenon or X-ray evidence of acroosteolysis (Dinman et al, 1971). The conditions were clearly associated with the cleaning of reactors, in which a heavy exposure to VC occurred. Only one case of Raynaud's phenomenon occurred among 557 workers employed in PVC fabrication.

During the early 1970's, VC liver disease was described in detail by Marsteller et al (1973, 1975). Observations on selected workers showed hepato- and splenomegaly to be common. Peritoneoscopy and guided liver biopsy identified severe portal hypertension in some, generally without cirrhotic fibrosis, although perisinusoidal and focal or diffuse capsular fibrosis were commonly seen. The portal hypertension could lead to bleeding esophageal varices, with possible fatal consequences. In heavily exposed individuals, the portal hypertension and hepatic fibrosis often progressed after cessation of exposure (Martin et al, 1974). The histology of malignant and nonmalignant liver disease has been well described by Popper and Thomas (1975; Thomas et al, 1975), who suggested the possibility of an interrelationship between hemangiosarcoma and the proliferation of sinusoidal lining cells and hepatocytes seen in VC fibrosis. Lelbach and Marsteller (1981) have also noted that the vast majority of hemangiosarcoma cases have appeared on a background of some degree of hepatic fibrosis. The implications of these suggestions for a hemangiosarcoma dose-response relation are uncertain.

During 1974, extensive studies were undertaken by the Environmental Sciences Laboratory of the total workforces of three polymerization plants in the states of New York, Michigan and West Virginia. The results from the New York plant (Lilis et al, 1975) indicated the presence of acroosteolysis in heavily exposed individuals. Hepato- and splenomegaly or hepatic tenderness was commonly observed and associated with duration of exposure and elevated alkaline phosphatase levels. Sixty-four of 354 had an enlarged or tender liver or spleen and of these, 41% had elevated alkaline phosphatase. Liver function tests were not particularly revealing, except for a correlation of elevated alkaline phosphatase levels with duration of exposure. Additionally, carcinogenic embryonic antigen titers were slightly higher among vinyl chloride exposed groups than in a smoking matched control population (Anderson et al, 1978).

Tamburro and Greenberg (1981) have evaluated the effectiveness of federally mandated screening tests for vinyl chloride exposed workers. Figure 1 shows the results on specificity and sensitivity for 78 individuals with hepatic status determined by biopsy. ICG clearance had the highest combined sensitivity and specificity, with SGPT the second most useful test. Elevated alkaline phosphatase had the greatest specificity of all tests, particularly for chemically-induced liver injury, but was lacking in sensitivity. SGOT and GGPT were of limited use because of their low specificity for chronic liver disease. They recommended the use of ICG clearance for screening, to be followed with alkaline phosphatase determinations for those with altered clearance.

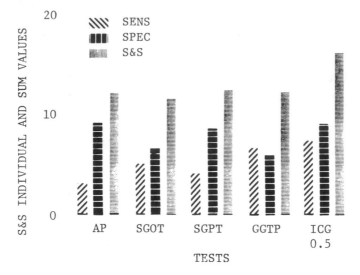

Figure 1: Sensitivity and specificity of various
biochemical screening tests and their sensitivity
and specificity sum values (S & S) based on 78
with biopsy documentation of their hepatic status.

Three of the seven individuals who died after 1974
with hemangiosarcoma in the previously described mortality
followup were examined in 1974. One, who died 22 months
after examination, had no noteworthy abnormalities on
examination (alkaline phosphatase was 88, slightly high).
A second, who died three years after examination, had a
slightly enlarged, palpable liver (11 x 6 cm) with normal
blood counts and chemistry. Only one of the above drank
alcohol at all and he only drank 2-3 beers/month. The
third, who died 22 months after examination, had a slightly enlarged liver (11 x 8 cm) and spleen (13 x 8 cm), and
slightly elevated alkaline phosphatase (93), SGOT (52) and
CEA (4.7). Thrombocytopenia was also present (75,000).
No data are available on later clinical parameters, but
the above results are clearly not sufficiently specific
for identification of a special risk.

Pulmonary abnormalities also have been associated
with VC/PVC exposure. Small opacities, predominantly
irregular, of profusion 1/0 or greater were found in 20 of
1,216 workers employed at PVC production in an Italian
plant (Mastrangelo et al, 1981). All had been exposed to
high levels of PVC dust (>10 mg/m^3). Lilis et al (1976)

reported that approximately 20% of VC/PVC workers with high exposures to PVC dust had abnormal X-rays, which correlated with duration of exposure and, also, with cigarette smoking. In contrast, only 4.7% of individuals in a PVC plant with low dust levels had abnormal X-rays. In addition to "typical pneumoconiosis," a granulatomous reaction to PVC dust has been reported (Arnaud et al, 1978). Miller et al (1975) have observed pulmonary function abnormalities (a reduction in the ratios FEV_1/FVC and MMF/predicted MMF) in both smokers and non-smokers heavily exposed to PVC dust (and also to VC monomer). Maltoni and Lodi (1981), observed greater percentage of abnormal sputum cytological results among VC exposed workers compared to several other groups of manufacturing workers or miners. Only workers in the chromium industry demonstrated a greater proportion of abnormal cells.

Ducatman et al (1975) have observed an increased frequency of chromosome abnormalities in the lymphocyte cultures of VC workers. Most of the abnormalities were "unstable" changes, such as fragments, dicentrics, and rings. This was confirmed by Purchase et al (1978), among others. Some of the group studied by Purchase was resampled 18 and 42 months later (Anderson et al, 1980). In those studied during January 1976, the frequency of abnormalities was increased in those who continued VC/PVC employment, but decreased in those who left the industry. In January 1978, no increased frequency was found in any worker. The authors attributed the decrease to the reduction in VC exposure.

HEALTH HAZARDS IN THE PVC PROCESSING INDUSTRY

Prior to identification of hemangiosarcoma in VC polymerization workers, little effort was made to control either the concentration of residual monomer in PVC dust or exposures to dust and VC that occurred in the various forming operations of the PVC fabricating industry. VC concentrations in excess of 10 ppm occurred frequently. While these concentrations were significantly lower than those of the polymerization industry, the much greater employment in the processing industry (hundreds of thousands vs. tens of thousands in the polymerization work) raised concern for population health effects, particularly for malignant disease for which no threshold was known. However, only two hemangiosarcomas have been documented in the PVC processing industry, one in an accountant in a plant making PVC fabric and one in an Italian plant making PVC sacks. A third case may have occurred in an electrical wire insulator, but the pathological diagnosis is uncertain (Lloyd, 1975). This is in contrast to 85 cases known to have occurred among polymerization workers (NIOSH, 1982). This is somewhat comforting and indicates a significantly lower total VC-related neoplastic risk among fabrication workers. However, it should be noted that case finding is likely to be poorer in this group than in polymerization workers.

A proportionate mortality study has been conducted of 4,341 deaths of former employees of 17 PVC fabricators (Chiazze Jr., et al, 1977). The direct PMR's suggested an excess in total cancer mortality among both white men and white women with the major excesses concentrated in cancers of the digestive organs. An excess of breast cancer was also seen in women, but not confirmed in a case-control study (which was of very low power and could only detect a threefold increased risk) (Chiazze, Jr. et al, 1980). The results of the proportionate mortality study must be considered cautiously. In such studies, elevated cancer risks and are typically seen because of a "healthy worker effect," which leads to a reduction in cardiovascular deaths relative to those of cancer. If PCMR's (proportionate cancer mortality ratios) had been calculated, rather than PMR's, digestive cancer would still be elevated but not at an 0.05 level of significance. Interestingly, an excess of stomach cancer was seen in the proportional mortality study of Baxter and Fox (1976).

SUMMARY

Overall, the results of the analysis of 12 studies of VC production and polymerization workers demonstrate an enormously elevated risk of liver malignancies, the possibility of a twofold increased risk of brain and central nervous system tumors and perhaps, also, of malignancies of the lymphatic and hematopoietic system. However, the role of other agents cannot be excluded in the etiology of nonhepatic malignancies. Bronchogenic carcinoma does not appear to be increased from exposures to VC monomer, although a relationship to PVC dust was suggested in one study. These conclusions must be considered in light of limited data on workers followed more than 25 years from onset of exposure. Considering the numbers of observed and expected deaths in all studies, it would appear that the excess of malignancies at nonhepatic sites is less than the excess of liver tumors. Data presented elsewhere in this volume (Nicholson and Henneberger, 1983) suggest that exposure reductions in 1974 may have virtually eliminated the VC-associated risk of liver cancer if the current U.S. standard is met. To the extent that VC exposure is associated with other cancers, a similar risk reduction would be expected.

Raynaud's phenomenon, acroosteolysis, sclerodermalike skin lesions, hepato- and splenomegaly with noncirrhotic hepatic fibrosis, and severe portal hypertension have been associated with past heavy exposures to VC. Evidence exists that the liver disease and portal hypertension may progress following cessation of exposure. However, all of the above syndromes were found largely in heavily exposed individuals. Their occurrence would be much less likely in workers exposed only to concentrations currently allowed. Pulmonary deficits, X-ray abnormalities, and, perhaps, lung cancer have been associated with VC/PVC exposure. Because of the possible contribution of PVC dust to these findings, engineering controls during polymer drying, bagging and usage are warranted.

REFERENCES

Anderson HA, Snyder MS, Lewinson T, Woo C, Lilis R, Selikoff IJ (1978). Levels of CEA among vinyl chloride and polyvinyl chloride exposed workers. Cancer 42:1560-1567.

Anderson D, Richardson CR, Weight TM, Purchase IFH, Adams WGF (1980). Chromosomal analyses in vinyl chloride exposed workers: Results from analysis 18 and 42 months after an initial sampling. Mutation Res 79:151-162.

Alexander V, Leffingwell SS, Lloyd JW, Waxweiler RJ, Miller RL (1980). Brain cancer in petrochemical workers: A case series report. Am J Ind Med 1:115-123.

Arnaud A, Pommier de Santi P, Garbe L, Payan H, Charpin J (1978). Polyvinyl chloride pneumoconiosis. Thorax 33:19-25.

Barnes AW (1976). Vinyl chloride and the production of PVC. Proc R Soc Med 69:277-281.

Baxter PJ, Fox AJ (1976). Angiosarcoma of the liver in P.V.C. fabricators. Lancet I:245.

Beaumont JJ, Breslow NE (1981). Power considerations in epidemiologic studies of vinyl chloride workers. Am J Epidem 114:725-734.

Bertazzi PA, Villa A, Foa V, Saia B., Fabbri L, Mapp C, Marcer C, Manno M, Marchi M, Mariani F, Bottasso F (1979). An epidemiological study of vinyl chloride exposed workers in Italy. Arh hig rada toksikol 30:379-397 (Suppl).

Buffler PA, Wood S., Eifler C, Suarez L, Kilian DJ (1979). Mortality experience of workers in a vinyl chloride monomer production plant. J Occ Med 21:195-202.

Byren D, Engholm G, Englund A, Westerholm P (1976). Mortality and cancer morbidity in a group of Swedish VCM and PCV production workers. Environ Health Persp 17:167-170.

Chiazze Jr L, Nichols WE, Wong O (1977). Mortality among employees of PVC fabricators. J Occ Med 19:623-628.

Chiazze Jr L, Wong O, Nichols WE, Ference LD (1980). Breast cancer mortality among PVC fabricators. J Occ Med 22:677-679.

Cordier JM, Fievez C, Lefevre MJ, Sevrin A (1966). Acro-osteolysis combined with skin lesions in two workers exposed in cleaning autoclaves. Cahiers Med Travail 4:14, 3-39.

Dinman BD, Cook VA, Whitehouse WM, Magnuson HJ, Ditcheck T (1971). Occupational Acroosteolysis. I. An epide-miological study. Arch Environ Health 22:61-73.

Ducatman A, Hirschhorn K, Selikoff IJ (1975). Vinyl chlo-ride exposure and human chromosome aberrations. Mutation Res 31:163-168.

Duck BW, Carter JT, Coombes EJ (1975). Mortality study of workers in a polyvinyl-chloride production plant. Lancet II:1197-1199.

Equitable Environmental Health, Inc, (1978). Epidemio-logical study of vinyl chloride workers. Equitable Environmental Health, Inc., 6000 Executive Blvd, Rockville MD 20852.

Filatova VS, Gronsberg ES (157). Hygienic working condi-tions in the production of polyvinyl chloride resins and measures for improvement. Gig Sanit 1:38-42 (Russian text).

Fox AJ, Collier PF (1976). Low mortality rates in indus-trial cohort studies due to selection for work and survival in the industry. Brit J Prev Soc Med 30:225-230.

Fox AJ, Collier PF (1977). Mortality experience of workers exposed to vinyl chloride monomer in the manufacture of polyvinyl chloride in Great Britain. Brit J Ind Med 34:1-10.

Greenwald P, Friedlander BR, Lawrence CE, Hearne T, Earle K (1981). Diagnostic sensitivity - an epidemiologic explanation for an apparent brain tumor excess. J Occ Med 23:690-694.

Jones JH (1981). Worker exposure to vinyl chloride and polyvinyl chloride. Environ Health Persp 41:129-136.

Lelbach WK, Marsteller HJ (1981). Vinyl chloride-associated disease. In: Ergebnisse der Inneren Medizin und Kinderheilkunde, Bd 47, Advances in Internal Medicine and Pediatrics. P. Frick et al Eds. Springer-Verlag, Berlin.

Lilis R, Anderson H, Nicholson W, Daum S, Fischbein AS, Selikoff IJ (1975). Prevalence of disease among vinyl chloride and polyvinyl chloride workers. Ann NY Acad Sci 246:22-41.

Lilis R, Anderson H, Miller A, Selikoff IJ (1976). Pulmonary changes among vinyl chlroide polymerization workers. Chest 69:299S-303S (suppl).

Lloyd JW (1975). Angiosarcoma of the liver in vinyl chloride/polyvinyl chloride workers. J Occ Med 17:333-334.

Maltoni C, Lodi P (1981). Results of sputum cytology among workers exposed to vinyl chloride monomer and poly(vinyl chloride). Environ Health Persp 41:85-88.

Maltoni C, Lefemine G, Ciliberti A, Cotti G, Carretti D (1981). Carcinogenicity bioassays of vinyl chloride monomer: A model of risk assessment on an experimental basis. Environ Health Persp 41:3-29.

Marsteller HJ, Lelbach WK, Muller R, Juhe S, Lange CE, Rohner HG, Veltman G (1973). Chronic toxic liver damage in workers of PVC producing plants. Deut Med Wochschr 98:2311-2314.

Marsteller HJ, Lelbach WK, Muller R, Gedigk P (1975). Unusual splenomegalic liver disease as evidenced by peritoneoscopy and guided liver biopsy among polyvinyl chloride production workers. Ann NY Acad Sci 246:95-134.

Mastrangelo G, Saiu B, Marcer G, Piazza G (1981). Epidemiological study of pneumoconiosis in the Italian poly(vinyl chloride) industry. Environ Health Persp 41:153-157.

Masuda Y (1979). Long-term mortality study of vinyl chloride and polyvinyl chloride workers in a Japanese plant. Arh hig rada toksikol 30:403-409 (suppl).

Miller A, Teirstein AS, Chuang M, Selikoff IJ, Warshaw R (1975). Changes in pulmonary function in workers exposed to vinyl chloride and polyvinyl chloride. Ann NY Acad Sci 246:42-52.

Monson RR, Peters JM, Johnson MN (1974). Proportional mortality among vinyl chloride workers. Lancet II:397-398.

National Institute for Occupational Safety and Health (U.S.)(October, 1982). Reported cases of angiosarcoma of the liver among vinyl chloride polymerization workers. (Unpublished).

Nicholson WJ, Hammond EC, Seidman H, Selikoff IJ (1975). Mortality experience of a cohort of vinyl chloride-polyvinyl chloride workers. Ann NY Acad Sci 246:225-230.

Nicholson WJ, Henneberger PK(1983). Trends in cancer mortality among workers in the synthetic polymers industry. This volume.

Ott MG, Langner RR, Holder BB (1975). Vinyl chloride exposure in a controlled industrial environment. Arch Environ Health 30:333-339.

Popper H, Thomas LB (1975). Alterations of liver and spleen among workers exposed to vinyl chloride. Ann NY Acad Sci 246:172-194.

Purchase IFH, Richardson CR, Anderson D, Paddle GM, Adams WGF (1978). Chromosomal analysis in vinvyl chloride exposed workers. Mutation Res 57:325-334.

Reinl W, Weber H, Greiser E (1979). The mortality of German vinyl chloride (VC) and polyvinyl chloride (PVC) workers. Arh hig rada toksikol 30:399-402 (suppl).

Rowe VK (1975). Experience in industrial exposure control. Ann NY Acad Sci 246:306-310.

Selikoff IJ, Hammond EC, Eds. (1982). Brain tumors in the chemical industry. Ann NY Acad Sci 381:1-364.

Smiranova NA (1961). On the question of bone lesions due to chronic intoxication by olefins and vinyl chloride. Vestn Renigenol Radiol 36:63-66 (Russian text).

Smith AH, Waxweiler RJ, Tryler HA (1980). Epidemiologic investigation of occupational carcinogenesis using a serially additive expected dose model. Am J Epidem 112:787-797.

Sucui I, Drejman I, Valaskai M (1963). Contribution to the study of vinyl chloride disease. Med Interna 15:967978.

Tamburro CH, Greenberg R (1981). Effectiveness of Federally required medical laboratory screening in the detection of chemical liver injury. Environ Health Persp 41:117-122.

Theriault G, Allard P (1981). Cancer mortality of a group of Canadian workers exposed to vinyl chloride monomer. J Occ Med 23:671-676.

Thomas LB, Popper H, Berk PD, Selikoff IJ, Falk II (1975). Vinyl-chloride-induced liver disease. From idiopathic portal hypertension (Banti's syndrome) to angiosarcomas. N Engl J Med 292:17-22.

Tribukh SR, Tikhomirova NP, Levina SV, Koslov LA (1949). Working conditions and measures for their sanitation in the production and utilization of vinyl chloride plastics. Gigiena Sanit 10:38-44.

Waxweiler RJ, Stringer W, Wagoner JK, Jones J (1976). Neoplastic risk among workers exposed to vinyl chloride. Ann NY Acad Sci 271:40-48.

Waxweiler FJ, Smith AH, Falk H, Tryoler HA (1981). Excess lung cancer risk in a synthetic chemicals plant. Environ Health Persp 41:159-165.

Wilson RH, McCormick WE, Tatum CF, Creech JL (1967). Occupational acroosteolysis. J Am Med Assoc 201:577-581.

Industrial Hazards of Plastics and Synthetic Elastomers, pages 177–189
© 1984 Alan R. Liss, Inc., 150 Fifth Ave., New York, NY 10011

PREVENTIVE MEASURES AGAINST OCCUPATIONAL HAZARDS IN THE PVC
PRODUCTION INDUSTRY

S. Tarkowski

World Health Organization, Regional Office for
Europe, Copenhagen, Denmark*

INTRODUCTION

For many years, the production of vinyl chloride monomer
and its polymerization to polyvinyl chloride resin were considered relatively safe industrial processes that posed no
significant occupational health hazard. As recently as
twelve years ago, handbooks of industrial hygiene still
stressed the fire and explosion hazard of vinyl chloride,
even though high concentrations of this chemical were known
to act as an anaesthetic. It was only the outbreak of acro-
osteolysis in the 1960's and the later discovery of the
carcinogenicity of vinyl chloride that focused attention on
the toxic hazard of this chemical associated with occupational
exposures and on the need to develop and implement preventive
measures at the workplace.

Only recently have experimental studies and clinical
observations shown that multiple systemic disorders can be
evoked by exposure to VCM. In direct contact, vinyl chloride
is a skin irritant. Such contact with the liquid may cause
frostbite after evaporation. Severe eye irritation may also
occur. Exposure to high concentrations of VCM causes depression of the central nervous system. Symptoms such as nausea
and disorders of vision and auditory response were observed
after acute exposure to VCM at concentrations from 10 000 ppm
to 25 000 ppm (Viola 1974).

* On leave from the Institute of Occupational Medicine,
Industrial Toxicology Branch, Lodz, Poland

Numerous authors have described the occurrence of acro-osteolysis in workers engaged in the polymerization of VCM. The disease is characterized by dissolution of the bone tips of the distal phalanges of the hands and feet. This disease is frequently associated with Raynaud's syndrome and sclero-dermatous changes. The incidence of cases does not exceed 5-6% of personnel during the polymerization of VCM, and the disease involves mainly autoclave cleaners (Harris and Adams 1967; Dinman et al. 1971; Lange et al. 1974). Liver damage has been claimed to be observed more frequently (about 30%). In addition, depression of the nervous system (Suciu et al. 1975) thrombocytopenia, spleno- and hepatomegalia and liver fibrosis (Smith et al. 1976) have been reported. Workers exposed to PVC dust showed fibrotic lung changes and altered pulmonary function tests which were more pronounced in those with long exposure (Lilis et al. 1976; Waxweiler et al. 1976).

Despite earlier evidence for induction of tumours in laboratory animals, more widespread acceptance of the carcino-genicity of VCM has occurred only since Creech and Johnson (1974) found that vinyl chloride exposure was associated with the development of cancer in humans. Excess cancer of the lung and nervous system has now been reported in addition to liver angiosarcoma, and the experimental data point to a multisystem oncogenic effect.

REDUCTION OF EXPOSURE LEVELS

Worker exposure may result from inhalation of gases of vinyl chloride monomer (VCM) or to polyvinyl chloride (PVC) dust. The level of exposure is process and job related. Workers are exposed to VCM during its manufacture and poly-merization, whereas exposure to PVC dust is encountered during PVC processing. At each stage of the production process in both VCM and polymerization plants, VCM is likely to escape into the air via random leaks in pump seals, valves, gaskets, storage tanks, etc. Controlling such leakage is a major problem.

The risk of exposure to VCM is greater during polymeriza-tion than during manufacture, and particularly high concen-trations of VCM may be experienced by workers employed in cleaning the reactor vessels where the polymerization occurs. There are two main types of process used for polymerization:

suspension and emulsion (dispersion). In both polymerization processes, some of the solid polymer deposits on the walls of the reactors. This deposition necessitates periodic cleaning of the reactors. Cleaning and unloading of the reactors are sources of significant levels of VCM exposures. In the past, this work was manually performed, with reactor cleaners spending hours daily inside the reactor, where concentrations of VCM reached several thousand ppm. In recent years, the cleaning has been done automatically, either by solvent washing or hydroblasting. As a result, workers enter the reactor for only short periods of time (10-15 min) every 20-30 polymerization processes.

These two polymerization processes are sources of relatively high levels of exposure to VCM as compared to other, much less frequently used processes, such as mass (bulk) or solvent polymerization. Concentrations of VCM in the air in PVC-producing factories vary from one place to another. Filatova and Gronsberg (1957) have reported that air concentrations of VCM in working places ranged from 40-312 ppm, with peak concentrations of up to 34 000 ppm. The highest exposure levels were found in the areas around the polymerization reactors. For example, values reported by Ott et al. (1975) reached 4 000 ppm, with concentrations of VCM varying over time, depending on the stage of polymerization (Krajewski et al. 1980). However, attempts to reduce leakages have led to significant reductions in VCM concentrations in the work atmosphere. Data presented by Jones (1981) provide examples. Surveys carried out in a VC polymerization plant in the United States have indicated that exposures could be reduced tenfold. In a mass polymerization plant, the percentage of personnel exposed to concentrations below 1 ppm increased from 2% in 1974 to 73% in 1976. A similar effect has been achieved in a suspension polymerization plant where exposure levels decreased twofold.

Improved hygienic conditions have been noted in other countries as well. Data presented by Krajewski et al. (1980) indicate similar trends in Poland, where introduction of mechanical cleaning of reactors and other technological improvements has resulted in reduced emissions of VCM into the workplace air.

PERMISSIBLE LEVELS OF EXPOSURE

Air and biological standards, commonly known as threshold limit values (TLVs), maximum allowable concentrations (MACs) and biological limit values, are only best estimates of safe levels of exposure, but they are extremely useful guidelines in the hands of occupational health and safety practitioners, helping them to judge the potential health hazards of industrial chemicals and to prevent health impairments in workers exposed to them. One possible exception may be chemical carcinogens, for which it is still questionable whether safe levels of exposure can be defined.

Vinyl chloride is a good example of a chemical with an evolving standard. Before its toxic potential was known, the primary aim in prevention was to avoid nauseous effects in workers. In 1962, the American Conference of Governmental Industrial Hygienists set a time-weighted average (TWA) limit recommendation of 500 ppm. In 1974 the B.F. Goodrich Chemical Company announced that deaths due to liver angiosarcoma among its workers could be linked with exposure to vinyl chloride. In view of these findings in 1974, the United States Occupational Safety and Health Administration (OSHA) made the emergency temporary standard of 50 ppm. This concept of an allowable "working level" was recommended by a consultant of the National Institute of Occupational Safety and Health (NIOSH) who based this recommendation on preliminary findings of Maltoni which showed no tumours in animals exposed to VCM at 50 ppm. However, Maltoni (1974) subsequently found that VCM induced angiosarcomas in both rats and mice at 50 ppm. The animal evidence by itself raises problems of interpretation in setting safe levels of exposure for humans. Maltoni's data clearly identified a dose-response effect, but these studies did not include concentrations of VCM below 50 ppm. It could be argued that tumours can occur below 50 ppm, providing enough animals are exposed for sufficiently long periods of time. Some confirmation of this argument came later when Maltoni et al. (1981) found effects of VCM at concentrations of 1 ppm. However, theoretical considerations do not permit a threshold to be established for chemical carcinogens, although very low concentrations may extend the latency period beyond the life expectancy.

Available data were not sufficient to set a standard which could be accepted without doubt. In proposing a permanent standard for VCM, OSHA considered both animal and human data and technological and economic feasibility. This led to setting the standard at 1 ppm as the TWA, with a ceiling value of 5 ppm.

The present standards for exposure to VCM in various countries, in terms of time-weighted averages (TWA) for an 8-hour working shift and ceiling concentrations of usually 10 or 15 minutes, are presented in Table 1.

Country	TWA (ppm)	Ceiling Concentration ppm	min
Canada	10	25	15
Finland	5	10	10
France			
existing factories	5 (week)	15	15
new factories	1	5	15
FRG			
existing factories	5	15	15
new factories	2 (year)	15	60
Japan	10		
Netherlands	10		
Norway	1	5	15
Sweden	1	5	15
Poland	12		
USA	1	5	15
USSR	12		

Table 1. Threshold limit values for VCM in the work environment

The adopted standards vary country to country and are the result of compromise between the recognized health risks and technological feasibilities. Some countries accept somewhat higher risk than do others. Because a no-response level cannot be established, it is generally recommended that permissible concentrations should not be exceeded and that every effort be made to further reduce levels of exposure to VCM.

CONTROL OF THE WORK ENVIRONMENT

To avoid excess exposure to VCM, exposure should be controlled by measuring concentrations of VCM in the work environment. The primary purpose of the control measurements is to assess actual levels of worker exposure. Therefore, measurements must be made in all areas where workers are likely to be in contact with VCM. These strategic sampling points, established by a professional industrial hygienist, will serve as locations for environmental monitoring.

Assessment of worker exposure is based on air sampling from personal sampling pumps. To check the intensity and frequency of peak exposures, the sampling period should not exceed 15 minutes. Using the presently available gas chromatography technique for determination of VCM in the sampled air, the 15-minute period of sampling is sufficient to collect representative samples of the work atmosphere. The lower limit of detection of VCM in the air for the present method is below 1 ppm.

Apart from the above, it is also important that periodic measurements are performed to check possible process or equipment leaks and emissions of VCM resulting from work practices.

In many industries, environmental monitoring has been significantly improved by introducing automatic monitoring techniques. Sequential samplers and automatic analyzers are being used for continuous monitoring of selected locations or operations and are most useful for exposure control, partic-ularly in emergencies.

BIOLOGICAL MONITORING OF EXPOSURE

Assessment of atmospheric concentrations of pollutants
and estimation of TWA concentrations do not always reflect
true levels of occupational exposure during the workshift.
Furthermore, they do not allow assessment of a dose of a
chemical absorbed during the exposure period, particularly
when routes of absorption other than inhalation may contribute
to overall intake of pollutants. Many successful attempts
have been made to support environmental monitoring with
biological monitoring, which would more precisely reflect
individual exposure. Attempts to develop a method of biological monitoring of exposure to VCM have also been made,
including both breath analysis and urine analysis.

In 1969 Baretta et al. presented results of their study
of human exposure to VCM. The authors constructed post-
exposure breath VCM decay curves from data obtained during
experimental exposures to relatively uniform concentrations
of VCM in an exposure chamber. They subsequently compared
these curves with those derived from TWA equal but broadly
fluctuating concentrations in the air in a chemical plant.
The decay curves for the breath VCM concentrations were
constructed by stepwise multiple regression.

There was a very close relationship between post-exposure
breath concentrations of VCM and TWA concentrations in the
atmosphere of the exposure chambers. In spite of broad
variation in TWA concentrations of VCM encountered in industry,
a remarkable correlation between breath concentrations and
corresponding TWA concentrations was found. The authors have
postulated that the close agreement between post-exposure
breath concentrations of VCM at the corresponding TWA air
concentrations, obtained in both experimental and industrial
conditions, suggests that breath analysis is valid for
estimating the worker's individual daily exposure to VCM.

However, this postulated method has not been widely
applied. One reason for this reluctance could be that detec-
tability of VCM in the exhaled air by the method used by
Baretta et al. becomes limited when air concentration of VCM
in the workplace has been reduced below 50 ppm. In the
authors' study, the absolute breath levels of VCM ranged from
about 20 ppm in one sample taken less than 1 hour after an
8-hour TWA of 250 ppm to barely detectable levels in samples
taken after exposures at TWA below 50 ppm.

A different approach to biologically monitor exposure to VCM has been made during recent years. The basic assumption has been to base biological monitoring of exposure on the excretion of a metabolite of VCM in the urine. VCM is metabolized by microsomal mixed-function oxidases and in conjugation with glutathione. The end-products of this metabolic activity are N-acetyl-S-(2-hydroxyethyl)cysteine and thiodiglycolic acid. The latter one constitutes a major fraction of metabolites excreted in the urine (Green and Hathway 1977). These metabolites have also been found by German researchers in the urine of persons exposed to VCM (Müller et al. 1978). The authors subsequently developed an analytical method for measuring urinary thiodiglycolic acid (Müller et al. 1979) which was applied in a study on monitoring of excretion of this VCM metabolite in urine from persons exposed to VCM in relation to its air concentrations (Heger et al. 1982).

The latter study was performed on 15 workers exposed to different air concentrations of VCM in a PVC-producing plant. Some correlation ($P=\propto<0.01$) was found when 12 h TWA concentrations of VCM were compared with the thiodiglycolic acid content(μg) in the portion of urine collected after the end of the workshift. However, individual variations in the excretion of thiodiglycolic acid were high enough that identification of workers exposed to VCM concentrations exceeding the TLV and those below this value could not be made with absolute certainty. These variations could have resulted from fluctuations in the levels of external exposure during the workshift. Better correlation ($P=\propto<0.005$) was found when the same air concentrations of VCM were correlated with the content of thiodiglycolic acid in the urine collected during the 12-hour period after the end of the workshift (Fig. 1). Again, however, the correlation was not strong enough to enable exact assessment of external exposure, for which peak concentrations during the period of exposure were not measured.

In conclusion, there appears to be no reliable method currently available for biological monitoring of exposure to VCM. Measurement of excretion of thiodiglycolic acid in the urine of exposed persons may provide, at best, a means of exposure surveillance which would only allow a guess as to whether or not permissible levels of exposure to VCM could have been exceeded. However, it does not include the possibility of assessing TWA air concentrations of VCM during the workshift. Reliance on measurements of excretion of thiodiglycolic acid for exposure assessment is also limited due to the influence of numerous factors on the rate of metabolism of VCM. Therefore, excretion of thiodiglycolic acid can be used as an indicator of intake of VCM as long as individual variability due to such factors as liver disease, use of drugs, alcohol intake, etc., is taken into account.

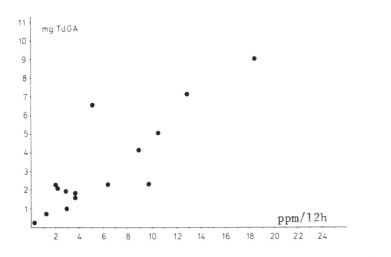

Fig 1.   The content of TdGA in the urine collected during 12 h after the workshift from persons exposed to different air concentrations of VCM expressed as 12 h TWA (Heger, et al. 1982).

MEDICAL SURVEILLANCE

Health effects of vinyl chloride identified by clinical and experimental observations have clearly indicated the need for health surveillance programmes for workers in polyvinyl chloride production and processing plants.

Towards this goal, the emphasis of programmes in most countries has now shifted towards the prevention of different health effects. This approach has resulted from the progress made in recent years in medical diagnosis and industrial technology. Due to the latter, high concentrations of VCM which in the past caused development of acro-osteolysis, Raynaud's syndrome and depression of the central nervous system have been significantly reduced. The health effects of exposure to lower concentrations have now been identified, the most important being liver damage. It is therefore desired and recommended that specific tests be conducted for early detection of liver damage.

Routine health screening programmes should include preplacement and periodic medical examinations and recording of past medical history. The latter should include alcoholic intake, hepatitis, exposure to hepatotoxic agents, drugs and other chemicals, blood transfusions and hospitalizations. Long-term follow-up of exposed persons should be carried out on at least an annual basis, when results of screening are normal.

Provision and availability of medical services to all workers exposed to VCM is a primary duty of industry. Even though small plants have more difficulty balancing the need for medical services with the costs for such service, the large proportion of the exposed population employed at these facilities must be considered.

Participation of workers in preventive programmes for diseases which may result from exposure to VCM is very important. The workers should be directly informed of the type of harmful effects that may be caused by VCM, the effective ways in which exposure can be controlled and the best methods of individual protection.

The informed worker will be more likely to be sensitive to adverse health effects possibly related to occupational exposure. The worker will also be more likely to discuss this possibility with his doctor. One very positive result will be improved record keeping of occupationally related health effects.

REFERENCES

Baretta ED, Steward RD, Mutchler JE (1969). Monitoring exposures to vinyl chloride vapor: breath analysis and continuous air sampling. Am Ind Hyg Ass J 30: 537

Creech JL, Johnson MN (1974). Angiosarcoma of liver in the manufacture of polyvinyl chloride. J Occup Med 16: 150.

Dinman BD, Cook WA, Whitehouse WM, Magnuson HJ and Ditchek T (1971). Occupational acro-osteolysis I. An epidemiological study. Arch environ hlth 22: 61.

Filatova VS, Gronsberg ES (1957). Sanitary-hygienic conditions of work in the production of polychlorvinylic tar and measures of improvement. Gig Tr Prof Zabd 1: 3. (in Russian)

Green T, Hathway DE (1977). The chemistry and biogenesis of the S-containing metabolites of vinyl chloride in rats. Chem Biol Interact 17: 137.

Harris DK, Adams WGF (1967). Acro-osteolysis occurring in men engaged in polymerization of vinyl chloride. Brit Med J 3: 172.

Heger M, Müller G, Norpoth K (1982). Untersuchungen zur Beziehung zwischen Vinylchlorid (VCM) - Aufnalsine und Metabolitenausscheidung bei 15 VCM-Exponierten. Int Arch Occup Environ Hlth 50: 187.

Jones JH (1981). Worker exposure to vinyl chloride and poly (vinyl chloride). Environ Hlth Persp 41: 129.

Krajewski J, Dobecki M, Mielczarska A (1980). Assessment of work environment in plants producing polyvinyl chloride. Med Pracy 31: 149 (in Polish).

Lange CE, Juhe S, Stein G, Veltman G (1974). The so-called vinyl chloride disease - an occupational systemic sclerosis. Int Arch Occup Hlth 32: 1.

Lilis R, Anderson H, Miller A, Selikoff IJ (1976). Pulmonary changes among vinyl chloride polymerization workers. Chest 69: 299.

Maltoni C, Lefemine G, Chieco P, Carretti D (1974). Vinyl chloride carcinogenesis: current results and perspectives. Med Lav 65: 421.

Maltoni C, Lefemine G, Ciliberti A, Cotti G, Caretti D (1981). Carcinogenicity bioessays of vinyl chloride monomer: a model risk assessment on an experimental basis. Environ Hlth Persp 41: 3.

Mastrangelo G, Saia B, Marcer G, Piazza G (1981). Epidemiological study of pneumoconiosis in the Italian poly (vinyl chloride) industry. Environ Hlth Persp 41: 153.

Müller G, Norpoth K, Kusters E, Herweg K, Versin E (1978). Determination of thiodiglycolic acid in urine specimens of vinyl chloride exposed workers. Int Arch Occup Environ Hlth 41: 199.

Müller G, Norpoth K, Wicknamasinghe RH (1979). An analytical method using GC-MS for the quantitative determination of urinary thiodiglycolic acid. Int Arch Occup Environ Hlth 44: 185.

Ott MG, Langner RR, Holder BB (1975). Vinyl chloride exposure in a controlled industrial environment. A long-term mortality experience in 594 employees. Arch Environ Hlth 30: 333.

Smith PM, Crosley IR, Williams DHJ (1976). Portal hypertension in vinyl chloride production workers. Lancet ii: 602.

Suciu I, Prodan L, Ilea E, Paduraru A, Pascu L (1975). Clinical manifestation in vinyl chloride poisoning. Ann NY Acad Sci 246: 53.

Viola PL (1974). La Malatia da Cloruro di Vinile. Med Lav 65: 81.

Waxweiler RJ, Stringer W, Wagoner JK, Jones J, Falk H, Carter C (1976). Neoplastic risk among workers exposed to vinyl chloride. Ann NY Acad Sci 271: 40.

CHAPTER III. STYRENE-CONTAINING POLYMERS

Industrial Hazards of Plastics and Synthetic Elastomers, pages 193–202
© 1984 Alan R. Liss, Inc., 150 Fifth Ave., New York, NY 10011

PRODUCTION AND USES OF STYRENE CONTAINING POLYMERS

Pentti Kalliokoski

University of Kuopio, Finland

The commercial production of polystyrene (PS) started in
Germany in 1930. The worldwide production capacity of this
first polymer containing styrene, which is still the most
important one, has increased from 60 t to 7.5 million t annually in the course of 50 years (Anon., 1981 a). More than
half of the styrene manufactured in the world is used in the
production of PS (Tossavainen, 1978). In addition to polystyrene, important styrene containing polymers include acrylonitrile-butadiene-styrene (ABS), styrene-acrylonitrile
(SAN), unsaturated polyester resins, and styrene-butadiene
rubber (SBR).

POLYSTYRENE

Table 1 lists the major applications or PS and their production rates in the United States and western Europe.
Packaging and construction are the areas were the most PS is
used.

Table 1. The main areas of use of polystyrene (solid and expandable) in the United States and western Europe in 1981.

	Production			
	United States		Western Europe	
Area of use	1,000 t	%	1,000 t	%
Packaging	666	40	620	43
Construction	124	8	229	16
Housewares	97	6	133	9
Refrigerators, freezers	18	1	94	6
Appliances, electronics	16	1	85	6
Furniture, furnishings	50	3	60	4
Toys, recreation	113	7	50	3
Miscellaneous	569	34	187	13
Total	1,653	100	1,458	100

In 1981 expandable polystyrene (EPS) comprised 26 % of the total demand in western Europe. The production of PS in Japan was 510,000 t in 1981 (Anon., 1982 a). The global consumption of PS totalled 5.15 million t in 1981 (Anon., 1981 a). The average operating rate was thus only 67.8 %. However, this already reflects growth; due to the economic recession, the global operating rate was as low as 62.6 % in 1980. This was not a unique phenomenon in the plastics world; a similar decrease also took place in the production of the other main plastics: polyethylene, polypropylene, and polyvinylchloride. The recovery is thought to continue, and the global operating rate in the production of PS is estimated to reach the level of 80 % in 1985 (Anon., 1981 a). The annual global growth rate in the concumption of polystyrene is predicted to be 5.2 % in the 1980s (Anon., 1982 b).

The residual monomer content was as high as 2 % in the early polystyrenes, and in the beginning of 1960s it was still about 1 %. Since then PS grades with low concentrations, 500 ppm,or less of residual styrene have been developed (Anon., 1981 b).

The impact strength of pure PS is poor. Butadiene is most

often used to enhance the impact properties of PS. The use of high-impact PS grades that meet Underwriters Laboratories flammability criteria V-0 is rapidly increasing in the United States. The production of these new PS grades, which can be used for television, radio, and business machine housings, is estimated to be 52,000 t this year in the U.S., where consumption is predicted to exceed 100,000 t in 1985 (Anon., 1982 c). Butadiene is also used together with special additives to manufacture other new modified PS grades called SSP. These resins have physical properties that fit between the properties of PS and ABS (Anon., 1980 a).

New applications of EPS include the insulation spray which can be injected into cavities where it forms a dense foam mass. The market estimate for this product is 50,000 t per year (Anon., 1981 c).

ACRYLONITRILE-BUTADIENE-STYRENE

The statistics for the consumption of ABS in the U.S. and western Europe, for 1981 are given in Table 2. Important applications include construction, appliances, transportation, and electronics (Anon., 1982 a).

Table 2. The main areas of use of acrylonitrile-butadiene-styrene resins in the United States and western Europe in 1981.

	Production			
	United States		Western Europe	
Area of use	1,000 t	%	1,000 t	%
Construction	100	23	10[1]	3
Appliances	85	19	74	24
Electrical	60	14	39	13
Transportation	56	13	70	23
Recreation	25	6	13	4
Furniture, luggage	19	4	7	2
Miscellaneous	96	21	97	31
Total	441	100	310	100

[1] includes only pipes and fittings

The market for ABS is rather poor at this moment. The operating rate in the U.S. was as low as 51 % in 1980 (Anon., 1981 d). Polypropylene is replacing ABS in some of its traditional areas of use. Thus the new applications of flame-retardant grades, including housings for telecommunication and business machines, are becoming relatively more important (Anon., 1980 b).

The typical level of residual acrylonitrile in ABS ranges from 6 to 8 mg/kg after processing. The British Ministry of Agriculture, Food and Fisheries regards this as safe and has recently approved ABS for food packaging (Anon., 1982 d).

STYRENE-ACRYLONITRILE AND OTHER STYRENE COPOLYMER RESINS

The market for SAN in the U.S. is presented in Table 3. Housewares were the most important application. However, SAN accounted for only 2 % of all the consumption of plastics in this area of use (Anon., 1982 a).

Table 3.   The main areas of use of acrylonitrile-styrene in the United States in 1981.

	Production	
Area of use	1000 t	%
Housewares	9	18
Compounding	8	16
Appliances	8	16
Packaging	6	12
Transportation	4	8
Miscellaneous	16	30
Total	51	100

SAN is relatively more important in Japan, where it had a market of 90,000 t in 1979 (Anon., 1980 c).

Methacrylate-butadiene-styrene (MBS) polymers are used as impact modifiers for polyvinylchloride. The consumption of these polymers in the U.S. was 20,000 t in 1981 (Anon., 1982 e).

New styrenic copolymers are being developed continually. One of the new, commercially interesting polymers is acrylonitrile-olefin rubber-styrene. It is reported to have good chemical resistance, good impact strength, and good outdoor retention properties (Anon., 1981 e).

STYRENE-BUTADIENE RUBBER

Styrene-butadiene rubber became important during World War II; for example, in the U.S. production increased from 5,000 t in 1941 to 750,000 t in 1944. It is still the most important synthetic rubber. The present worldwide production of SBR, 5.2 million t per year exceeds even the 4 million t per year supply of natural rubber. (Mullins, 1981). Tire industry is the major user of SBR.

The manufacture of PS, ABS, SAN, and SBR takes place in closed processes. Therefore only low concentrations of styrene are found in the air of polymerization factories nowadays. Styrene concentrations above 1 ppm were seldom found in areas where workers were frequently present in a West-German styrene and polystyrene production plant (Thiess and Friedheim, 1978). In a styrene and polystyrene factory in the U.S., most of the measured levels of styrene in the breathing zone were less than 5 ppm. The highest concentrations found were 20 ppm (Maier et al., 1974). The concentrations of styrene were usually below 1 ppm in a Soviet factory manufacturing polystyrene and copolymers (Ponomareva and Zlobina, 1972). Butadiene, acrylonitrile, and organic solvents can also be found in the air of polymerization plants. Due to the unreacted residual monomer or thermal degradation, some styrene is expelled into the atmosphere also in polymer processing but the exposure to styrene rarely exceeds 1 ppm in these operations (Tossavainen, 1978).

UNSATURATED POLYESTER RESINS

Unsaturated polyester resins are mainly used in the manufacture of fiberglass reinforced plastics. These resins
usually contain about 40 % styrene as a reactive diluent.
When the products are manufactured in open molds, as much as
10 % of the styrene volatilizes into the workplace air
(Kalliokoski and Jantunen, 1981). Therefore, the exposure
of employees to styrene is by far the heaviest in the reinforced plastic industry, where a number of investigations
have indicated exposure levels of 20 - 300 ppm typical in
open mold work (see Table 4).

Table 4.   Environmental concentrations of styrene in
           reinforced plastic plants.

Reference		Country	Exposure level (ppm)
Bardodej et al.	(1960)	Czechoslovakia	15-200
Zielhuis et al.	(1964)	Netherlands	25-90
Simko et al.	(1966)	Czechoslovakia	5-195
Huzl et al.	(1967)	Czechoslovakia	50-100
Matsuhita et al.	(1968)	Japan	up to 600
Götell et al.	(1972)	Sweden	20-290
Bodnei et al.	(1974)	U.S.A.	45-550
Kalliokoski	(1976)	Finland	0.1-300
Rosensteel et al.	(1977)	U.S.A.	35-110
Bergman	(1977)	Sweden	10-170
Brooks et al.	(1979)	U.S.A	40-230
Kjellberg et al.	(1979)	Sweden	3-14
Crandall	(1981)	U.S.A	2-180
Schumacher et al.	(1981)	U.S.A	up to 300

Hand lamination is the most common working practice. The
resin is applied to the surface of the mold either by the
hand lay-up or the spray-up method. These methods closely
resemble roller and spray painting. First, a layer of
pigmented polyester resin, the so called gel coat, is
applied. This is then covered with layers of resin and
fiberglass. The laminate must be manually rolled out also
when the spray-up method is used to remove air pockets.

Injection, rotation, and pressing are more mechanized and

automated manufacturing methods. In the first method, resin is injected into the gap between the mold halves filled with fiberglass. In the rotation method, resin is sprayed onto a rotating product. In the pressing method, resin is forced onto the fiberglass mat surface in a press.

The main uses of reinforced plastics in the U.S. and Western Europe in 1981 are presented in Table 5. It should be noted that this table includes the reinforcements. The consumption of the mere resin in the U.S. was 452,000 t in 1981 (Anon., 1982 a).

Table 5.   The main areas of use of unsaturated polyester resins in the United States and western Europe in 1981 (incl. reinforcements).

	Production			
	United States		Western Europe	
Area of use	1,000 t	%	1,000 t	%
Transportation	139	19	105	19
Construction	138	19	105	19
Boats	132	18	70	12
Industrial parts[1]	120	16	125	22
Electrical	112	15	115	20
Miscellaneous	99	13	45	8
Total	740	100	565	100

[1] incl. pipes, ducts, and tanks

LIST OF REFERENCES

Anon. (1980 c).   Japanese plastics export dip; stockpiling cuts into domestic sales. Mod. Plast Int 6(11):6.

Anon. (1981 a).   World competition will get tougher. Plastics World 39(2):14.

Anon. (1981 b). Polystyrene - half a centery of development and innovation. Plast Rubber Int 6(4):158.

Anon. (1981 c). EPS insulation system is performance alternative to urea-formaldehyde. Mod Plast Int 11(9):14.

Anon. (1981 d). What's happening in ABS market. Plastic World 39(6):18.

Anon. (1981 e). Sun heats pools via styrenic panels. Mod Plast Int 11 (6):22.

Anon. (1982 a). Materials 1982. Mod Plast Int 12 (1):39.

Anon. (1982 b). Gains in Latin America will pace world plastics growth. Mod Plast Int 12 (1):8.

Anon. (1982 c). Why so many new V-O PS grades. Mod Plast Int 12 (9):50.

Anon. (1982 d). British Ministry approves ABS for food packaging. Mod Plast Int 12 (8):4.

Anon. (1982 e). Impact modifiers/processing aids. Mod Plast Int 12 (9):55.

Bardodej A, Malek B, Volfova B, Zelena E (1960). The hazard of styrene in the production of glass laminates. Cesk Hyg 5:541 (in Czechoslovakia).

Bergman K (1977). Exposure to styrene in plastic boat industry. I. Technical-hygienic study. Arbete och Hälsa 3:1 (in Swedish).

Bodnei AH, Butler GJ, Okawa MT (1974). Health Hazard evaluation/Toxicity Determination Report No. 73-103-128-American Standard Fiberglass Inc., Stockton, California, Cincinnati: National Institute for Occupational Safety and Health, 10 pp.

Brooks SM, Anderson LA, Tsay J-Y, Carson A, Buncher CR, Elia V, Emmett EA (1979). Investigation of workers exposed to styrene in the reinforced plastic industry - health and psychomotor status, toxicological and industrial hygiene data and effect of protective equipment as it relates to exposures through lung and skin routes. Cincinnati: University of Cincinnati, 330 pp.

Crandall MS (1981). Worker exposure to styrene monomer in reinforced plastic boat-making industry. Amer Ind Hyg Assn J 42:499.

Götell P, Axelson O, Lindelöf B (1972). Field studies on human styrene exposure. Work Environ Health 9:76.

Huzl F, Sykova J, Mainevova J, Jankova J, Srutek J, Junger V, Lahn V (1967). Prac Lek 19:121 (in Czechoslovakian).

Kalliokoski P (1976). The reinforced plastic industry - a problematic work environment. Työterveyslaitoksen tutki-muksia 122 (in Finnish).

Kalliokoski P, Jantunen M (1981). Control of airborne styrene in reinforced plastics industry. 2nd World Congress of Chemical Engineering. Montreal Oct. 4-9. 1981.

Kjellberg A, Wigaeus E, Engström J, Åstrand I, Ljungquist E (1979). Long term effects of styrene exposure in a polyester plant. Arbete och Hälsa 18:30 (in Swedish).

Maier A, Ruhe R, Rosensteel R, Lucas JB (1974). Health Hazards Evaluation/Toxicity Determination Report No. 72-90-107. Arco Polymer Inc., Monaco, PA. Cincinnati: National Institute for Occupational Safety and Health, 28 pp.

Matsuhita T, Matsumoto T, Miyagaki J, Maeda K, Takeuchi Y, Katajima J (1968). Nervous disorders considered to be symptoms of chronic styrene poisoning. Saigai Igaku 11:173 (in Japanese).

Mullins L (1981). Changing horizons for rubber-like materials. Plast Rubber Int 7 (5):89.

Ponomareva NI, Zlobina NS (1972). Working conditions and the state of the upper respiratory tract in workers engaged in the production of block and emulsion polystyrene and its copolymers. Gig Tr Prof Zabol 15:22 (in Russian).

Rosensteel RE, Mayer CR (1977). Health Hazard Evaluation Determination Report No. 75-150-378, Reinell Boats Inc., Poplar Bluff, Missouri. Cincinnati: National Institute for Occupational Safety and Health, 52 pp.

Schumacher RL, Breysse PA, Carlyon WR, Hibbard RP,

Kleinman GD (1981). Styrene exposure in the fiberglass fabrication industry in Washington State. Amer Ind Hyg Assn J 42:143.

Simko A, Jindrichova J, Pultarova H (1966). The effect of styrene on the health state of workers employed in laminate production. Prac Lek 18:348 (in Czechoslovakian).

Thiess AM, Friedheim M (1978). Morbidity among persons employed in styrene production, polymerization and processing plants. Scand J Work Environ Health 4 (suppl. 2):203.

Tossavainen A (1978). Styrene use and occupational exposure in the plastics industry. Scand J Work Environ Health 4 (suppl. 2):7.

Zielhuis RL, Hartogensis F, Jongh J, Kalsbeek JW, van Rees H (1964). The Health of workers processing reinforced polyesters. In XIV International Congress pf Occupational Health, Madrid, Spain, 16-21 September, 1963, volume III, p. 1092.

Industrial Hazards of Plastics and Synthetic Elastomers, pages 203–213
© 1984 Alan R. Liss, Inc., 150 Fifth Ave., New York, NY 10011

THERMODEGRADATION OF STYRENE–CONTAINING POLYMERS

Pirkko Pfäffli

Institute of Occupational Health

Haartmaninkatu 1, 00290 Helsinki 29, Finland

Introduction

The widespread and increasing use of plastics
warrants the evaluation of possible health hazards in
their production, processing and consumption. In the 80's
the annual growth in the consumption of plastics has
averaged 3,5 % and presently the world production totals
about 50 million tons per year (Modern Plastics International, 1982). Toxicological research on the different
raw materials and additives used in plastics has increased
during the last few years. Research has also focused on
the hazards caused by different residues in consumer
goods. Little attention has been given to the possible
health hazards of the thermodegradation of plastics.
During the processing of plastics to end products, they
are subjected to relatively high temperatures, which
induce breakdown of polymer or additives and give rise to
a complex mixture of compounds, which vaporize into the
air. This mixture of compounds presents a potential risk
to workers who process plastics in the consumer goods
industry. The most interest for the toxic effects of the
thermodegradation of plastics has been shown by the fire
toxicologists. The origin of this type of toxicological
research has been the fact that casualities in fires are
often caused by toxic compounds from burning materials. It
is, however, impossible to infer circumstances of industrial hygiene from the data obtained in fire research as
in fires the temperatures are much higher than in processing, and in the case of fire the appearance of carbon
monoxide, hydrogen cyanide, and nitrogen oxides, as strong

acute toxicants, is important (Chaigneau M. et al. 1974,
Michail J. 1982).

The toxicological effects of the thermodegradation
products of styrene-containing plastics at temperatures
near to processing temperatures have been studied by
Zitting et al. (1978 and 1980). They have been able to
show that the degradation products of these plastics can
cause metabolical and biochemical imbalance in exposed
organisms with sub-chronic exposure.

Styrene-containing plastics

The styrene plastics are a diverse group of polymers
varying from simple homopolystyrene resins (PS) through
styrene-acrylonitrile copolymers (SAN) to the various
types of acrylonitrile-butadiene-styrene terpolymers
(ABS). These are thermoplastics processed with methods
which utilize heat under controlled conditions. About 6
million tons of these styrene thermoplastics are produced
in the world annually (Modern Plastics Intern. 1982).
Another type of styrene-containing plastics are the
unsaturated polyester plastics. These are formed through
condensation of a polyol, glycol, with unsaturated
carboxylic acids. Styrene is used as a reactive solvent
for these polyesters. It produces cross-linking bonds
between the polyester chains, thus transforming the
polyester into a rigid, insoluble thermoset resin.
Styrene-butadiene (SB) copolymer is also used as a blend
with natural rubber to make an elastomer (SBR) (Berg et al
1982). The addition of the suitable amounts of additives
transforms the normally rigid, unstable polymer into a
plastic. The styrene plastics usually contain only up to 5
% additives; fire retardants form an exception where this
figure can reach 28 % in some cases (Brighton et al. 1979,
Gächter et al. 1979).

Processing

In the processing of polystyrene thermoplastics, heat
is applied to give the plastic the degree of fluidity
required for shaping it into the particular products. The
processing temperatures range from 150 to 335°C. The
shaping techniques of styrene-containing thermoplastics

consist of extrusion, form molding, injection molding, blow molding, thermoforming, and fabrication of expanded beds. The other types of processes are hot wire cutting and mechanical machining; the latter generates local heat. Polyester thermosets, on the other hand, cannot be processed with heat. But they also emit, as organic material, degradation and combustion products at elevated temperatures (Brighton et al. 1979).

Thermal degradation

At low temperatures the polystyrene thermoplastics are solid. When polystyrene is slowly heated, it reaches a transitional state. The originally very rigid structure becomes soft, and then fluidy. This takes place at temperatures between 80° and 150°C. Further raising of the temperature causes gradual breakdown of the macromolecules and degradation of the polymer.

Figure 1.   POLYMER DEGRADATION

Polymer $\sim$ C–C–C–C–C–C–C–C–C–C $\sim$
                          $\downarrow$ energy (heat)
Fragments $\sim$ C=C–C–C$\sim$ $\sim$C–C=C$\sim$ $\sim$C–C $\sim$
of polymer

The activation energy for the thermal degradation reaction of polystyrene has been reported to be about 30 kcal/mol. It can vary depending on the original method of polymerization used and it is also affected by molecular weight (Kishore et al. 1976, Kokta et al. 1973, Malhotra et al. 1975, Dickens 1980). The thermal degradation of polystyrenes begins at lower temperature in air than in inert gas (Hoff et al. 1982).

The degradation reactions involve: random scission of "weak links" (which are randomly distributed within the polymer chains), depolymerization (yields monomer(s)), intramolecular transfer (produces oligomers, i.e., dimers, trimers, etc.) and intermolecular transfer (reduces the molecular weight and yields short–chain fragments) (Cameron et al. 1978). The exact mechanism of degradation of styrene polymers is still debated in the literature.

When styrene–containing thermoplastics are degraded,

styrene and styrene dimers are the main products among the
volatiles identified and quantified followed by the other
volatile oligomers and the short-chain fragments of high
boiling point as an oily product. There remains a polymer
residue of lower molecular weight than the starting polymer. In addition to these compounds, small quantities of
other aromatic hydrocarbons also evolve from polystyrene
plastics: These include benzene, toluene, ethylbenzene,
isopropylbenzene, n-propylbenzene, allylbenzene, and
α-methylstyrene. Small quantities of lowmolecular-weight
aliphatic hydrocarbons are also generated (Hoff et al.
1982, Pfäffli et al. 1978). Commercial plastics sometimes
contain minor amounts of hydrocarbons trapped inside the
polymer as residues from the original polymerization.

The thermal degradation of a copolymer or blend of
styrene and butadiene resembles that of PS itself (McNeil
1978), excluding the fact that more oligomeric compounds
arise from BS (Fig. 2).

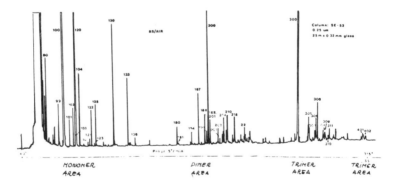

Figure 2. A gas chromatogram of thermal degradation
products from BS in air (A. Hesso, unpublished).

Often ABS and SAN show a degradation rate similar to that
of polystyrene (Hoff et al. 1982). On the other hand, ABS
can also differ from PS in degradation behavior, especially in the styrene monomer and fragment contents of
the degradation products. The differences in degradation
rate and quantity and quality of degradation products can
be assumed to be related to differences in relative, orig-

inal monomer content.

The chemical composition of copolymer and terpolymer fragments resembles that of the original polymers. Thus, polystyrene fragments contain styrene units as dimers, trimers, etc. Fragments of styrene-butadiene copolymers contain, besides styrene units, also aliphatic chains originating from polybutadiene. Similarly, fragments of styrene-acrylonitrile contain, besides styrene units, even nitrile units. Fragments also hold isomeric variations (Blazso et al. 1980). Thus, nitrogenous polymers, like SAN and ABS, also emit nitriles under thermal degradation; these emissions consist of small amounts of acetonitrile, acrylonitrile, and vinylacetonitrile. The processing temperatures are usually not high enough, however, to give rise to hydrogen cyanide, ammonia, or nitrogen oxides (Edgerley 1980). These compounds can appear in cases of fire (Chaigneau et al. 1974, Michail 1982).

Other polymers, either as blends or as copolymers, may influence polystyrene degradation. Polyvinyl chloride (PVC) and polyvinylacetate (PVA) in blend with polystyrene cause a more rapid degradation of polystyrene than is the case with the pure PS. Monomer production (styrene) is, however, retarded and an increase in the rate of chain scission has been observed (Dodson 1976, Jamieson 1976, Grassie 1977).

Also some additives, such as fire retardants, have been noticed to accelerate polymer degradation. Especially bromine flame retardant additives, in spite of their being good flame retardants, are detrimental to the polymer structure in terms of polymer fragments and smoke emitted (Brauman 1981, Brauman 1981, Brighton 1979).

Thermo-oxidative degradation

When thermal degradation takes place in the presence of air, the degradation is thermo-oxidative (Fig. 3). In normal machining, it is difficult to avoid the presence of air. The air enters the hot machining cylinder at first with and on the plastic granules. The plastic encounters fresh air anew when the ready-made hot pieces (articles) fall out of the working machine. During processing, thermal and thermo-oxidative degradation occur simul-

taneously in the plastic. The oxidation reaction begins
with the formation of hydroperoxide groups along the polymer chain. The hydroperoxides decompose via alkoxy radicals to give carbonyl and hydroxyl groups (Brighton
1979). Thus the degradation products of plastics contain,
besides hydrocarbons, also oxidized compounds such as
alcohols, ketones, aldehydes, organic acids, and carbon
oxides (Hoff 1982, Pfäffli 1978)

The thermo-oxidative degradation of styrenated polyesters, prepared, e.g., from phthalic anhydride, maleic
anhydride, and propylene glycol, is a complex reaction
with the formation of, i.a., phthalic acid and anhydride,
esters of glycol, aliphatic and aromatic hydrocarbons,
carbon oxides as well as fragments of polymer chain with
high molecular weight (Sugimura 1979).

Figure 3.   OXIDATIVE POLYMER DEGRADATION

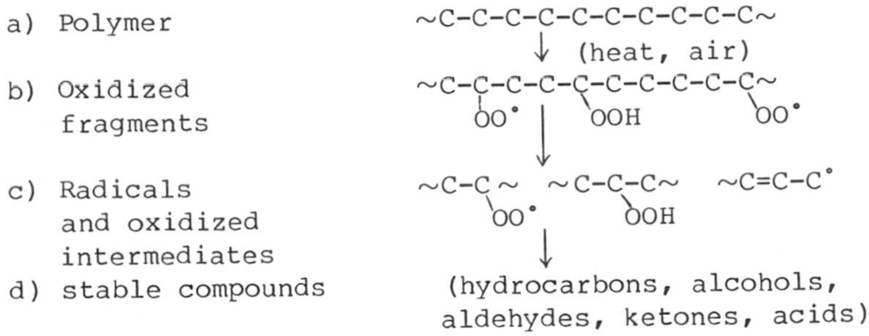

There are very few reports on concentrations of air-
borne substances during the processing of plastics.
Brighton et al. (1979) give a range of 1-7 ppm for styrene
concentration during an injection molding process.
Edgerley (1980) simulated thermal degradation of several
thermoplastics in laboratory scale. He also carried out
some measurements in factory conditions. He measured
carbon monoxide concentrations developed at various temperatures with several plastics. With styrene-containing
plastics, he also measured benzene and styrene in the case
of polystyrene, and hydrogen cyanide, ammonia, nitrogenous
fumes, and styrene in the case of acrylonitrile-
butadiene-styrene. He found all the concentrations very
low compared with the accepted threshold limit values,
although the degradation temperatures applied were at the
upper limit of the corresponding processing temperatures.

A Swedish-Finnish group (Hoff et al. 1982) published
a report on the thermal degradation of polyethylene and
styrene-containing thermoplastics. Preliminary work was
done in the laboratory. The work place measurements were
then planned on the basis of the laboratory experiments.
Measurements were carried out in several factories.
Thermal degradation of styrene thermoplastics yielded
styrene only as a minor product. Thus, a considerable part
of the volatiles were compounds other than styrene. The
yield of these compounds was higher from PS and BS (impact
PS) than from ABS and SAN. The products with a high
boiling point structurally resembled the original polymer.
They were dimers, trimers, tetramers, etc. of the styrene
polymer. They were probably such fragments of polymer
chain that are volatile at the processing temperature but
involatile at the ambient temperature. In the atmosphere,
this oily fraction forms tiny droplets and particles with
an average size below 0.2 µm. These particles easily
aggregate (Fig. 4). The oxidized products found both in
the laboratory and in the working atmosphere were
benzaldehyde, benzoic acid, acetophenone, and small
amounts of aliphatic aldehydes and acids, e.g., formaldehyde, acetaldehyde, formic acid, and acetic acid. Oxidized
compounds were also found in the aerosol fraction. Small
amounts of oxidation intermediates, i.e., hydroperoxides
and alkoxy radicals (Westerberg et al. 1982), were detected in factory air. Nitrogen containing plastics, ABS
and SAN, also emitted small amounts of organic nitriles,
e.g., acetonitrile and trace concentrations of acrylonitrile.

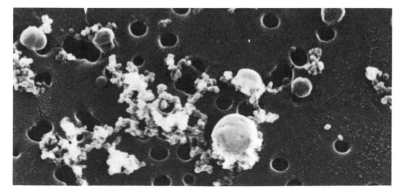

Figure 4.   Scanning electron micrograph. PS fume on a
Nuclepore filter (0.2 µm pore; 20,000 x)

The measured atmospheric concentrations were fairly low in comparison with existing TLV's (ACGIH 1982), with the exception of aerosol (fume) which seemed to form the main hygienic problem. The calculated combined effect of the air impurities for the processing of styrene-thermo-plastics averaged 0.5. When the combined effect was calculated so that fumes were not included, the resulting combined effect was only 0.1. The calculation of the combined effect was not, however, entirely adequate as threshold limit values were available neither for many of the individual compounds nor for compound groups, e.g., radicals and peroxides. For fumes the TLV of respirable dusts, 5 mg/m^3, was used. As a matter of fact the TLV for fumes should be lower because the fumes contain irritative functional groups. These groups occur both as chemically bonded in their molecules and as dissolved compounds. The results were variable to some extent because of differences in raw materials, ventilation, etc. Process-dependent variations can be seen in figure 5.

In this study the industrial hygienic measurements were carried out during "normal process". In cases of malfunction of processes there may exist much higher concentrations of impurities in the air.

Conclusions

The characteristic features of air-borne impurities during "normal" processing of polystyrene thermoplastics are: the appearance of
- a mixture of very many individual compounds with a wide range of molecular weights
- relatively low concentration levels of single compounds or compound groups when compared with existing TLV's
- oxidized compounds with irritative effects
- reactive intermediate compounds (peroxides, radicals)
- fumes with very tiny particles and oxidized groups both as chemically bonded and as dissolved compounds

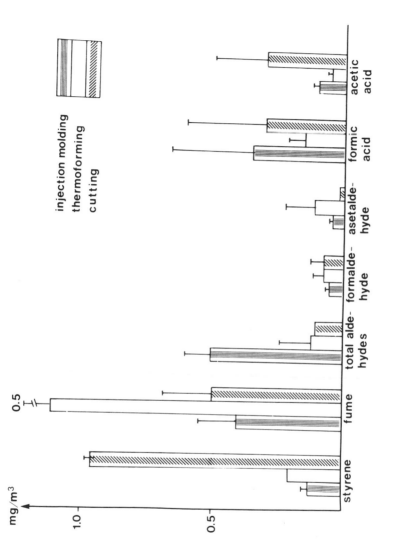

Figure 5. Air-borne concentrations of thermal degradation products during processing of styrene-thermoplastics. Medians of average concentrations in the various workplaces (Pfäffli & Vainiotalo, unpublished).

ACGIH, American Conference of Governmental Industrial Hygienists. Threshold Limit Values for 1982.

Berg H, Olsen H, Pedersen E (1982) Gas dannelse ved vulkanisering of gummi. Arbeijdsmiljøfondet, 110 p.

Blazso M, Ujszaszi K, Jakab E (1980). Isomeric structure of styrene-acrylonitrile and styrene-methylacrylate copolymer pyrolysis products. Chromatographia 13:3:151-156.

Brauman SK (1981). Influence of fire retardant cyagard RF-1/ammonium polyphosphate on degradation of polypropylene and polystyrene. J Fire Retard Chem 8:1:8-23.

Brauman SK, Chen IJ (1981). Influence of the fire retardant decarbomodiphenyl oxide-antimony trioxide on degradation of polystyrene. J Fire Retard Chem 8:1:28-36.

Brighton CA, Pritchard G, Skinner GA (1979). Styrene Polymers: Technology and Environmental Aspects. Applied Science Publishers Ltd, London, 284 p.

Cameron GG, Meyer JM, McWalter JT (1978) Thermal degradation of polystyrene. 3. A Reappraisal. Macromolecules 11:4:696-700.

Chaigneau M, Le Moan G (1974). Etude de la pyrolyse des matèriaux en matières plastiques. VIII Polyacrylonitrile et copolymères. Annales Pharm Franc 32:9110:485-490.

Dickens B (1980). Thermal degradation and oxidation of polystyrene studied by factor-jump thermogravimetry. Polym Degradat Stabil 2:4:249-268.

Dodson B, McNeil IC (1976) Degradation of polymer mixtures VI. Blends of poly(vinylchloride) with polystyrene. J Polymer SC 14:353-364.

Edgerley GP (1980). The oxidative pyrolysis of plastics. Fire and Materials 4:2:77-82.

Grassie N (Ed) (1977). Developments in Polymer Degradation-1. Applied Science Publishers Ltd, London, p 165, 196, 199, 201.

Gächter R, Müller H (1979). Taschenbuch der Kunststoff-Additive. Carl Hanser Verlag, München, 578p.

Hoff A, Jacobsson S, Pfäffli P, Zitting A, Frostling H (1982). Degradation products of plastics. Polyethylene and styrene-containing thermoplastics - Analytical, occupational and toxicological aspects. Scand J Work Environ Health 8:suppl. 2, 60p.

Jamieson A, McNeil IC (1976) Degradation of polymer mixtures VII. Blends of polyvinyl acetate with polystyrene. Polym Sci Polym Chem 14:3:603-608.

Kishore K, Pai Verneker VR, Nair MNR (1976). Thermal degradation of polystyrene. J Appl Polymer Sc 20:2355-2365.

Kokta BV, Valade JL, Martin WN (1973). Dynamic thermogravimetric analysis of polystyrene: effect of molecular weight on thermal decomposition. J Appl Polymer Sc 17:1-19.

Malhotra SL, Hesse J, Blanchard LP (1975). Thermal decomposition of polystyrene. Polymer 16:81-93.

McNeil IC, Ackerman L, Gupta SN (1978). Degradation of polymer mixtures. IX. Blends of polystyrene with polybutadiene. J Polymer Sci Polym Chem 16:9:2169-2181.

Michail J (1982). Determination of hydrogen cyanide in thermal degradation products of polymeric materials. Fire and Materials 6:1:13-15.

Modern Plastics International (1982)

Pfäffli P, Zitting A, Vainio H (1978). Thermal degradation products of homopolymer polystyrene in air. Scand J Work Environ & Health 4:suppl.2:22-27.

Sugimura Y, Tsuge S (1979). Studies on thermal degradation of aromatic polyesters by pyrolysis-gas chromatography. J Cromatog Sc 17:269-272.

Westerberg L-M, Pfäffli P, Sundholm F (1982). Detection of free radicals during processing of polyethylene and polystyrene plastics. Am Ind Hyg Assoc J 43:544-546.

Zitting A, Pfäffli P, Vainio H (1978). Effect of thermal degradation products of polystyrene on drug biotransformation and tissue glutathione in rat and mouse. Scand J Work Environ & Health 4:suppl.2:60-66.

Zitting A, Heinonen T (1980). Decrease of reduced glutathione in isolated rat hepatocytes caused by acrolein, acrylonitrile, and the thermal degradation products of styrene copolymers. Toxicol 17:333-341.

Zitting A, Heinonen T, Vainio H (1980). Glutathione depletion in isolated rat hepatocytes caused by styrene and the thermal degradation products of polystyrene. Chem-Biol Interactions 31:313-318.

Zitting A, Savolainen H (1980). Effects of single and repeated exposures to thermo-oxidative degradation products of poly(acrylonitrile-butadiene-styrene) (ABS) on rat lung, liver, kidney, and brain. Arch Toxicol 46:295-304.

Industrial Hazards of Plastics and Synthetic Elastomers, pages 215–225
© 1984 Alan R. Liss, Inc., 150 Fifth Ave., New York, NY 10011

METABOLISM AND MUTAGENICITY OF STYRENE AND STYRENE OXIDE

H. Vainio, H. Norppa and G. Belvedere[1]

Institute of Occupational Health,
Haartmaninkatu 1, Helsinki 29, Finland and
[1]Istituto di Ricerche Farmacologiche "Mario
Negri", Milan, Italy

CYTOCHROME P-450 DEPENDENT METABOLISM OF STYRENE

The main urinary metabolites of styrene have been
known for over 20 years (Fig. 1). The key intermediate in
the formation of most of these metabolites is styrene-
7,8-oxide which has been identified in microsomal preparations of rat liver after short periods of incubation
(Leibman and Ortiz, 1969). Styrene has been shown to be
oxidized to (R)- and (S)-styrene-7,8-oxides in the ratio
1:1.3 by rat liver microsomal cytochrome P-450 (Watabe et
al., 1981).

Styrene binds to cytochrome P-450 with a type I
difference spectrum in control, phenobarbital and
3-methylcholanthrene induced rat liver microsomes (Vainio
and Zitting, 1978; MacKenzie and Vainio, 1981). The activity of styrene monooxygenase depends on NADPH, NADH
having a slight synergistic effect (Salmona et al., 1976).
Styrene-7,8-oxide is readily hydrated to styrene glycol by
the microsomal epoxide hydrolase (Oesch, 1973). The (R)-
and (S)-oxides are hydrolyzed in a regiospecific manner by
epoxide hydrolase to (R)- and (S)-styrene glycols at relative rates of 1:4, respectively (Watabe et al., 1982). The
affinity (K_m) for styrene monooxygenase shows broad
interspecies variability (Belvedere et al., 1977).

Styrene-7,8-oxides are also conjugated with
glutathione (Oesch, 1973; Pacheka et al., 1979). Rat liver
cytosolic glutathione S-transferase convert (R)- and

(S)-oxides stereoselectively to corresponding glutathione conjugates (Fig. 1).

Rats given styrene i.p. also excrete 2-vinylphenol and 4-vinylphenol in conjugated form in the urine (Hiratsuka et al., 1982). Styrene-1,2- and 3,4-oxides administered i.p. to rats are recovered in the urine as 2-vinylphenol and 4-vinylphenol, respectively. The excretion of 4-vinylphenol has also been found among workers exposed to styrene (Pfäffli et al., 1981). The excretion of this metabolite is 0.3 % in respect to mandelic acid, and different subjects' excretion of 4-vinylphenol correlates well with their excretion of mandelic acid.

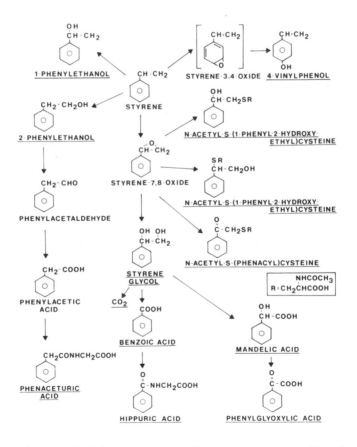

Fig. 1. The metabolic pathways of styrene. Underlined metabolites are excreted in the urine.

METABOLISM NOT DEPENDENT ON CYTOCHROME P-450

Washed human erythrocytes have been found to oxidize styrene to styrene-7,8-oxide in the absence of NADPH. This reaction, almost completely inhibited by carbon monoxide, is not catalyzed by methaemoglobin (Belvedere and Tursi, 1981). The amount of styrene-7,8-oxide formed is proportional to the molar fraction of oxyhaemoglobin present (Tursi et al., 1982). The affinity of styrene for oxyhaemoglobin is, however, much lower than the affinity for rat liver microsomal cytochrome P-450 (cf. Vainio et al., 1982b).

The flavoprotein ferredoxin reductase can also catalyse the oxidation of styrene to styrene-7,8-oxide in the presence of NADPH. This reaction is inhibited by superoxide dismutase and catalase, and H_2O_2 can be used instead of NADPH (Belvedere et al., 1982). The formation of styrene-7,8-oxide is most likely catalyzed by free reactive oxygen intermediates, because superoxide dismutase and catalase inhibit this reaction. Oxyhaemoglobin catalyzed styrene oxidation is not inhibited by these two enzymes (Belvedere et al., 1982, Tursi et al., 1982).

REACTIVE INTERMEDIATES AND MUTAGENICITY OF STYRENE

Studies in experimental animals and in man indicate that the major route for the metabolism of styrene proceeds via the formation of styrene-7,8-oxide. Styrene-7,8-oxide binds covalently to nucleic acid bases (Hemminki et al., 1980); is a direct base-pair type mutagen in bacteria (Table 1); is positive in several non-mammalian eukaryotic mutagenicity assays (Table 2); produces point mutations and chromosome damage in mammalian cell cultures (Table 3); and is carcinogenic in animals (Maltoni et al., 1979). It is therefore reasonable to assume that this epoxide metabolite is responsible for the (geno)toxicity of styrene. Other reactive compounds (such as styrene-3,4-oxide and phenylacetaldehyde) have not been detected in vitro or in vivo and are presumed to be minor metabolic intermediates.

Table 1. Mutagenicity of styrene and styrene-7,8-oxide in bacteria. For references see Norppa (1981), Norppa and Vainio (1982) and Vainio et al. (1982a).

Test system	Styrene		Styrene-7,8-oxide	
	Metabolic activation:		Metabolic activation:	
	None	Present	None	Present
Salmonella typhimurium				
Base-pair mutations	−	+[a]	+	+
Frameshift mutations	−	−	−	−
Escherichia coli	−[b]		+	
Klebsiella pneumoniae			+	
Bacillus subtilis	+[c]			

[a]Both positive and negative results reported. [b]Greim et al. (1977). [c]Mentioned in a review by Kawachi et al. (1980); rec assay.

　　Styrene has been reported to be mutagenic only after metabolic activation (Tables 1-3). The conflicting results on the in vitro mutagenicity of styrene may reflect differences in activation and inactivation capacities between the metabolizing systems used. In rodents exposed to styrene or styrene-7,8-oxide in vivo, chromosome damage has been observed in some studies with mice (and one in rats), but not in Chinese hamsters (Table 4). This may be explained by the presence of higher levels of epoxide hydrolase in Chinese hamsters as compared to mice (Norppa et al., 1979).

　　Styrene is able to increase chromosome aberrations and sister chromatid exchanges (SCEs) in vitro - without exogenous metabolizing systems - in human lymphocytes cultured with whole blood (Linnainmaa et al., 1978; Norppa et al., 1980). No clear effect on SCEs is seen in isolated lymphocyte cultures which contain only 0.01 % of the erythrocytes present in whole blood cultures (Norppa et al., 1982). When erythrocytes are added into these purified cultures SCEs increase. Thus, erythrocytes seem to be important in the metabolic activation of styrene in the lymphocyte assay. This is also supported by the findings of Belvedere and Tursi (1981) and Tursi et al. (1982) on styrene oxidation to styrene-7,8-oxide by erythrocytes and oxyhaemoglobin. Conner et al. (1982) have pointed out the possible importance of extrahepatic metabolism in their studies on mice exposed to styrene by inhalation. The increase of SCEs is similar in bone marrow, alveolar macrophages and regenerating liver cells of hepatectomized or normal (bone marrow and macrophages studied) mice after styrene exposure. Chromosome aberrations are increased in the peripheral lymphocytes of styrene exposed workers of the reinforced plastics industry (Meretoja et al., 1977; Norppa et al., 1981; Camurri et al., 1982). The role of red blood cells in the formation of this chromosome damage is presently unknown.

Table 2. Mutagenicity of styrene and styrene-7,8-oxide in nonmammalian eukaryotic systems. For references see Norppa (1981), Norppa and Vainio (1982) and Vainio et al. (1982a).

Test system	Styrene	Styrene-7,8-oxide
Drosophila melanogaster		
Recessive lethals	+	+
X/Y chromosome loss	−	
Silkworm		
Point mutations	−a	
Schizosaccharomyces pombe		
Forward mutations:		
No activation		+
With S-10 mix	−	
Host-mediated assay	−b	+b
Saccharomyces cerevisiae		
Gene conversions:		
No introduced activation	+c	+
With S-10 mix	−	
Host-mediated assay	−b	+b
Allium cepa		
Chromosome aberrations	+	−
Micronuclei	+	+
Anaphase fragments and bridges	−	+
Vicia faba		
Chromosome aberrations		+
Micronuclei		+

aMentioned in a review by Kawachi et al. (1980).
bLoprieno, personal communication. cDel Carratore et al. (1982); yeast strain containing cytochrome P-450.

Table 3. Mutagenicity of styrene and styrene-7,8-oxide in mammalian cell cultures. For references see Norppa (1981), Norppa and Vainio (1982) and Vainio et al. (1982a).

Test system	Styrene		Styrene-7,8-oxide	
	Metabolic activation:		Metabolic activation:	
	None	Present	None	Present
Chinese hamster cell lines				
V79:				
HGPRT mutations	−	+[a]	+	−[a]
Chromosome aberrations			+	
Micronuclei			+	
Anaphase bridges			+	
CHL:				
Chromosome aberrations	−	+	+[b]	
CHO:				
Sister chromatid exchance	−	+[c]	+	+
Mouse cell line				
L5178Y:				
TK-mutations			+	−
Wistar rat hepatocytes				
Unscheduled DNA synthesis				−
Human cells				
Lymphocytes				
(whole blood cultures):				
Chromosome aberrations	+		+	
Micronuclei	+		+	
Sister chromatid exchange	+		+	
Unscheduled DNA synthesis	−[d]			
EUE:				
Unscheduled DNA synthesis	−	+		

[a]Liver perfusion system used for metabolic activation (Beije and Jenssen, 1982). [b]Ishidate et al. (1981). [c]Only with cyclohexene oxide present (De Raat, 1978). [d]Pero et al. (1982); purified lymphocytes.

Table 4. Chromosome damage induced by styrene and styrene-7,8-oxide in rodents in vivo. For references see Norppa (1981), Norppa and Vainio (1982) and Vainio et al. (1982a).

Test system	Styrene	Styrene-7,8-oxide
Rat		
Bone marrow:		
Chromosome aberrations	±	
Mouse		
Bone marrow:		
Chromosome aberrations	-[a]	+
Micronuclei	+	-[b]
Sister chromatid exchange	+	
Alveolar macrophages:		
Sister chromatid exchange	+	+[b]
Regenerating liver cells:		
Sister chromatid exchange	+	+[b]
Primary spermatocytes:		
Translocations		-
Post-meiotic germ cells:		
Dominant lethals		-
Chinese hamster		
Bone marrow:		
Chromosome aberrations	-	-[c]
Micronuclei	-	-
Sister chromatid exchange		(+)[d]

[a]Sbrana et al. (1983). [b]Conner et al. (1982). [c]A positive effect in fatal treatment (Norppa et al., 1979). [d]A slight effect at 500 mg/kg body weight (Norppa et al., 1979).

The importance of epoxide metabolites is a complex issue, because the activating and inactivating pathways may intermingle. A major factor in determining the relative and sequential significance of these routes is certainly the chemical structure of the substrates. For example, after activation by epoxidation, styrene is mainly inactivated by the formation of a glutathione conjugate or a dihydrodiol, whereas the covalent binding of the epoxide could result in toxic effects. The availability of an intermediate epoxide metabolite is not only determined by the relative activities (V_{max}) of the enzymes which form and inactivate epoxides, but also by the apparent K_m's of these enzymes. Activities determined in vitro, which are close to the V_{max} values, may even give a misleading impression about the relative importance of the enzymes at the rather low concentrations encountered in living animals. At low concentrations of styrene, because of its low K_m, the contribution of the epoxide forming monooxygenase attack is likely to be greater than the role of epoxide hydrolase.

As styrene most probably exerts its mutagenic effects via its metabolites, the non-linear clearance observed at high levels of exposure suggests that the toxic effects seen at high exposures will be disproportionally smaller than those seen at lower exposure levels. Therefore, the extrapolation of toxicity data obtained at saturating levels of exposure may lead to the underestimation of hazards encountered in exposure to low concentrations of styrene.

REFERENCES

Beije B and Jenssen D (1982). Investigation of styrene in the liver perfusion/cell culture system. No indication of styrene-7,8-oxide as the principal mutagenic metabolite produced by the intact rat liver. Chem-Biol Interact 39: 57.
Belvedere G, Blezza D and Cantoni L (1982). Ferredoxin reductase catalyzes styrene oxidation to styrene oxide. Experientia, in press.
Belvedere G, Cantoni L, Facchinetti T and Salmona M (1977). Kinetic behaviour of microsomal styrene monooxygenase and styrene epoxide hydratase in different animal species. Experientia 33: 708.
Belvedere G and Tursi F (1981). Styrene oxidation to styrene oxide in human blood erythrocytes and lymphocytes. Res Comm Chem Pathol Pharmacol 33: 273.
Camurri L, Codeluppi S, Pedroni C and Scarduelli L (1982). Chromosomal aberrations and sister chromatid exchanges in styrene exposed workers. Mutat Res, in press.
Conner MK, Alarie Y and Dombroske RL (1982). Multiple tissue comparisons of sister chromatid exchanges induced by inhaled styrene. In Tice R, Costa DL and Schaich KM (eds): "Genotoxic Effects of Airborne Agents", New York:

Plenum Publishing Corporation, p 433.

Del Carratore R, Bronzetti G, Bauer C, Corsi C, Nieri R, Paolini M and Giagoni P (1982). Study of cytochrome P_{450} in yeast D_7 strain. An alternative model to microsomal assay. Mutagenicity of styrene. In 12th Annual Meeting of European Environmental Mutagen Society, Espoo, Finland, 20-24 June, 1982, Abstracts, p 160.

De Raat WK (1978). Induction of sister chromatid exchanges by styrene and its presumed metabolite styrene oxide in the presence of rat liver homogenate. Chem-Biol Interact 20: 163.

Greim H, Bimboes D, Egert G, Göggelman W and Krämer M (1977). Mutagenicity and chromosomal aberrations as an analytical tool for in vitro detection of mammalian enzyme-mediated formation of reactive metabolites. Arch Toxicol 39: 159.

Hemminki K, Paasivirta J, Kurkirinne T and Virkki L (1980). Alkylation products of DNA bases by simple epoxides. Chem-Biol Interact 30: 259.

Hiratsuka A, Aizawa T, Ozawa N, Isobe M, Watabe T and Takabatake E (1982). The role of epoxides in the metabolic activation of styrene to mutagens. Eisei Kagaku 28: P-34.

Ishidate M Jr, Sofuni T and Yoshikawa K (1981). Chromosomal aberration tests in vitro as a primary screening tool for environmental mutagens and/or carcinogens. GANN Monograph on Cancer Research 27: 95.

Kawachi T, Yahagi T, Kada T, Tazima Y, Ishidate M, Sasaki M. and Sugiyama T (1979). Cooperative programme on short-term assays for carcinogenicity in Japan. In Montesano R, Bartsch H, Tomatis L (eds): "Molecular and Cellular Aspects of Carcinogen Screening Tests". IARC Scientific Publications No. 27, Lyon, IARC, p 323.

Leibman KC, Ortiz E (1969). Oxidation of styrene in liver microsomes. Biochem Pharmacol 18: 552.

Linnainmaa K, Meretoja T, Sorsa M, Vainio H (1978). Cyto-genetic effects of styrene and styrene oxide. Mutat Res 58: 277.

MacKenzie P, Vainio H (1981). Interactions of styrene and styrene oxide with partially purified cytochrome P-450 and P-448 from rat liver microsomes. Toxicol Lett 9: 183.

Maltoni C, Failla G, Kassapidis G (1979). First experimen-tal demonstration of the carcinogenic effects of styrene oxide. Med Lavoro 5: 358.

Meretoja T, Vainio H, Sorsa M, Härkönen H (1977). Occu-pational styrene exposure and chromosomal aberrations. Mutat Res 56: 193.

Norppa H (1981). "Chromosome Damage Induced by Styrene, Styrene Oxide and Some Analogues", Academic dissertation, Helsinki, Institute of Occupational Health and University of Helsinki.

Norppa H, Elovaara E, Husgafvel-Pursiainen K, Sorsa M, Vainio H (1979). Effects of styrene oxide on chromosome aberrations, sister chromatid exchange and hepatic drug biotransformation in Chinese hamsters in vivo. Chem-Biol Interact 26: 305.

Norppa H, Sorsa M, Pfäffli P, Vainio H (1980). Styrene and styrene oxide induce SCEs and are metabolised in human lymphocyte cultures. Carcinogenesis 1: 357.

Norppa H, Vainio H (1982). Genetic toxicity of styrene and some of its derivatives. Scand J Work Environ Health, in press.

Norppa H, Vainio H, Sorsa M (1981). Chromosome aberrations in lymphocytes of workers exposed to styrene. Am J Ind Med 2: 299.

Norppa H, Vainio H, Sorsa M, Belvedere G (1982). Metabolic activation of styrene by erythrocytes in human lymphocyte cultures. In 12th Annual Meeting of European Environmental Mutagen Society, Espoo, Finland, 20–24 June, 1982, Abstracts, p 148.

Oesch F (1973). Mammalian epoxide hydrases. Inducible enzymes catalysing the inactivation of carcinogenic and cytotoxic metabolites derived from aromatic and olefinic compounds. Xenobiotica 3: 305.

Pachecka J, Gariboldi P, Cantoni L, Belvedere G, Mussini E, Salmona M (1979). Isolation and structure determination of enzymatically formed styrene oxide glutathione conjugates. Chem-Biol Interact 27: 313.

Pero RW, Bryngelsson T, Högstedt B, Åkeson B (1982). Occupational and in vitro exposure to styrene assessed by unscheduled DNA synthesis in resting human lymphocytes. Carcinogenesis 3: 681.

Pfäffli P, Hesso A, Vainio H, Hyvönen M (1981). 4-Vinylphenol excretion suggestive of arene oxide formation in workers occupationally exposed to styrene. Toxicol Appl Pharmacol 60: 85.

Salmona M, Pachecka J, Cantoni L, Belvedere G, Mussini E, Garattini S (1976). Microsomal styrene mono-oxygenase and styrene epoxide hydrase activities in rats. Xenobiotica 6: 585.

Sbrana I, Lascialfari D, Rossi AM, Loprieno N, Bianchi M, Tortoreto M, Pantarotto C (1983). Bone marrow cell chromosomal aberrations and styrene biotransformation in mice given styrene on a repeated oral schedule. Chem-Biol Interact, in press.

Tursi F, Samaia M, Salmona M, Belvedere G (1982). Styrene oxidation to styrene oxide in human erythrocytes is catalyzed by oxyhemoglobin. Experientia, in press.

Vainio H, Norppa H, Hemminki K, Sorsa M (1982a). Metabolism and genotoxicity of styrene. In Snyder R, Parke DV, Kocsis J, Jollow DJ, Gibson GG, Witmer CM (eds): "Biological Reactive Intermediates-II: Chemical Mechanisms and Biological Effects, Part A", New York: Plenum Publishing Corporation, p 257.

Vainio H, Tursi F, Belvedere G (1982b). What are the significant toxic metabolites of styrene? In Hietanen E, Laitinen M, Hänninen O (eds): "Cytochrome P-450, Biochemistry, Biophysis and Environmental Implications", Elsevier Biomedical Press, p 679.

Vainio H, Zitting A (1978). Interaction of styrene and acetone with drug biotransformation enzymes in rat liver. Scand J Work Environ Health 4: suppl 2, 47.

Watabe T, Ozawa N, Yoshikawa K (1982) Studies on metabolism and toxicity of styrene. V. The metabolism of styrene, racemic, (R)-(+)-, and (S)-(-)-phenyloxiranes in the rat. J Pharm Dyn 5: 129.

Industrial Hazards of Plastics and Synthetic Elastomers, pages 227–238
© 1984 Alan R. Liss, Inc., 150 Fifth Ave., New York, NY 10011

Styrene, Styrene Oxide, Polystyrene, and
β-Nitrostyrene/Styrene Carcinogenicity in Rodents

J.E. Huff
National Toxicology Program
National Institute of Environmental Health Sciences
Research Triangle Park, NC   USA

Propelled by portals whose only shame
is a zeppelin's shadow crossing a field
of burning bathtubs,
I ask myself: There must be more to life
than this?

Richard Brautigan
Rommel Drives on Deep into Egypt (1970)

Most styrene is used to produce plastics and resins:
polystyrene, acrylonitrile-butadiene-styrene terpolymer,
and styrene-acrylonitrile copolymer. Lesser amounts are
used to make styrene-butadiene rubber. Styrene oxide is
used as a reactive diluent in epoxy resins to reduce the
viscosity of mixed systems prior to curing, as an intermediate in the preparation of agricultural and biological
chemicals, of cosmetics, and of surface coatings in the
treatment of textiles and fibers. Polystyrene is used in
packaging (containers, sheet and film, and loose fill), furnishings and consumer products, construction, electrical
parts, and others (IARC, 1979). Three major fabricated
forms of polystyrene are used in the US: injection molding
(50%), extrusions (35%), and beads (10%), with the remainder
in packaging. β-Nitrostyrene, a chain stopper in styrene
type polymerization reactions, possesses antibacterial,
antifungal, and insecticidal activities and has been proposed as a repellant for bats and other rodents (NCI,
1979a).

This paper summarizes the available literature on the carcinogenicity studies of styrene (Dow 1978; Ponomarkov and Tomatis, 1978; NCI, 1979b; CMA, 1980), styrene oxide (Weil et al., 1963; Van Duuren, et al., 1963; Maltoni et al., 1979; Maltoni, 1982; Ponomarkov et al., 1983), polystyrene (Nothdruft, 1956; Oppenheimer et al., 1958; Riviere et al., 1960), and β-nitrostyrene/styrene (NCI, 1979a). Additionally, for each chemical certain select identifier data are given: Chemical Abstracts Service Registry Number (CAS RN), chemical formula, synonyms, physical appearance, method of preparation, discovery/year of first commercial introduction, and current production figures.

STYRENE -- Production amounts (USA) for styrene and some styrene polymers are listed in Table 1. The latest monthly production figures published by the US International Trade Commission (USITC 1983a) report 266×10^6 kg styrene for May 1983; this projects to an annual amount of 2650×10^6 kg. Imports into the US for 1982 totaled 9.5×10^6 kg (USITC, 1983b). Profile information on styrene is given in Table 2.

Table 1. US Styrene and Select Styrene Polymer Production (C&EN, 1983)

	1982	1981	1980	1979
		(10^6 kg)		
Styrene	2700	3000	3100	3400
Polystyrene	1450	1700	1600	1800
Styrene-Acrylonitrile	41	50	50	56
Styrene-Butadiene Copolymers	290	380	360	410
Acrylonitrile-Butadiene-Styrene	336	410	420	540

Table 2. Styrene Properties

CAS RN: 100-42-5   Formula: $C_6H_5-CH=CH_2$; C_8H_8
Synonyms: Ethenylbenzene; Styrole; Tyrolene; Vinyl Benzene
Appearance: Volatile liquid (colorless, viscous)
Preparation: Catalytic dehydrogenation of ethylbenzene
Discovery: 1831 by Bonastre during distillation of Storax
            Resin
Commercial Production: 1938
1981 US Production: 3,005,000,000 kg (21st top chemical)
1982 US Production: 2,696,000,000 kg (22nd top chemical)

Styrene Carcinogenesis -- Four reports on 14 studies are available for review: Dow, 1978; Ponomarkov and Tomatis, 1978; NCI, 1979b; and CMA, 1980. Summary information in Table 3 is given to compare routes of administration and doses. The studies and results are described in chronological order.

Table 3. Styrene Carcinogenesis Studies

Laboratory	Route	Dose	Reference
Dow Chemical USA	Vapor Inhalation	Rats: 0, 600, 1000 ppm	Dow, 1978
International Agency for Research on Cancer	Gavage in Olive Oil	Mice: 0, 1350 mg/kg for 16 weeks Rats: 0, 500 mg/kg for 120 weeks	Ponomarkov & Tomatis, 1978
National Cancer Institute	Gavage in Corn Oil	Rats: 0, 500, 1000, 2000 mg/kg Mice: 0, 150, 300 mg/kg	NCI, 1979b
Chemical Manufacturers Association	Drinking Water	Rats: 0, 125, 250 ppm M: 0, 7.7, 14 mg/kg F: 0, 12, 21 mg/kg	CMA, 1980

In the Dow Study (Dow, 1978) male and female Sprague-Dawley rats were exposed to air containing styrene vapor (0, 600, or 1200 ppm) for 6 hours per day, 5 days per week, for 18.3 months (males) or for 20.7 months (females). The 1200 ppm level was reduced to 1000 ppm after 2 months because of reduced weight in males. The experiment was terminated at 2 years. While the results did not show particularly significant effects, in female rats the incidence of adenocarcinomas of the mammary glands in the 600-ppm group was increased (1/85, 7/85, 0/85). More compelling were the increases in leukemia-lymphosarcoma in female rats (1/85, 6/85, 6/85); the rates in the 600-ppm males were elevated as well (1/85, 5/86, 1/84). The report rightly concludes that these findings are "suggestive of an association between the exposure of these female rats to styrene vapor and an increased incidence of tumors of the leukemia-lymphosarcoma type. In males, the results are even more inconclusive but tend to support the suggestive association found in the females".

The IARC Study (Ponomarkov and Tomatis, 1978) involved pre-
natal exposure by a single intubation of styrene on day 17
of gestation to O_{20} mice (1350 mg/kg), C57 B1 mice (300
mg/kg), and BD IV rats (1350 mg/kg).   The progeny were
administered styrene weekly beginning at weaning: the O_{20}
mice received 1350 mg/kg for 16 weeks; the C57 B1 mice
received 300 mg/kg and the BD IV rats 500 mg/kg until 120
weeks of age.

Table 4. Lung Neoplasms in O_{20} Progeny
(Ponomarkov and Tomatis, 1978)

Lung Neoplasm	Male Mice (mice with lesion/effective number, %)		Female Mice	
	Control	Treated	Control	Treated
Adenoma	4/19,21%	12/23,52% P = 0.039[a]	10/21,48%	14/32,44% P = 0.50
Carcinoma	4/19,21%	8/23,35% P = 0.264	4/21,19%	18/32,56% P = 0.007
Combined	8/19,42%	20/23,89% P = 0.003	14/21,67%	32/32,100% P = 0.0008

[a] [These calculated P values are unadjusted Fisher
exact pairwise comparisons].

The incidences observed both in treated males and females
were increased compared to controls. For males the overall
response reflects mainly increases in adenomas, whereas for
females the carcinomas contribute mostly.   The combined
incidences   are   the   appropriate   comparisons   for   lung
neoplasms (McConnell, 1983); for each sex the increases were
significant and should be considered chemically induced.

In the C_{57} B1 mice progeny some numerical (though not statistically significant) increases in lymphomas were observed
for males (3/12, 25% versus 9/24, 38%) and for females
(5/13, 38% versus 13/24, 54%).

For the BD IV rat progeny the incidences of neoplasms were
not different among groups.

The US National Cancer Institute Study (NCI, 1979b) used F344 rats and B6C3F$_1$ mice, 50 of each sex in treated groups and 40 rats and 20 mice of each sex as vehicle controls. Doses given by gavage in corn oil 5 times per week were for rats 0, 500, 1000, 2000 mg/kg and for mice 0, 150, and 300 mg/kg. All groups except the 500 mg/kg rats were treated for 78 weeks and observed for 27 weeks (rats) or 13 weeks (mice); the 500 mg/kg male and female rats were treated for 103 weeks. These studies were conducted at Litton Bionetics.

No styrene-related neoplasms were seen in male or female rats. For male mice increases in alveolar/bronchiolar neoplasms occurred (Table 5); adenomas were seen in 0/20, 1/44, and 3/48 females.

Table 5. Lung Neoplasms in Male B6C3F$_1$ Mice (NCI, 1979b)

Alveolar/ Bronchiolar Neoplasm	Vehicle Control	150 mg/kg	300 mg/kg
Adenoma	0/20	3/45	4/49
Carcinoma	0/20	3/45	5/49
Combined	0/20	6/45	9/49

The dose-response trend for the combined tumors is significant ($P = 0.023$) and the 300 mg/kg group had a significantly ($P = 0.024$) higher incidence of neoplasms than the vehicle controls.

The CMA (1980) study was conducted at Litton Bionetics using male and female Charles River COBS CD (SD) BR rats (Sprague-Dawley). Styrene in drinking water was made available continuously at concentrations of 0, 125, and 250 ppm for 104 weeks. The daily styrene intakes were estimated to have been 0, 7.7, and 14.0 mg/kg for male rats and 0, 12, and 21 mg/kg for female rats. A dose-related decrease in water consumption was reported (g/day/kg bw) for males (control, 78; 125 ppm, 69; 250 ppm, 63) and for females (118; 107; 93). The report concluded that ingestion of styrene in the drinking water under the conditions of this study produced "no deleterious dose-related effects in the rat". According to the data a marginal increase in neoplasms of the mammary glands in female rats were observed (Table 6).

Table 6. Mammary Gland Neoplasms in Female
Sprague-Dawley Rats (CMA, 1980).

Mammary Gland Neoplasm	Control	125 ppm (12 mg/kg/day)	250 ppm (21 mg/kg/day)
Fibroadenoma	45/96, 49%	15/30, 50%	37/60, 62%
Adenoma	1/96, 1%	0/30	0/60
Adenocarcinoma	8/96, 8%	5/30, 17%	8/60, 13%
Combined	49/96, 51%	18/30, 60%	40/60, 67%

For the combined incidence rates, the trend statistic for a
dose-response relation was significant ($P = 0.032$) and the
incidence in the 250-ppm group was significantly ($P = 0.039$)
higher than observed in the controls.

Styrene Carcinogenicity Summary -- Using the gavage route of
exposure both the IARC and the NCI studies showed increases
in lung neoplasms in mice. Ponomarkov and Tomatis (1978)
reported higher incidences of adenomas in male O_{20} mice and
of carcinomas in female O_{20} mice; the combined incidences of
neoplasms were elevated in both sexes compared to controls
(Table 4). In the male $B6C3F_1$ mice (NCI, 1979b), alveolar/
bronchiolar adenomas/carcinomas were observed with a significant dose-related increase in the treated groups compared
to vehicle controls (Table 5). Female $B6C3F_1$ mice showed a
slight (not statistically significant) increase in lung adenomas (0/20, 1/44, 3/48). For the C_{57} Bl mice (Ponomarkov
and Tomatis, 1978) some trends were seen for lymphomas in
male (25% versus 38%) and in females (38% versus 54%).
Likewise, the Sprague-Dawley rats (Dow, 1978) showed a
marginal increase in leukemia-lymphosarcoma. Mammary gland
neoplasms were increased with dose in female Sprague-Dawley
rats (CMA, 1980).

STYRENE OXIDE -- General information about this epoxide
appears in Table 7.

Styrene Oxide Carcinogenesis -- Two dermal studies were
reported (Weil et al., 1963; Van Duuren et al., 1963);
neither showed any positive response. Weil et al. (1963)
applied 0, 5%, or 10% styrene oxide in acetone three times
weekly to groups of C3H mice for up to 17 months. No skin

tumors were seen. Van Durren et al. (1963) delivered by brush 0.1 ml of a 10% solution of styrene oxide in benzene onto male Swiss-Millerton mice three times weekly. Skin tumors were seen on 3/30 treated and on 11/150 "benzene controls". Both studies are considered inadequate.

Table 7. Styrene Oxide Properties

CAS RN: 96-09-3   Formula: $C_6H_5 - \overset{\displaystyle CH - CH_2}{\underset{O}{\diagup}}$; C_8H_8O

Synonyms: 1,2-Epoxyethylbenzene; Epoxystyrene; Phenyl-oxirane; Styrene Epoxide

Appearance: Liquid (colorless to pale yellow)

Discovery: 1905 by Fourneau and Tiffeneau (a-Phenyl-b-iodoethanol with Potassium Hydroxide)

Preparation: 1967 by Epoxidation of Styrene with Peroxyacetic Acid

Commercial Production: 1974

1981 US Production: 1,000,000 kg (estimated)

Maltoni et al. (1979) and Maltoni (1982) gave styrene oxide in olive oil by gavage once daily (4-5 days weekly) to male and female Sprague-Dawley rats at doses of 0, 50, or 250 mg/kg. Treatment was discontinued after 52 weeks and the animals observed until natural death. Both reports are preliminary; only stomach lesions are recorded. Because the results do not correspond between these two reports, data were taken only from the Maltoni 1982 reference; also a footnote to the results table indicates that one or more tumors may have been present in the same animal.

Table 8. Forestomach Neoplasms in Sprague-Dawley Rats (Maltoni, 1982)

Forestomach Neoplasm	Male Rat			Female Rat		
	0	50 mg/kg	250 mg/kg	0	50 mg/kg	250 mg/kg
Papillomas & Acanthomas	0/39	3/39	7/39	0/40	2/37	5/38
Squamocellular Carcinomas	0/39	9/39	16/39	0/40	7/37	20/38

Dose-related increases in forestomach benign and malignant neoplasms were seen in both male and female rats. Apparently other organs/tissues have not been yet examined or reported.

Ponomarkov, Cabral, and Wahrendorf (1983) administered styrene oxide in olive oil at weekly doses of 0 and 100-150 mg/kg to male and female BD IV rats for two years. Mothers of these progeny received a single 200 mg/kg dose at gestation day 17. Forestomach lesions were increased in male and female progeny (Table 9) as were lung tumors in females (1/55, 2% versus 7/60, 12%).

Table 9. Forestomach Neoplasms in BD IV Rats
(Ponomarkov, Cabral, and Wahrendorf, 1983)

Forestomach Neoplasm	Male		Female	
	0	100-150	0	100-150
	(mg/kg)		(mg/kg)	
Papilloma	0/49	7/42	2/55	2/60
Carcinomas	0/49	10/42	1/55	16/60

The US National Cancer Institute Study on styrene oxide was conducted by the Frederick Cancer Research Center. The NCI asked the National Toxicology Program to evaluate these data and report the results. The NTP policy states that no results will be reported without having the actual experimental data verified and the subsequent interpretations peer reviewed. Because this process has not been finished for styrene oxide only the experimental protocols and vague notions will be given here.

Groups of 50-52 male and female F344/N rats and B6C3F$_1$ mice received 0, 275, or 550 mg/kg styrene oxide in corn oil 3 times/week for 104 weeks. Unverified findings in both sexes of both species lend support to the forestomach lesions observed by Maltoni et al. (1979), Maltoni (1982), and Ponomarkov, Cabral, and Wahrendorf (1983).

Styrene Oxide Carcinogenesis Summary -- Both benign and malignant lesions of the rodent forestomach were caused by the gavage administration of styrene oxide in oil (Tables 8 and 9). This reactive, irritant chemical caused lesions apparently confined to the site of local application. These findings suggest that styrene oxide appears to be toxic to stratified squamous epithelium by direct contact. The forestomach of rats and mice is often a target organ for chemical carcinogens, particularly when the chemical is administered by oral intubation. The squamous-lined forestomach (nonglandular portion) is the proximal two-thirds of the stomach, immediately adjacent to the esophagus, and is demarcated from the distal glandular stomach. The latter is composed of columnar secretory epithelium similar to that of many (or most) higher mammals, including humans. In common with many other non-rodent mammals, humans have no direct counterpart of the rodent forestomach, except possibly the squamous epithelium at the squamocolumnar junction of the cardiac portion of the stomach, which can be a site for either squamous cell carcinoma or adenocarcinoma.

The glandular portion of the rodent stomach is rarely a site of carcinogenesis in untreated animals or those given chemical carcinogens. Gastric adenocarcinomas are, however, frequent in humans in certain countries of the world, including Japan and Chile, for example. During the past 50 years the incidence of these neoplasms has decreased for as yet unknown reasons both in the US and in some European countries. Conversely, carcinoma of the human esophagus is increasing in the US. In many respects, the rodent forestomach more closely resembles the human esophagus than the human stomach.

POLYSTYRENE -- Select information is given in Table 10. Molecular weights extend from about 10,000 to 1,000,000.

Table 10. Polystyrene Properties

CAS RN: 9003-53-6	Appearance: Solid (Transparent, Hard)

Formula: $-CH-CH_2-$
           $\quad\vert\qquad (C_8H_8)n$
           C_6H_5

Synonyms: Ethenylbenzene Homopolymer; Oligostyrene; Polystyrene Latex; Polystyrol; Styrene Polymers; Vinylbenzene Polymer

Discovered: 1839 by Simon after allowing styrene to stand in air (gelatinous); 1845 by Blyth and Hoffman by heating styrene in air (solid).

Preparation: Continuous by a modified mass polymerization process of styrene using a peroxide initiator and ethyl benzene reacted at 120-160°C.

1980 US Production: 1,600,000,000 kg
1981 US Production: 1,645,000,000 kg
1982 US Production: 1,450,000,000 kg

Polystyrene Carcinogenesis -- Subcutaneous implantations (embedding) of discs (smooth or perforated), rods, spheres, or fibers made of polystyrene into the abdominal wall of Wistar rats induced local sarcomas (IARC, 1979; Nothdruft, 1956; Oppenheimer et al., 1958; Riviere et al., 1960). Oppenheimer et al. (1958) reported that a plastic sheet or film causes malignant tumors whereas other forms such as textiles, sponges, or powders ordinarily do not. If the plastic is removed from the connective tissue sheath or "pocket" by six months, no tumors are induced. Removing the complete pocket at any time seems to presage no tumors. The meaningfulness of these "solid state" tumors remains obscure.

β-NITROSTYRENE AND STYRENE -- This combination is usually supplied as a 30:70 mixture. This solution was tested because workers exposed to β-nitrostyrene during manufacture or fabrication would also likely be exposed simultaneously to styrene.

β-Nitrostyrene (30%) and Styrene (70%) Carcinogenesis -- Groups of 50 male and 50 female F344 rats and B6C3F$_1$ mice were exposed to the mixture in corn oil by gavage three times weekly for 79 weeks (rats, untreated for 29 weeks) or for 78 weeks (mice, untreated for 14 weeks) (NCI, 1979a). Dosages based on β-nitrostyrene in the styrene solution were 150 or 300 mg/kg for male rats, 75 or 150 mg/kg for female rats, and 87.5 or 175 mg/kg for male and female mice. Vehicle control groups of 20 male and 20 female rats and mice received corn oil alone on the same dosing schedule. The report concludes that "there was no convincing evidence for the carcinogenicity of a solution of β-nitrostyrene and styrene in Fischer 344 rats or in B6C3F$_1$ mice". For the 87.5 mg/kg male mice, however, increases were observed in the incidences of alveolar/bronchiolar adenomas/carcinomas (0/20; 11/49, 22%: P=0.016; 2/36, 6%). Most of this effect was due to lung adenomas (0/20, 8/49, 1/36) whereas lung carcinomas were 0/20, 3/49, 1/36. This effect becomes more significant by correspondence with increased lung tumors in mice exposed to styrene alone (Ponomarkov and Tomatis, 1978; NCI, 1979b) and given that in the 175 mg/kg male mice group 14 were killed at week 36 (2 others later) with hemorrhage or hemorrhagic necrosis of the liver (explained as "a handling accident").

CARCINOGENESIS SUMMARY -- Styrene appears to induce lung tumors in mice (NCI, 1979b; Ponomarkov and Tomatis, 1978), while the evidence for a similar response is less convincing for a mixture of β-nitrostyrene/styrene (NCI, 1979a). Styrene in vapor results in hematopoietic lesions in rats and in drinking water associates with a marginal increase in mammary neoplasms in female rats (CMA, 1980). Styrene oxide causes neoplasms locally in the rodent forestomach (Maltoni, 1982; Ponomarkov, Cabral, and Wahrendorf, 1983) at the site of gavage administration. Polystyrene implanted sub-cutaneously gives rise to "protective" sarcomas (Nothdruft, 1956; Oppenheimer et al., 1958; Riviere et al., 1960).

REFERENCES

1. C&EN (1983). Top 50 Chemicals, Chem. Eng. News 13 June 1983.

2. CMA (1980). Toxicological Study on Styrene Incorporated in Drinking Water of Rats for Two Years in Conjunction with a Three-Generation Reproduction Study. Chemical Manufacturers Association, Washington, DC (LBI Projects No. 2612 and No. 21175) 995 pages.

3. Dow (1978). Two Year Chronic Inhalation Toxicity and Carcinogenicity Study on Monomeric Styrene in Rats. Final Report. CMA sponsored study. 150 pages.

4. IARC (1979). IARC Monographs on the Evaluation of the Carcinogenic Risk of Chemicals to Humans, Volume 19: Some Monomers, Plastics and Synthetic Elastomers, and Acrolein, 513 pages, International Agency for Research on Cancer, Lyon, France.

5. Maltoni, C. (1982). Early Results of the Experimental Assessments of the Carcinogenic Effects of One Epoxy Solvent: Styrene Oxide, Chapter 7, 97-110, In: England, A., Ringen, K., and Mehlman, M. (Editors). Occupational Health Hazards of Solvents. Princeton Sci. Pub. Inc., Princeton, NJ. 259 pages.

6. Maltoni, C., Failla, G., and Kassapidis, G. (1979). First Experimental Demonstration of the Carcinogenic Effects of Styrene Oxide. Long-Term Bioassays on Sprague-Dawley Rats by Oral Administration. Med. Lavoro 5:358-362.

7. McConnell, E.E. (1983). Guidelines for Combining Benign and Malignant Neoplasms as an Aid in Determining Evidence of Carcinogenicity (NTP Draft Document).

8. NCI (1979a). Bioassay of a Solution of β-Nitrostyrene and Styrene for Possible Carcinogenicity, CAS NO. 102-96-5, CAS NO. 100-42-5, TR 170, National Cancer Institute, Bethesda, MD, 81 pages.

9. NCI (1979b). Bioassay of Styrene for Possible Carcinogenicity, CAS NO. 100-42-5, TR 185, National Cancer Institute, Bethesda, MD, 90 pages.

10. Nothdruft, H. (1956). Experimentelle Sarkomanslosung durch Eingeheilte Fremdkorper. Strahlentherapie 100:192-210.

11. Oppenheimer, B.S., Oppenheimer, E.T., Stout, A.P., Wilhite, M., and Danishefsky, I. (1958). The Latent Period in Carcinogenicity by Plastics in Rats and its Relation to the Presarcomatous Stage. Cancer 11(1):204-213.

12. Ponomarkov, V., Cabral, J.R.P., and Wahrendorf, J. (1983). A Carcinogenicity Study of Styrene Oxide in Rats (Draft IARC Document).

13. Ponomarkov, V. and Tomatis, L. (1978). Effects of Long-Term Oral Administration of Styrene in Mice and Rats. Scand. J. Work Environ. Health 4(Suppl. 2):127-135.

14. Riviere, M.R., Chouroulinkov, I., and Guerin, M. (1960). Sarcomes Produits par Implantation de Polystyrene chez le Rat: Resultats Sensiblement Differents Suivant les Souches D'animaux Utilisees C.R. Soc. Biol. 154:485-487.

15. USITC (1983a). Preliminary Report on U.S. Production of Selected Synthetic Organic Chemicals (including Synthetic Plastics and Resin Materials) April, May, and Cumulative Totals, 1983, Series C/P-83-5. United States International Trade Commission, Washington, DC.

16. USITC (1983b). Imports of Benzenoid Chemicals and Products 1982, USITC Pub. No. 1401, 99 pages, United States International Trade Commission, Washington, DC.

17. Van Duuren, B.L., Nelson, N., Orris, L., Palmes, E.D., and Schmitt, F.L. (1963). Carcinogenicity of Epoxides, Lactones, and Peroxy Compounds. J. Nat. Can. Instit. 31(1):41-55.

18. Weil, C.S., Condra, N., Haun, C., and Striegel, J.A. (1963). Experimental Carcinogenicity and Acute Toxicity of Representative Epoxides. Amer. Indust. Hyg. Assoc. J. 24:305-325.

Industrial Hazards of Plastics and Synthetic Elastomers, pages 239–262
© **1984 Alan R. Liss, Inc., 150 Fifth Ave., New York, NY 10011**

TOXICITY OF THE COMPONENTS OF STYRENE POLYMERS: POLYSTYRENE,
ACRYLONITRILE-BUTADIENE-STYRENE (ABS) AND STYRENE-BUTADIENE-
RUBBER (SBR).   REACTANTS AND ADDITIVES

Lawrence Fishbein

Department of Health and Human Services, Food
and Drug Administration, National Center for
Toxicological Research, Jefferson, Arkansas

INTRODUCTION

Styrene containing polymers such as polystyrene,
styrene-butadiene and styrene-acrylonitrile copolymers,
styrene-butadiene-rubber (SBR) and acrylonitrile-butadiene
styrene (ABS) are among the most widely produced materials
which are employed for a large variety of applications
(IARC, 1979a,b; Tossavainen, 1978).

The principal objective of this overview is to highlight the toxicological features of a number of the more
important additives and reactants (except styrene) that are
employed in the production and processing of polystyrene,
acrylonitrile-butadiene-styrene (ABS) and styrene-butadiene-
rubber (SBR). However, initially it is instructive to
examine features of the scope of production and use applications of these polymers as well as some typical recipes for
their production.

The morbidity among persons employed in styrene production, polymerization and processing plants (Thiess and
Friedheim, 1978), mortality experience of styrene-polystyrene polymerization workers (Nicholson et al, 1978),
styrene use and occupational exposure in the plastics
industry (Tossavainen, 1978), epidemiological investigations
of styrene-butadiene rubber production and reinforced
plastics production (Meinhardt et al, 1978) and a hematolo-
gic survey of workers at a styrene-butadiene synthetic
rubber manufacturing plant (Checkoway and Williams, 1982)
have all been recently reported. Additionally, the International Agency for Research on Cancer (IARC) has recently

reviewed the carcinogenic risk of styrene, polystyrene and styrene-butadiene copolymers (IARC, 1979a) as well as acrylonitrile-butadiene-styrene copolymers (IARC, 1979b).

POLYSTYRENE

Polystyrene resins are produced by a modified mass polymerization process in a continuous manner. Typically, the liquid styrene monomer is diluted with a relatively small amount of a diluent, e.g., 5-15% of ethylbenzene. (In some cases more diluent is employed and the process is then referred to as a solution process.) The heated mixture of styrene, solvent and initiator is reacted at 120-160°C and the unreacted monomer and solvent are removed after polymerization is complete (IARC, 1979a).

Initiators which have been employed for styrene polymerization include: benzoyl and lauroyl peroxides, dibenzoyl and dicumyl peroxides, tert.-butyl peroxypivalate, di-tert.-butylperoxide, cumene hydroperoxide, tert.butyl hydroperoxide, and p-menthane hydroperoxide.

Polystyrene is an excellent example of a material which can be fabricated into cellular form by utilization of expandable formulations. Particles in the size range of 0.2 to 3.0 mm are prepared either by heating polymer particles in the presence of a blowing agent and allowing the blowing agent to penetrate the particle, or by polymerizing the styrene monomer in the presence of a blowing agent so that the blowing agent is entrapped in the polymerized bead. Typical blowing agents used in such processes are the various isomeric pentanes and hexanes, halocarbons, or mixtures of these materials (Nicholson et al, 1978; Skochdopole, 1966).

In 1976, the United States, Western Europe and Japan produced 1453 million kg polystyrene (49% straight polystyrene and 51% impact modified polystyrene), 1700 million kg and 582 million kg respectively. Worldwide production of polystyrene in 1976 was estimated to have been approximately 4200 million kg (IARC, 1979a).

In 1981 production of polystyrene in the United States was estimated to reach 3.8 billion pounds, less than 10% above 1980 production and 5% below 1979 production of 4.0

billion pounds. The major fabricated forms of polystyrene in the United States currently are: injection molding 50%, extrusions, 35%, beads 10%, and one-third of end products are inpackaging (Anon, 1981). The use of polystyrene resins in the United States in 1977 was as follows: packaging (35%); toys, sporting goods and recreational articles (1%); housewares, furnishings and consumer products (10%); appliances and TV cabinets (9%); disposable serviceware and flatware (8%); construction (7%); electrical and electronic parts (6%); miscellaneous chemical and industrial moulding (4%); furniture (2%); and other uses (8%) (IARC, 1979a). For polystyrene construction-related applications, especially foam products used for insulation is considered to be a major growth market in the future once building activity resumes. The principal single product category in this area is insulation board for roofing and laminates for sheathing products (SRI, 1982).

In 1977, Western Europe use pattern of polystyrene was as follows: packaging (48%), appliances (10%), housewares (9%), refrigerators (85%), furniture (6%), toys (4%), and other uses (15%).

The working temperatures for polystyrene vary from 150° to 300°C, and in some operations, like wire-cutting they may rise considerably higher. At these elevated temperatures the polymer is degraded to some extent and the degradation products can be emitted to the work environment (Pfäffli et al, 1978).

ACRYLONITRILE-BUTADIENE-STYRENE (ABS)

Acrylonitrile-butadiene-styrene copolymers are a blend of two phases: (1) graft polymers of acrylonitrile and styrene on a polybutadiene substrate, and (2) styrene-acrylonitrile copolymers with the acrylonitrile content ranging from 20-35%. Some acrylonitrile-butadiene-styrene resins may also contain α-methylstyrene-acrylonitrile copolymers. A typical medium impact grade of acrylonitrile-butadiene-styrene copolymer is derived from 57.4% styrene, 13.3% butadiene and 29.3% acrylonitrile, and has a softening point of 108°C and a specific gravity of 1.05 (IARC, 1979b).

Acrylonitrile-butadiene-styrene copolymers are produced commercially in the United States by graft polymerization of acrylonitrile and styrene on a polybutadiene substrate by three processes: emulsion, suspension and bulk (Morneau et al, 1978). The emulsion process consists of three distinct polymerizations, e.g., a polybutadiene substrate latex is prepared; styrene and acrylonitrile are grafted onto the polybutadiene substrate; and styrene-acrylonitrile copolymers are formed. The following formulation is a typical redoxysystem for the initial step for the production of a polybutadiene substrate (parts by weight): butadiene (175.0), cumene hydroperoxide (initiator) (0.30), sodium pyrophosphate (activator) (2.50), sodium oleate (emulsifier) (4.0), dextrose (activator) (1.0), ferrous sulfate (activator) (0.05) and water (200.00).

In the next step in the emulsion process, styrene and acrylonitrile are grafted into the polybutadiene substrate. A recipe for a 40% rubber graft reaction (parts by weight) is as follows: polybutadiene-acrylonitrile (93.7), latex (50% solids) (900), water (1055.0), rubber reserve soap (2.0), 2% aqueous potassium persulfate (initiator) (240.0), styrene (455.0), acrylonitrile (235.0) and terpinolene (chain-transfer agent) (4.8). This graft latex containing 40% rubber substrate is blended with emulsion SAN copolymer latex prepared separately to produce typical ABS resins containing 10-30% rubber. The thermal stability of the resin may be improved by adding such antioxidants as di-tert.butyl-p-cresol and tris (nonylphenyl)phosphate to the latex (Morneau et al, 1978).

In contrast to the emulsion process, the suspension process begins with a polybutadiene rubber which is so highly crosslinked, that it is soluble in the monomers. The following formulation (parts by weight) is representative of suspension ABS: soluble polybutadiene rubber (14.0), styrene (62.0), acrylonitrile (26.0), tert.butyl peracetate (free radical initiator) (0.07), di-tert.butyl peroxide (free radical initiator) (0.05), terpinolene (chain-transfer agent) (0.90), water (120.0) and acrylic acid-2-ethyl hexyl acrylate copolymer (0.30).

In 1976, the United States, western Europe and Japan produced 420-450, 290 and 232 million kg respectively of acrylonitrile-butadiene-styrene copolymers. Worldwide production of acrylnitrile-butadiene-styrene copolymers in 1976 was estimated to have been approximately 1000 million kg (IARC, 1979b).

The 476.7 million kg of ABS used in the United States in 1977 were in: piping (22%), automotive components (20%), appliance components (15%), components of business machines, telephones and electrical and electronic equipment (10%), pipe fittings (7%), recreational vehicle components (7%) and other uses (19%) (IARC, 1979b). The use pattern for the 180 million kg of ABS used in western Europe in 1975 was as follows: appliances (31%), automotive components (22%), electrical/electronic applications (13%), furniture (7%), recreation equipment (including luggage and boats) (5%), piping (3%), and other uses (19%) (IARC, 1979b).

STYRENE-BUTADIENE-RUBBER (SBR)

SBR was the first elastomer used in polyblends for toughening polystyrene. This random copolymer represents the most widely used synthetic rubber amounting to 60% of United States consumption and 63% of worldwide consumption. Most SBR made by emulsion polymerization contains 23-25% styrene randomly dispersed with butadiene in the polymer chain. The microstructure of the butadiene units is 14-19% cis-1,4; 60-68% trans-1,4 and 17-21% 1,2-configuration (Platter, 1977). SBR that is made in solution contains about the same amount of styrene as that made in emulsion and both the random and block copolymers have been produced in solution. They have lower trans, slightly lower vinyl and higher cis contents than emulsion SBR (Bauer, 1979).

SBR is available in dry and latex form. There are over 100 different types of dry SBR rubber and about 25 different latexes; but only a portion of these are available from any one manufacturer (Saltman, 1973).

Two methods of polymerization are in widespread use today and both are used for SBR. The emulsion process is used for standard SBR and the solution process for the other varieties (Bauer 1979; Saltman, 1965, 1973).

Three basic types of styrene-butadiene copolymers are available in the United States, viz., 1) styrene-butadiene elastomers (commonly called SBR or styrene-butadiene rubber), 2) styrene block copolymers with butadiene, and 3) styrene-butadiene copolymer latexes. Dry SBR made by emulsion polymerization contains about 23-25% styrene units and 75-77% butadiene units on a polymer basis. When SBR is made by solution polymerization the composition varies, with typical grades containing about 10–25% styrene units and 75–90% butadiene units (IARC, 1979a).

An emulsion system normally contains water, monomers, initiator and foam (Saltman, 1973). An initiator system which usually contains paramenthane hydroperoxide and sodium formaldehyde sulfoxylate is used in combination with a mercaptan to polymerize butadiene and styrene. A typical SBR recipe is as follows (parts per 100 parts monomer): butadiene (75.0), styrene (25.0), n-dodecyl mercaptan (0.5), paramenthane hydroperoxide (0.10), soap flakes (5.0) and water (180.0). The termination of the reaction is achieved by the addition of a "short stop" (e.g., sodium diethyldi-thiocarbamate or sodium polysulfide) which reacts rapidly with free radicals and oxidizing agents destroying any re-maining initiator and prevents the formation of new chains. After removal of unreacted monomers, an antioxidant such as a mixed alkylated diphenylamine is added to protect the product from oxidation (Meinhardt et al, 1978).

Another typical recipe for SBR made by cold emulsion polymerization at $5^{\circ}C$ is as follows (parts per 100 monomer): butadiene (70), styrene (30), water (180), fatty acid soap (2.25), disproportionated rosin soap (2.25), potassium chloride (0.3), sodium formaldehyde–2–naphthalene sulphonate (0.04), tetrasodium ethylenediamine tetraacetate (0.025) and ferrous sulphate heptahydrate (0.013).

A typical recipe for SBR made by hot emulsion polymeri-zation is as follows (parts per 100 monomers): butadiene (75), styrene (25), water (180), fatty acid or rosin soap (5), n-dodecyl mercaptan (0.5) and potassium persulfate (0.3).

An additional recipe for cold SBR is as follows (parts per 100 monomer): butadiene (71.0), styrene (29.0), tert.-dodecyl mercaptan (0.18), paramenthane hydroperoxide (0.03), ferrous sulfate (0.03), trisodium phosphate 90.50), EDTA (0.035), sodium formaldehyde sulfoxylate (0.08), rosin acid soap (4.50), water (200.0) (Saltman, 1965). A more recent cold SBR recipe (number 1500) (parts per 100 monomer, phm) is as follows: butadiene (71), styrene (29), p-menthane hydroperoxide (100%) (0.12), tert.dodecylmercaptan (0.2); emulsifier makeup (phm): water (final monomer–water ratio adjusted to 1:2) (190), disproportioned rosin acid soap

(4.5-5), trisodium phosphate dodecahydrate (buffer electrolyte) (0.50), sodium salt of naphthalenesulfonic acid-formaldehyde condensate (secondary emulsifiers (0.02-0.1), disodium salt of EDTA (iron complexing agent) (0.01), sodium dithionite (oxygen scavenger) (0.025); sulfoxylate activator makeup (phm): ferrous sulfate heptahydrate (0.040, disodium salt of EDTA (0.05), sodium formaldehyde sulfoxylate (0.10), water (10); short stop makeup (phm): sodium dimethyldithiocarbamate (0.10), sodium nitrite (0.02), sodium polysulfide (0.05), water (8.0) (Bauer, 1979).

In hot polymerization, processes are similar to that for cold polymerization, except that the reaction temperature is about 50°C and the reaction is stopped after 12 hours with about 75% conversion. Styrene and butadiene are copolymerized in a typical solution polymerization, in a hexane or cyclohexane solution in the presence of a small amount of n-butyl lithium at 50°C; 98% conversion is achieved after 4 hours and the reaction is terminated by adding a fatty acid deactivator followed by addition of a stabilizer. Solvent residues are removed when the elastomer is coagulated (IARC, 1979a).

An illustrative hot SBR recipe (number 1000) (phm) is as follows: butadiene (71), styrene (29), potassum peroxydisulfate (0.3), n-dodecylmercaptan (0.5); emulsifier makeup (phm): water (final monomer-water ratio adjusted to 1:2) (190), sodium stearate (soap flakes) (5); short stop makeup (phm): hydroquinone (0.10) (Bauer, 1979).

Recipes for SBR made by solution polymerization vary greatly depending upon the properties desired. SBR is vulcanized (typically 1.5 - 2.0 parts sulfur per 100 parts of polymer are used) and accelerators, antioxidants, activators, fillers (e.g., carbon black) and softeners may be employed all depending on the desired properties of the finished rubber. Additionally, SBR is extended with aromatic and naphthenic oils to improve handling and processing (IARC, 1979a).

In a typical solution polymerization, styrene and butadiene are copolymerized in a hexane or cyclohexane solution in the presence of a small amount of n-butyl lithium at 50°C. Approximately 98% conversion is achieved after 4 hours, and the reaction is terminated by adding a fatty acid deactivator followed by addition of a stabilizer; solvent residues are then removed when the elastomers coagulate (IARC, 1979a).

Styrene block copolymers with butadiene can be produced by anionic solution polymerization with sec-butyl lithium or n-butyl lithium in a solvent such as cyclohexane, isopentane, n-hexane or mixtures. The styrene is homopolymerized, followed by addition of butadiene. More styrene is then added and the polymer is finally coagulated from the solution with water (IARC, 1979a).

In 1976 the production of SBR latex in the United States, western Europe and Japan amounted to 141, 1075 and 558 million kg, respectively. In 1976 the use pattern for SBR in the United States was as follows: in the production of tires and tire products, 65% (passenger car tire, 44%, truck and bus tires, 10%, tread rubber, 7%, and other tires, 4%); mechanical goods (non-automotive), 17%; latex applications, 10%; automotive applications, 6%; and miscellaneous applications 2%. In western Europe, SBR is used primarily for the production of tires (68% of total production). The use of SBR in Japan in 1976 was as follows: in automotive applications (including tires), 56%; industrial goods, 8%; footwear, 7%; and other uses, 29% (IARC, 1979a).

In 1981, the consumption of SBR in the United States was approximately 996,000 metric tons with a predicted increase of less than 3% for 1982. SBR is the major synthetic elastomer which makes up some 53% of the total elastomer industry in the United States with the major end uses in 1981 as follows: tire and tread rubber, 73%; mechanical goods, 15% and latex, 9% (Anon, 1982).

TOXICITY OF REACTANTS AND IMPURITIES IN STYRENE POLYMERS AND COPOLYMERS

It is initially important to briefly outline the steps in the production of styrene per se as some of the reactants may be encountered as impurities of styrene in subsequent polymerizations and copolymerizations of this monomer. The process steps include: 1) alkylation of benzene with ethylene to form ethylbenzene; 2) reaction of ethylbenzene with oxygen; 3) reaction of ethylbenzene hydroperoxide with propylene; 4) recovery of unreacted propylene; 5) hydrogenation of unreacted ethylbenzene hydroperoxide to 2-phenylethanol;

6) recovery of unreacted ethylbenzene and propylene oxide; 7) dehydration of 2-phenyl carbinol to styrene and 8) purification of styrene. Inhibitors such as sulfur and tert.-butyl catechol are added during the purification step to prevent polymerization (Tossavainen, 1978).

In one plant's operation for the production of styrene, ethylbenzene is received at the plant in tank cars then reacted with superheated steam and a catalyst of iron oxide to produce styrene monomer. The crude styrene so produced is purified with the removal of unreacted ethylbenzene, benzene, toluene and xylene. These impurities are hydrogenated in another operation and distilled to separate the components for later use or distribution (Nicholson et al, 1978).

Ethylbenzene

Despite the fact that ethylbenzene is produced in enormous amounts as an intermediate in the commercial production of styrene by dehydrogenation, there is a paucity of data on its' mutagenicity, carcinogenicity, reproductive and chronic effects. Ethylbenzene is primarily an irritant to the skin and less markedly of mucous membranes. At high concentrations, ethylbenzene is a narcotic. Subchronic studies on rats, rabbits, guinea pigs and monkeys exposed to ethylbenzene by various routes have been reported (Wolf et al, 1956). Female rats exhibited histopathological changes in the liver and kidneys when a total dose of 408 mg/kg ethylbenzene was administered by gavage over 6 months. Inhalation of 600 ppm (2.6 g/m^3) ethylbenzene adversely affected the livers, kidneys, and testes of all animals tested and no adverse effects in the hematopoietic system were observed in any of the animals.

Benzene

The toxicity of benzene has recently been extensively reviewed (EPA, 1980; IARC, 1982; Laskin and Goldstein, 1977; Snyder and Kocsis, 1975). Chronic exposures to commercial benzene or benzene-containing mixtures can cause damage to the hematopoietic system including pancytopenia. The relationship between benzene exposures and the development of acute myelogenous leukemia has been established in epidemiological studies in a number of countries and different

industries (IARC, 1982). IARC (1982) in their evaluation of
benzene concluded that reports linking exposure to benzene
with other malignancies are considered to be inadequate to
permit an evaluation. Additionally, there is limited evidence that benzene is carcinogenic in experimental animals.
For example, oral administration of benzene to rats resulted
in an increase in the incidence of Zymbal-gland carcinomas.
Anemia, lymphocytopenia and bone-marrow hyperplasia and an
increased incidence of lymphoid tumors occurred in male mice
exposed by inhalation to benzene. IARC (1982) also concluded that there is sufficient evidence that benzene is
carcinogenic to man.

Benzene induced cytogenetic abnormalities (chromosomal
aberrations and sister chromatid exchanges) in mammalian
cells in vitro; and the micronucleus test in mice and rats
has been consistently positive. Additionally, numerous
studies have shown that benzene exposure to experimental
animals in vivo leads to induction of chromosomal aberrations in the bone-marrow cells. Benzene does not induce
specific gene mutations in bacterial systems (S. typhimurium, E. coli, B. subtilis) or in Drosophila melanogaster
(IARC, 1982).

Butadiene

1,3-Butadiene is a major industrial chemical which is
made in very large quantities (e.g., production in the
United States in 1981 was estimated at 3.4 billion pounds
(Anon, 1980) and used mainly for the production of synthetic
elastomers such as acrylonitrile-butadiene-styrene (ABS),
styrene-butadiene-rubber (SBR) and polybutadiene. Although
many workers can be potentially exposed to butadiene, surprisingly little is known about its toxicity or its metabolism (Malvoisin and Roberfroid, 1982). Its acute toxic
effects are narcosis, respiratory mucous membrane and eye
irritation. However, little is known regarding the long-
term consequences of chronic exposures to butadiene. Hematopoietic dysfunction and increased cancer morbidity has
been associated with exposures to chloroprene (monochlori-
nated butadiene) in the rubber industry (Sanotskii, 1976).

Earlier reports indicated the apparent direct mutagenic
activity of gaseous butadiene toward S. typhimurium strains
TA 1530 and TA 1535, whereas the mutagenic effects when a

fortified S-9 rat liver fraction was added were not unambiguous (deMeester et al, 1978). Recent studies clearly demonstrate that the presence of a liver microsomal activating system is an essential prerequisiste for the transformation of butadiene into mutagenic intermediate(s) such as epoxide(s). For example, gaseous butadiene was mutagenic toward S. typhimurium TA 1530 when the incubation mixture was supplemented with a NADPH fortified rat liver microsomal preparation, with mutagenicity increasing with the dose (deMeester et al, 1980).

In rat liver microsomes, 1,3-butadiene was metabolized to butadiene monoxide, which was subsequently transformed into 3-butene-1,2-diol by microsomal epoxide. In the metabolism of butadiene oxide in microsomes, four metabolites were found, e.g., two stereoisomers of DL-diepoxybutane, and two stereoisomers of 3,4-epoxy-1,2-butanediol; no meso-diepoxybutane was detected (Malvoisin and Roberfroid, 1982).

Acrylonitrile

Acrylonitrile (CH_2=CHCN, vinyl cyanide, VCN) is the most extensively used aliphatic nitrite, ranking 42nd on the list of chemicals produced in the United States. The production of VCN in 1979 was over 900 million kg. Total western European and Japanese production of VCN in 1976 amounted to 915 and 633 million kg, respectively (IARC, 1979b). It has been estimated that in the United States approximately 125,000 workers are potentially exposed to acrylonitrile, which is mainly used in the manufacture of synthetic rubbers, plastics and acrylic fibers (Finklea, 1977).

VCN has long been recognized as a potent acute toxin. The primary site of toxicity is the central nervous system and with higher doses, adrenal necrosis and congestive lung edema is evident (Abreu and Ahmed, 1979; Szabo et al, 1976).

Recently, a statistically significant enhanced incidence of lung and large intestine cancer was noted among workers occupationally exposed to this monomer (IARC, 1979b; O'Berg, 1980). Animal experiments involving oral administration and inhalation exposure have demonstrated that tumors are induced in the brain, forestomach and Zymbal gland in male and female rats (Maltoni et al, 1977; U.S.

Consumer Product Safety Commission, 1978; U.S. Department of Labor, 1978).

Acrylonitrile, in aqueous or gas phases, induces reverse mutations in S. typhimurium TA 1530, TA 1535, TA 1590, TA 100, TA 1538, TA 98 and TA 1978 in the presence of a 9000 xg supernatant of liver from mice or rats (deMeester et al, 1978b,c; Milvy, 1978; Milvy and Wolff, 1977; Venitt, 1978). It is also a direct acting mutagen for E. coli strains WP_2, WP_2uvrA, and WP_2uvrApolA (Venitt, et al, 1977).

The effect of several factors in the liver extract mediated mutagenicity of acrylonitrile in S. typhimurium TA 1530 and identification of four new in vitro metabolites (e.g., cyanoacetic acid, cyanoethanol, acetic acid and glycolaldehyde) have been reported. These studies suggested the intermediate formation of a radical species and an epoxide [H_2C-CH-CN] (Duverger-Van Bogaert et al, 1981).

Acrylonitrile did not produce significant increases in cytogenetic aberrations in the mouse-bone marrow when given orally for 4, 21 or 30 days at doses equal to 7, 14 and 21 mg/kg/day respectively or by i.p. for the same periods at doses of 0, 15 and 20 mg/kg/day. Rats treated orally with 16 daily doses of acrylonitrile (40 mg/kg/day) showed no increase of aberrant metaphases in the bone marrow over controls (Rabellow-Gay and Ahmed, 1980). Additionally, it was reported that acute treatment with acrylonitrile has no clastogenic effects on male mouse cells in vivo (Leonard et al, 1981).

Application of acrylonitrile to primary Syrian golden hamster embryo cells (HEC) in culture produced foci of morphologically transformed cells and caused enhancement of virus transformation (simian adenovirus, SA7) (Parent and Casto, 1979). The transformed foci of cells produced by VCN were indistinguishable from foci induced with known carcinogens (Casto et al, 1977).

Teratogenic effects have been reported in the offspring of rats treated during pregnancy either by gavage (Murray et al, 1976), by inhalation (Murray et al, 1978) or in the drinking water (Beliles and Mueller, 1977).

TOXICITY OF SOLVENTS AND ADDITIVES

Hexane

n-Hexane is one of many solvents (e.g., heptane, benzene, cyclohexane, chlorobenzene, etc.) used in the synthetic elastomer industry as the reaction medium in both the solution and emulsion procedures. Although n-hexane is used in large volumes, extensive data are lacking as to its carcinogenicity, mutagencity or teratogenicity. n-Hexane can cause a peripheral neuropathy in exposed workers (Wada et al, 1965). Similar clinical and pathological evidence of peripheral neuropathy is produced in rats by injection or inhalation of n-hexane (Schaumberg and Spencer, 1976). n-Hexane is metabolized to 2-hexanol, 2,5-dimethylfuran, 5-hydroxy-2-hexanone, 2,5-hexanediol, 2,5-hexanedione and gamma-valerolactone (Perbellini et al, 1981). [Methyl-n-butyl ketone, a widely used solvent, also produces the same metabolites and also produces a polyneuropathy in experimental animals (Krasavage et al, 1980)].

Methylethyl Ketone (MEK)

MEK, an extensively used solvent, is an agent to which those employed in the production of SBR can be exposed (Checkoway and Williams, 1982). It is an irritant of the eyes, mucous membranes and skin and at high concentrations causes narcosis. MEK has been found to enhance the neurotoxicity of n-hexane (Altenkirch et al, 1977) as well as that of methyl n-butyl ketone (Saida et al, 1976). 2-Butanol, 3-hydroxy-2-butanone and 2,3-butanediol were identified as metabolites in the sera of guinea pigs dosed with MEK (DiVencenzo et al, 1980).

MEK (1000 or 3000 ppm) inhaled by pregnant rats for 7 hr/day on days 6 through 15 of gestation was shown to be embryotoxic, fetotoxic and potentially teratogenic (Schwetz et al, 1974). In a more recent study, pregnant Sprague-Dawley rats inhaled 400, 1000 or 3000 ppm MEK for 7 hr/day on days 6 through 15 of gestation. The results of this study (Deacon et al, 1981) duplicated the increased incidence of skeletal variants observed above (Schwetz et al, 1974) but did not indicate that inhaled MEK causes either an embryotoxic or teratogenic response in rats at exposure levels up to 3000 ppm.

Phenyl-2-naphthylamine (PBNA)

PBNA has been employed (until recently) as an antioxidant for SBR often at levels of 1-2% (IARC 1978, Meinhardt
et al, 1978). It has been reported that commercial samples
of PBNA previously produced in the United States were contaminated with 20-30 mg/kg of the carcinogen 2-naphthylamine
(NIOSH, 1976). While PBNA samples in the United Kingdom in
the past contained 15-50 mg/kg 2-naphthylamine, levels of
less than 1 mg/kg have been reported in recent years in one
commercial product (Veys, 1973). Commercial samples of PBNA
in Japan have been found to contain aniline, 2-naphthol and
2-naphthylamine as impurities (IARC, 1978).

PBNA is a suspect carcinogen because of its possible
breakdown to 2-naphthylamine, a potent human bladder carcinogen (IARC, 1978). A statistically significant increase in
the incidence of all tumors, particularly hepatomas in male
mice, has been found in animals administered PBNA orally or
by single subcutaneous injection (Innes et al, 1969). 2-
Naphthylamine has been found in the urine of workers who
have been exposed to PBNA, and in the urine of healthy human
volunteers to whom PBNA was administered (Kummen and Tordoir
1975) as well as in urine of dogs administered PBNA orally
(Batten and Hathway, 1977).

Earlier reports suggested that PBNA did not cause cancer in exposed workers. These reports were based on epidemiological evidence that there was no increase in cancer
incidence in workers engaged in the large-scale production
of PBNA in a number of countries for many years (IARC, 1978;
NIOSH, 1976).

Di-n-butylphthalate (DBP)

The most important industrial use of plasticizers with
polystyrene has been in latexes where dibutylphthalate aids
film formation and flexibility (Sears and Touchette, 1982).
DBP, as well as di-s-ethylhexyl phthalate (DEHP) produce
testicular injury when administered orally to rats at relatively high dose levels (200-2800 mg/kg) (Foster et al,
1980; Gray et al, 1977; Oishi and Hiraya, 1980).

In more recent studies, oral administration of DBP produced uniformly severe seminiferous tubular atrophy in rats

and guinea pigs but caused only focal atrophy in mice. Hamsters showed no testicular changes with DBP and only minor changes in response to DEHP and di-n-pentyl phthalate (DPP) (Gray et al, 1982).

Metabolic studies with phthalate diester have indicated that this effect is mediated via the corresponding monoesters and/or alcohols which are formed by partial hydrolysis in the G.I. tract (Albro et al, 1973; Lake et al, 1977; Rowland et al, 1977). The rate of intestinal monohydrolysis of DEHP was significantly slower in hamster than in rats and mono-(2-ethylhexyl)-phthalate (MEHP) did cause focal seminiferous tubular atrophy in hamsters. However, mono-n-butyl phthalate (MBP) had no such effect. The decrease in testicular zinc concentration and enhancement of urinary zinc excretion produced in rats by DEHP and DPP was not found in hamsters. Hence, species differ widely in their sensitivity to the testicular toxicity of phthalate esters (Gray et al, 1982).

Peroxide Initiators and Cross-Linking Agents

A relatively large number of peroxides and mixtures of peroxides have been employed as initiators for styrene polymerization as well as cross-linking agents for styrene-butadiene-rubbers. These include: tert.butyl peroxy pivalate; dibenzoyl peroxide; tert.butyl peroxy-2-ethyl hexanoate; dicumyl peroxide, di.tert.butyl peroxide; cumene hydroperoxide; tert.butyl hydroperoxide, and p-menthane hydroperoxide.

Despite the broad utility and extensive employment in polymerization reactions, there is a paucity of data relating to their chronic and carcinogenic potential (Kotin and Falk, 1963; Swern et al, 1970; Van Duuren et al, 1963). For example, in an early study in 1963, tert.butyl ester of peroxybenzoic acid administered to mice by inhalation gave rise to malignant lymphomas (Kotin and Falk, 1963). Other peroxides and hydroperoxides examined in this study and which were found to possess carcinogenic properties include: acetyl pereoxides, lauroyl peroxide, sodium peracetate, diisopropyl benzene hydroperoxide, p-methane hydroperoxide, cumene hydroperoxide, meta- and para-tert.butyl isopropylbenzene hydroperoxide, methylethyl ketone peroxide, and di-tert.butyl peroxide. Benzoyl peroxide and lauroyl peroxide have been recently reported to be effective skin tumor promoters (Slaga et al, 1982). In earlier skin-painting studies, benzoyl peroxide was found to be a non-carcinogen (Van Duuren et al, 1963).

A number of organic peroxides have been reported to be mutagenic in earlier studies. For example, cumene hydroperoxide is mutagenic in E. coli (Chevallier and Luzatti, 1960) and Neurospora (Jensen et al, 1951; Latarjet et al, 1958); di-tert.butyl peroxide is mutagenic in Neurospora (Jensen et al, 1951) and tert.butyl hydroperoxide is mutagenic in Drosophila (Altenberg, 1940, 1958), E. coli (Chevallier and Luzatti, 1960) and Neurospora (Dickey et al, 1949).

Flame Retardants

A variety of chlorinated and brominated organo derivatives, as well as antimony trioxide, have been employed as flame retardants singly or in combination for polystyrene as well as ABS and SBR elastomers (Larsen, 1980). These include for polystyrene: 1) 2,3-dibromo-1-propanol and 2) hexabromocyclododecane (HBCD) (for expanded foam polystyrene); 3) decabromodiphenyl oxide (DBDPO) and 4) 1,2-bis(2,4,6-tribromophenoxyethane (BTBPE) (for high impact polystyrene); 5) Dechlorane Plus ($C_{18}H_{12}Cl_{12}$) (prepared from hexachlorocyclopentadiene and 1,5-cyclooctadiene); 6) 2-chloroethanol phosphate (Fyrolcef) and 7) antimony trioxide (1, 2, and 7 are additive flame retardants). (Mixtures of antimony trioxide and halogen compounds are called synergistic flame retardants. For example, mixtures of 5% antimony trioxide and 10% bromine are equivalent to those containing 14% chlorine). DBDPO is widely employed with antimony oxide in high impact polystyrene polymers (Larsen, 1980).

Flame retardants used for acrylonitrile-butadiene-styrene (ABS) include: 1) hexabromocyclododecane; 2) BTBPE, 3) Firemaster 680 ($C_{14}H_8Br_6O_2$); 4) decabromodiphenyl oxide; 5) Dechlorane Plus and 6) antimony trioxide (1 and 2 are additive flame retardants). Flame retardants used for SBR include: 2,3-dibromo-1-propanol (for SBR latexes) and 1,3-chloro-2-propanol, which are additive flame retardants.

Data concerning the toxicity of the great majority of commercially available flame retardants are generally lacking. 2,3-Dibromo-1-propanol exhibits direct-acting base substitution mutagenic activity in S. typhimurium TA 100 and TA 1535 but not in TA 1538 (Tan and Rosenkranz, 1978). The mutagenicity of 2,3-dibromo-1-propanol was further increased in the presence of rat-liver microsomes. It should be noted that 2,3-dibromo-1-propanol is also a hydrolysis product and an impurity in the flame retardant tris(2,3-dibromopropyl)-phosphate (Tris-BP) and has been found in human and rat urine after Tris-BP administration (Blum et al, 1978; St. John et al, 1976). 2,3-Dibromopropanol is also activated to covalently protein-boundproducts, but at a much slower rate than Tris-BP (Söderlund et al, 1981). 2,3-Dichloro-1-propanol was also found to be much more active as a direct-acting mutagen in S. typhimurium TA 100 and TA 1535 than the corresponding 2,3-dibromopropanol (Carr and Rosenkranz, 1978).

Decabromodiphenyl oxide (DBDPO) fed in the diet at levels of 80 mg/kg/day to rats for 30 days produced liver enlargement and thyroid hyperplasia. The liver enlargement consisted of centrolobular hepatocellular cytoplasmic enlargement and vacuolation. This was enhanced at 800 mg/kg/day for 30 days and accompanied by renal hyaline degenerative cytoplasmic changes (Norris et al, 1974, 1975). A two year feeding study employing Sprague-Dawley rats dosed at 1.0, 0.1 and 0.01 mg/kg DBDPO in the diet did not result in any observed carcinogenic effects. The decabromodiphenyl oxide used in this study was 77.4% parent compounds, 21.8% monabromodiphenyl oxide and 0.8% octabromodiphenyl oxide) (Kociba et al, 1975; Ulsamer et al, 1980).

1,2-Bis(2,4,6-tribromophenoxy)ethane (BTBPE) is non-toxic orally or dermally in acute studies. Subacute feeding studies conducted on rats showed minimal effects and low accumulation in fat, liver, and muscle tissues. The small amounts that accumulated rapidly decreased on withdrawal of the test diet (Larsen, 1980).

Antimony trioxide (Sb_2O_3) is used as a flame retardant (generally in combination with chlorinated paraffins) for a broad range of polymers and plastics. Spleen and testicular damage were reported in male and female rats after administration of 60 to 1070 mg/kg/day in the diet for 30 days (Clement Associates, 1978). Rats fed 2% antimony trioxide in the diet for 8 months exhibited an accumulation of antimony in body tissues 40 days after exposure, especially in the thyroid gland (Clement Associates, 1978; Mischutin, 1977). Lifetime carcinogenicity studies for antimony trioxide have not been reported.

SUMMARY

The toxicity of the components of styrene polymers, e.g., polystyrene, ABS and SBR, were reviewed with primary focus on the reactive monomers (except styrene) (e.g., acrylonitrile, butadiene) as well as on impurities and solvents such as benzene, hexane and methylethyl ketone, and additives such as phenyl-2-naphthylamine, di-n-butyl phthalate, and a number of peroxide initiators and flame retardants (e.g., 2,3-dibromopropanol, decadibromodiphenyl oxide and antimony trioxide). It is stressed that toxicity data are generally lacking for the majority of additives employed in the production of styrene polymers. Information is also lacking as to the numbers of individuals at potential risk and the extent of their exposure to the large number of additives employed.

REFERENCES

Abreu ME, Ahmed AE (1979). Studies on the mechanism of acrylonitrile neurotoxicity. Toxicol Appl Pharmacol 48:A54.
Albro PW, Thomas R, Fishbein L (1973). Metabolism of diethylhexylphthalate in rats, isolation and characterization of the urinary metabolites. J Chromatog 76:321.
Altenberg LS (1940). The production of mutations in Drosophila by tert.butyl hydroperoxide. Proc Natl Acad Sci 40:1037.
Altenberg LS (1958). The effect of photoreacting light on the mutation rate induced by tert.butyl hydroperoxide. Genetics 43:662.
Altenkirch HJ, Mager G, Stoltenburg G, Helmbrecht J (1977). Toxic polyneuropathies after glue sniffing. J Neurology 214:137.
Anon (1980a). Butadiene. Chem Eng News, Nov 17, p 18.
Anon (1980b). Production of chemicals in the United States. Chem Eng News 58[18]:35.
Anon (1981). Polystyrene. Chem Eng News, Aug 31, p 22.
Anon (1982). SBR. Chem Eng News, March 8, p 20.
Bauer RG (1979). Elastomers, synthetic (SBR). In: (ed) Grayson M "Kirk-Othmer Encyclopedia of Chemical Technology", Vol. 8, 3rd ed., New York, Wiley, pp. 608-625.

Batten PL, Hathway DE (1977). Dephenylation of N-phenyl-2-naphthylamine in dogs and its possible oncogenic implications. Brit J Cancer 35:342.

Beliles RP, Mueller S (1977). Three generation reproduction study of rats receiving acrylonitrile in drinking water. Acrylonitrile Progress Report, second generation. Submitted by Litton Bionetics, Inc. to the Manufacturing Chemists Association. LBI Project No. 2660, November.

Blum A, Ames BN (1977). Flame-retardant additives as possible cancer hazards. Science 195:17.

Carr HS, Rosenkranz HS (1978). Mutagenicity of derivatives of the flame retardant tris(2,3-dibromopropyl) phosphate: Halogenated propanols. Mutat Res 57:381.

Casto BC, Janosko N, DiPaolo JA (1977). Development of a focus assay model for transformation of hamster cells in vitro by chemical carcinogens. Cancer Res 37:3508.

Checkoway H, Williams TM (1982). A hematology survey of workers at a styrene-butadiene synthetic rubber manufacturing plant. Am Ind Hyg Assoc J 43:164.

Chevallier MR, Luzatti D (1960). The specific mutagenic action of 3 organic peroxides on reverse mutations of 2 loci in E. coli. Compt Rend 250:1572.

Clement Associates (1978). Dossier on Antimony Trioxide. Prepared for TSCA Interagency Testing Committee, Washington, DC, pp. 1-30

Deacon MM, Pilny MD, John JA, Schwetz BA, Murray FS, Yakel HO, Kunra RA (1981). Embryo- and fetotoxicity of inhaled methylethyl ketone in rats. Toxicol Appl Pharmacol 59:620.

deMeester C, Poncelet F, Roberfroid M, Mercier M (1978a). Mutagenicity of butadiene and butadiene monoxide. Biochem Biophys Res Commun 80:298.

deMeester C, Poncelet F, Roberfroid M, Mercier M (1978b). Mutagenic activity of acrylonitrile. A preliminary study. Arch Int Physiol Biochem 86:418

deMeester C, Poncelet F, Roberfroid M, Mercier M (1978c). Mutagenicity of acrylonitrile. Toxicology 11:97

deMeester C, Poncelet F, Roberfroid M, Mercier M (1980). The mutagenicity of butadiene towards Salmonella typhimurium. Toxicol Lett 6:125.

Dickey FH, Cleland GH, Lotz C (1949). The role of organic peroxides in the induction of mutations. Proc Natl Acad Sci 35:581.

Divencenzo GD, Hamilton ML, Kaplan CJ, Dedinas J (1980). Characterization of the metabolites of methyl-n-butyl ketone. In: (eds) Spencer PS, Schaumberg HH. Experimental and Clinical Neurotoxicology. Baltimore, Williams and Williams, pp. 846-855.

Duverger-VanBogaert M, Lambotte-Vandepaer M, DeMeester C, Rollmann B, Poncelet F, Mercier M (1981). Effect of several factors on the liver extract mediated mutagenicity of acrylonitrile and identification of four new in vitro metabolites. Toxicol Letters 7:311.

EPA (1980). Measurement of benzene body-burden for populations potentially exposed to benzene in the environment. Office of Pesticides and Toxic Substances. Environmental Protection Agency. Washington, DC.

Finklea JF (1977). Current Intelligence Bulletin: Acrylonitrile. NIOSH. July.

Foster PMD, Thomas LV, Cook MW, and Gangolli SD (1980). Study of the testicular effects and changes in zinc excretion produced by some n-alkyl phthalates in the rat. Toxicol Appl Pharmacol 54:392.

Gray TJB, Butterworth KR, Gaunt IF, Grasso P, Gangolli SD (1977). Short term toxicity study of di(2-ethylhexyl)-phthalate in rats. FD Cosmet Toxicol 15:389

Gray TJB, Rowland IR, Foster PMD, Gangolli SD (1982). Species differences in the testicular toxicity of phthalate esters. Toxicol Letters 11:141.

IARC (1978). N-Phenyl-2-naphthylamine. In: IARC Monographs on the Evaluation of the Carcinogenic Risk of Chemicals to Man. Vol. 16. Some Aromatic Amines and Related Nitro Compounds – Hair Dyes, Colouring Agents and Miscellaneous Industrial Chemicals. International Agency for Research on Cancer, Lyon, France, pp. 325–341.

IARC (1979a). Styrene, Polystyrene and Styrene-Butadiene Copolymers. In: IARC Monographs on the Evaluation of the Carcinogenic Risk of Chemicals to Man. Vol. 19. Some Monomers, Plastics and Synthetic Elastomers, and Acrolein. International Agency for Research on Cancer, Lyon, France, pp. 231–274.

IARC (1979b). Acrylonitrile, Acrylic and Modacrylic Fibers and Acrylonitrile-Butadiene-Styrene and Styrene-Acrylonitrile Copolymers. In: IARC Monographs on the Evaluation of the Carcinogenic Risk of Chemicals to Man. Vol. 19. Some Monomers, Plastics and Synthetic Elastomers and Acrolein. International Agency for Research on Cancer, Lyon, France, pp. 73–113.

IARC (1982). Benzene. In: IARC Monographs on the Evaluation of the Carcinogenic Risk of Chemicals to Humans. Vol. 29, Some Industrial Chemicals and Dyestuffs. International Agency for Research on Cancer, Lyon, France, pp. 93–148.

Innes JRM, Ulland BM, Valerio MG, Petrucelli L, Fishbein L, Hart ER, et al (1969). Bioassay of pesticide and industrial chemicals for tumorigenicity in mice: A preliminary note. J Natl Cancer Inst 42:1101.

Jensen KA, Kirk I, Kolmark G, Westergaard M (1951). Chemically induced mutations in Neurospora. Cold Springs Harbor quant Biol 16:245.

Kociba RJ, Frauson LO, Humiston CG, Norris JM et al (1975). Results of a two-year dietary feeding study with decabromodiphenyl oxide (DBDPO) in rats. Combust Toxicol 2:267.

Kotin P, Falk HL (1963). Organic peroxides, hydrogen peroxide, epoxides and neoplasia. Radiat Res (Suppl) 3:193.

Krasavage WJ, O'Donoghue JL, DiVincenzo GD, Terhaar CJ (1980). The relative neurotoxicity of methyl-n-butyl ketone, n-hexane and their metabolites. Toxicol Appl Pharmacol 52:433.

Kummer R, Tordoir WF (1975). Phenylbetanaphthylamine (PBNA), another carcinogenic agent? Tidjschr Geneeskd 5:415.

Lake BG, Phillips JC, Linnell JC and Gangolli SD (1977). The in vitro hydrolysis of some phthalate diesters by hepatic and intestinal preparations from various species. Toxicol Appl Pharmacol 39:239.

Larsen ER (1980). Halogenated flame retardants. In: (ed) Grayson M. Kirk-Othmer Encyclopedia of Chemical Technology. Vol. 10, 3rd ed., New York, Wiley, pp. 373-395.

Laskin S, Goldstein BD (1977). Benzene toxicity: A critical evaluation. J Tox Environ Hlth, Suppl 2:1-147

Leonard A, Garny V, Poncelet F, Mercier M (1981) Mutagenicity of acrylonitrile in mouse. Toxicol Letters 7:329.

Maltoni C, Ciliberti A, DiMaio V (1977). Carcinogenicity bioassays in rats of acrylonitrile administered by inhalation and by ingestion. Med Lav 68:401.

Malvoisin E, Roberfroid M (1982). Hepatic microsomal metabolism of 1,3-butadiene. Xenobiotica 12:137.

Meinhardt TJ, Young RJ, Hartle RW (1978) Epidemiologic investigations of styrene-butadiene rubber production and reinforced plastics production. Scand J Work Environ Hlth 4(Suppl2):240.

Mercier M (1981). Effect of several factors on the liver extract mediated mutagenicity of acrylonitrile and identification of four new in vitro metabolites. Toxicol Lett 7:314.

Milvy P (1978). Letter to the editor. Mutat Res 57:110.

Milvy P, Wolff M (1977). Mutagenic studies with acrylonitrile. Mutat Res 48:271.

Mischutin V (1977). Safe flame retardant. Am Dyest Ep 66:51.

Morneau GA, Pavelich WA, Roettger LG (1978). ABS resins. In: (ed) Grayson M "Kirk-Othmer Encyclopedia of Chemical Technology" Vol 1, 3rd ed., New York, Wiley-Interscience, pp. 442-456.

Murray RJ, Nitschke KD, John JA, Smith JF et al (1976). Teratologic evaluation of acrylonitrile monomer given to rats by gavage. Report from Dow Chemical Co. (USA).

Murray FJ, Nitschke KD, John FA, Smith JF et al (1978). Teratologic evaluation of inhaled acrylonitrile monomer in rats. Report from Dow Chemical Co. (USA). May 31.

Nicholson WJ, Selikoff IJ, Seidman H (1978). Mortality experience of styrene-polystyrene polymerization workers. Scand J Work Environ Hlth 4 (Suppl 2):247.

NIOSH (1976). Current Intelligence Bulletin: Metabolic Precursors of a known human carcinogen, beta-naphthyl-amine. US Dept. of Health Education and Welfare, Rockville, MD, pp. 1-3.

Norris JM, Ehrmantraut JW, Gibbons CL, Kociba RJ, Schwetz BA, et al (1974). Toxicological and environmental factors involved in the selection of decabromodiphenyl oxide as a fire retardant chemical. J Fire Flammability/Combust Toxicol 1:51.

Norris JM, Kociba RJ, Schwetz BA, Rose JQ, et al (1975). Toxicology of octabromobiphenyl and decabromodiphenyl oxide. Environ Hlth Persp 11:153.

O'Berg MT (1980). Epidemiologic studies of workers exposed to acrylonitrile. J Occup Med 22:245.

Oishi S, Hiraga K (1980). Testicular atrophy induced by phthalic acid esters: effect on testosterone and zinc concentrations. Toxicol Appl Pharmacol 53:35.

Parent RA, Casto BC (1979). Effect of acrylonitrile on primary Syrian Golden Hamster embryo cells in culture: Transformation and DNA fragmentation. J Natl Cancer Inst 62:1025.

Perbellini L, Brugnone F, Faggionato G (1981). Urinary excretion of the metabolites of n-hexane and its isomers during occupational exposure. Brit J Ind Med 38:20.

Pfäffli P, Zitting A, Vainio H (1978). Thermal degradation products of homopolymer polystyrene in air. Scand J Work Environ Hlth 4 (Suppl2):22.

Platzer N (1977) Elastomers for toughening styrene polymers. Chem Tech October pp 634-641.

Rabello-Gay MN, Ahmed AE (1980). Acrylonitrile: In vivo cytogenetic studies in mice and rats. Mutat Res 79:249.

Rowland IR, Cottrell RC, Phillips JC (1977). Hydrolysis of phthalate esters by the gastro-intestinal contents of the rat. Food Cosmet Toxicol 15:17.

Saida K, Mendell JR, Weiss HS (1976). Peripheral nerve changes produced by methyl n-butyl ketone and potentiation of methyl ethyl ketone. J Neuropathol Expt Neurology 35:207.

Saltman WM (1965). Elastomers Synthetic. In: "Kirk-Othmer 'Encyclopedia of Chemical Technology", Vol. 7, 2nd ed., New York, Wiley, pp. 678-705.

Saltman WM (1973). Styrene-butadiene rubbers. In: (ed) Merton M, Rubber Technology, New York, Van Nostrand Reinhold, pp. 178-198.

Sanotskii IV (1976). Aspects of the toxicology of chloroprene: Immediate and long-term effects. Environ Hlth Persp 17:85.

Schaumberg HH, Spencer PS (1976). Central and peripheral nervous system degeneration produced by pure n-hexane: An experimental study. Brain 99:183.

Schwetz BA, Leong BKJ, Gehring PJ (1974). Embryotoxicity and fetotoxicity of inhaled carbon tetrachloride, 1,2-dichloroethane and methylethyl ketone. Toxicol Appl Pharmacol 28:452.

Sears JK, Touchette NW (1982). Plasticizers. In: (ed) Grayson, M "Kirk-Othmer's Encyclopedia of Chemical Technology", Vol. 18, 3rd ed., Wiley, New York, pp. 111-183.

Skochdopole RE (1966). Foamed Plastics. In: "Kirk-Othmer Encyclopedia of Chemical Technology" Vol. 9, 2nd ed., New York, Wiley, pp. 847-884.

Slaga TG, Klein-Szanto AJP, Triplett, LC, Yottt LP, Trosko JE (1981). Skin tumor-promoting activity of benzoyl peroxide, a widely used free radical generating compound. Science 213:1023.

Snyder R, Kocsis JJ (1975). Current concepts of chronic benzene toxicity. CRC Crit Rev Toxicol 3:265.

Söderlund EJ, Nelson SD, Dybing E (1981). In vitro and in vivo covalent binding of the kidney toxicant and carcinogen tris(2,3-dibromopropyl phosphate). Toxicology 21:291.

SRI (1982). CEH Report Developments in the Polyolefins Industry. Chemical Industries Division Newsletter. Stanford Research International, Menlo Park, CA. Jan-Feb., pp. 5-6.

St. John LE Jr, Eldefrawi ME, Lisk DJ (1976). Studies of possible absorption of a flame retardant from treated fabrics worn by rats and humans. Bull Environ Contam Toxicol 15:192.

Swern D, Wieder R, Donough M, Meranze DR, Shimkin MB (1970). Investigation of fatty acids and derivatives for carcinogenic activity. Cancer Res 30:1037.

Szabo S, Reynolds ES, Kovacs K (1976). Animal model: Acrylonitrile-induced adrenal apoplexy. Am J Pathol 82:653.

Thiess AM, Friedheim M (1978). Morbidity among persons employed in styrene production, polymerization and processing plants. Scand J Work Environ Hlth 4(Suppl2):203.

Tossavainen A (1978). Styrene use and occupational exposure in the plastics industry. Scand J Work Environ Hlth 4 (Suppl 2):7.

Ulsamer AG, Osterberg RE, McLaughlin J (1980). Flame-retardant chemicals in textiles. Clin Toxicol 17:101.

US Consumer Product Safety Commission (1978). Assessment of Acrylonitrile Contained in Consumer Products. Final Report, Contract No. CPSC–C–77–0009, Task No. 1014K, H.1.A/Economic Analysis, Washington, DC.

US Department of Labor (1978). Occupational exposure to acrylonitrile (vinyl cyanide). Proposed standard and notice of hearing. Fed Regist 43:2586–2621.

VanDuuren BL, Nelson N, Orris L, Palmes ED, Schmitt FL (1963). Carcinogenicity of epoxides, lactones and peroxy compounds. J Natl Cancer Inst 31:41.

Veys CA (1973). A study of the incidence of bladder tumours in rubber workers (Thesis for Doctorate of Medicine). Faculty of Medicine, University of Liverpool, Liverpool, United Kingdom.

Venitt S, Bushell CT, Osborne M (1977). Mutagenicity of acrylonitrile (cyanoethylene) in E. coli. Mutation Res 45:283.

Venitt S (1978). Letter to the editor. Mutation Res 57:107.

Wada Y, Okamoto S, Takaji S (1965). Intoxication polyneuropathy following exposure to n-hexane. Clin Neurol 5:591.

Wolf MA, Rowe VK, McCollister DD, Hollingsworth RL, Oyen F (1956). Toxicological studies of certain alkylated benzenes and benzene. Arch Ind Hlth 14:387.

Industrial Hazards of Plastics and Synthetic Elastomers, pages 263–277
© 1984 Alan R. Liss, Inc., 150 Fifth Ave., New York, NY 10011

OCCUPATIONAL HAZARDS IN PRODUCTION AND PROCESSING
OF STYRENE POLYMERS - EPIDEMIOLOGIC FINDINGS

William J. Nicholson and Diane Tarr

Environmental Sciences Laboratory, Mount Sinai
School of Medicine of CUNY, New York 10029

INTRODUCTION

Of the major plastic monomers, styrene is exceeded only
by ethylene, propylene and vinyl chloride in terms of production. In 1977, approximately 3×10^6 metric tons were
produced in the United States and 7×10^6 metric tons worldwide (IARC, 1979). Approximately 60% of the monomer produced was used in homopolymers, largely for the packaging
industry. Other important uses of styrene are in the production of copolymers with acrylonitrile (SAN) and acrylonitrile and butadiene (ABS). Styrene also finds widespread
use as a copolymer with butadiene in the production of the
synthetic elastomer, styrene-butadiene rubber (SBR), which
forms the basis of approximately 80% of U.S. rubber products.
Finally, it is extensively used as a solvent and cross-linking agent for polyester resins in the fiber reinforced
plastic (FRP) industry. Estimates of the number of workers
employed in the various industries using styrene-based
polymers are given in Table 1 along with typical exposure
levels (Tossavainen, 1978). In addition to occupational
exposure, low-level environmental contamination can occur
from combustion of styrene-based products, as the thermal
decomposition of polystyrene leads to evolution of the
monomer, in contrast to other polymer materials.

MORTALITY STUDIES OF STYRENE-EXPOSED WORKERS

Studies on the mortality of populations exposed to
styrene are fraught with difficulty because of confound-

Table 1
Occupational exposures to styrene

Process	Percent of styrene production	Number of workers involved	Typical exposure (ppm)
Monomer production	100	10,000	1 - 20
Polymer production (PS,ABS,SBR)	60 - 70	50,000	1 - 20
Reinforced plastics production (FRP)	20 - 30	200,000	20 - 300
Polymer processing (PS,ABS,SBR)	5 - 10	1,000,000	0.01 - 1

From Tossavainen, 1978

ing exposures to other known carcinogenic materials. Benzene exposures can occur in the production of styrene monomer, as the principal production process utilizes benzene to produce ethylbenzene, which, in turn, is dehydrogenated to form styrene. Copolymerization involves exposures to acrylonitrile and/or butadiene, which are carcinogenic in animals (Huff, 1983). Finally, some of the additives used in styrene products may be carcinogenic, as well as other chemicals used in facilities producing styrene or polystyrene.

Only three studies provide data on possible human carcinogenicity of styrene, each having some of the confounding exposures mentioned above. All were of groups of individuals employed in monomer production, polymerization or polymer fabrication, where exposures were relatively limited. No data exist on the mortality of individuals employed in the FRP industry, where styrene concentrations were (and are) commonly ten times higher. Table 2 lists the cohorts observed and some of the characteristics of the three studies.

As can be seen from Table 2, the large number of individuals lost to follow-up in the study by Frentzel-Beyme et al (1978) severely limits its usefulness. Of those exposed, 7% of the German workers and 71% of foreign "guest workers" were untraced. Of those traced, 74 had died, 12 from cancer. Only 37 deaths occurred in those with five or more years of exposure. The overall result did not demonstrate excess mortality for any cause of death. However, the limitations of the study are clear.

Table 2

Population and follow-up characteristics of
three studies of styrene exposed workers

Study	Country	Analysis cohort size	Percent traced	Number of deaths analyzed	Percent of total
Ott et al. 1980	USA	2904	97.4	303	10.4
Nicholson et al. 1978	USA	560	100.0	83	14.8
Frentzel-Beyme et al. 1978	GER	1960	93.0Ger 29.0For	73	3.7

	Minimum exposure (years)	Minimum latency (years)	Years follow-up	Exposures
Ott et al. 1980	1	1	1940-1975	<10 ppm (3)[a]
Nicholson et al. 1978	5	10	1960-1975	<20 ppm (5)
Frentzel-Beyme et al. 1978	1 mo.	1 mo.	1956-1976	< 1 ppm[b]

[a] ( ) = Estimated average exposure
[b] Current measurement after installation of controls

The study of Nicholson et al (1978) successfully
traced all of 563 men employed in styrene production,
polymerization and polymer processing who had 5 years of
employment on May 1, 1960 and were 10 years from onset of
work in a large U.S. production facility. The basic
mortality data are shown in Table 3 and demonstrate no
excess mortality from any cause of death. Analyses
according to years from onset of exposure and calendar
years of observation did not reveal any pattern of excess
mortality. However, because of the limited number of
deaths, the data can be used only to establish upper
limits of risk. For example, the data are only sufficient to indicate that the SMR for lymphoma or leukemia is
less than 280 at the 0.05 level of significance and that
of lung cancer, less than 220. While no excess mortality
was identified in the cohort observed, the above publication mentioned the existence of 7 deaths from leukemia
and 5 of malignancy of the lymphatic system among 444
deaths known to have occurred in the plant workforce.
While the ages of death were not available for exact
proportionate mortality calculations, the number of
lymphomas is in line with expectations, while leukemia
appears to be in excess by as much as a factor of two.
However, the possibility of high exposures to benzene in
the facility during earlier years weakens the likelihood

Table 3

Expected and observed mortality experiences of
560 individuals employed in styrene production
and polymerization prior to 1 May 1955, followed
ten years after onset of exposure
(1 May 1960 -- 31 December 1975)

Cause of Death	Expected	Observed	SMR
All causes	106.41	83	78
Cancer	21.01	17	81
Cancer of the lung	6.99	6	117
Leukemia	0.79	1	126
Lymphomas	1.25	1	80
Other cancer	11.98	9	75
Heart and circulatory dis.	56.35	52	92
Respiratory diseases	6.64	1	15
Other causes of death	22.41	13	58

From Nicholson et al. 1978

of association of any possible excess of leukemia with styrene exposure.

The final study of styrene mortality is that of Ott et al (1980) who described the experience of 2,904 individuals with potential exposure to styrene prior to January 1, 1976. Three hundred three deaths occurred, 292 among production and nonprofessional research employees. The mortality experience for this latter group is shown in Table 4 and is compared to the expected deaths calculated from U.S. white male rates and rates from observations on other company employees. As can be seen, the mortality of styrene-exposed individuals compares favorably with each group; the only excess of note being six leukemias compared to 2.9 expected using U.S. rates and 1.6 from company rates. Lymphomas were also elevated, but not at an 0.05 level of significance. The excess leukemia was further investigated in an analysis of cancer incidence in the styrene-exposed population compared to that expected from rates of the Third National Cancer Survey. In this analysis, it was found that a significant number of lymphatic leukemias (5 observed vs. 0.26 expected) occurred in individuals who were exposed to styrene (< 5 ppm), ethylbenzene (< 5 ppm), polystyrene extrusion fumes, and colorants. Indeed, four of the five cases worked in the same general area during the period of time, 1947-1948, although their dates of death and other exposures varied widely. Interestingly, there were no deaths of lymphocytic leukemia among 442 individuals exposed to higher

Table 4

Observed and expected deaths by cause for total production
and non-professional research employees (2310 men), 1940–1976

Causes	Observed deaths	Expected deaths US white males	SMR	Expected deaths, Company comparison	SMR
All causes	282	357.8	79	287.6	98
Malignant neoplasms	55	64.2	87	65.0	85
Respiratory system	14	20.8	67	23.9	59
Digestive system	16	18.0	89	21.2	67
Lymphatic and hematopoietic system except leukemia	6	4.5	133	2.6	230
Leukemia	6	2.9	207	1.6	375
Other sites	13	18.0	72	15.7	83
Cardiovascular disease	143	172.4	83	141.5	101
Nonmalignant respiratory dis.	12	14.3	84	10.0	120
All other causes	72	106.9	67	71.1	101

From Ott et al. 1980

concentrations of styrene (5-9 ppm). Because of a lack of a definitive exposure-response-relationship and the presence of possible confounding exposures, the authors refrained from drawing any conclusions on an etiological relationship.

In 1976, nine cases of various types of leukemia were identified in two SBR plants and reported to the U.S. National Institute for Occupational Safety and Health (Meinhardt et al, 1978). All occurred after 1971 in a population of 5,600 workers. No data were presented on the expected numbers of deaths from leukemia in the group. Concern generated by the findings in the two plants led to reports on the leukemias present in two large ongoing studies of rubber workers. McMichael et al (1976) reported a relative risk of 6.2 for lymphatic and hematopoietic malignancies among employees producing elastomers, including SBR. However, this was based upon only 6 cases, 3 leukemias and 3 lymphomas. In a subsequent case control study of the same plant, a relative risk of 2.4 was found for the same exposure group (Spirtas et al, 1976). The difference in the two values reflect the uncertainties associated with small numbers of cases. A similar investigation by Monson et al (1978) showed an excess of leukemia to be present in calendering, extrusion, tire building and rubberized fabrics. However, the excess was associated with exposure to solvents and not to styrene.

Since 1976, considerable interest has existed in potential carcinogenicity of styrene. This has been heightened by the data on mutagenicity and possible carcinogenicity of styrene and styrene oxide (Huff, 1983). Some suggestive human data are available from studies of polymerization workers and SBR production facilities. The irony of the situation is that no studies have been conducted of groups exposed to the enormously higher concentrations found in the FRP industry. The data available on styrene carcinogenicity are such that they can provide no assurance of safety at these higher exposures.

CARCINOGENIC RISK FROM COMONOMERS USED WITH STYRENE

The two principal comonomers used in styrene-based copolymers have each been shown to be carcinogenic in animals. Of these, acrylonitrile has also been associated with lung cancer in humans. In a group of 1,345 male employees, with potential exposure to acrylonitrile and followed from 1956 through 1976, 8 cases of lung cancer occurred compared with 4.4 expected from company rates (O'Berg, 1980). Further, there was a correlation of increased risk with intensity and duration of exposure. Among production workers employed between 1950 and 1952, 6 lung cancer deaths were observed vs. 1.5 expected ($p < 0.01$). A second study of 327 employees of a rubber chemicals plant with potential exposure to acrylonitrile identified 9 deaths of lung cancer compared to 5.9 expected, based on mortality rates for U.S. white males (4.7 based on mortality rates for other rubber workers from the same city) (Delzell and Monson, 1982). The excess was greatest among those who had worked for more than 5 years in the facility. These data, while limited by the small numbers, strongly suggest that acrylonitrile is carcinogenic for humans.

While butadiene has been demonstrably carcinogenic in animals (Huff, 1983), no data are available from human exposures.

CLINICAL FINDINGS AMONG STYRENE-EXPOSED WORKERS

A variety of symptoms have been reported from styrene exposure, dating from the rapid increase in production during World War II. Eye, nose and throat irritation, various respiratory symptoms, and gastrointestinal disturbances, such as nausea, vomiting and loss of appetite, are commonly reported by styrene-exposed workers. Other abnormalities reported include headaches, tiredness and sleep disturbance. After a period of exposure, adaptation may occur and there can be a decrease in the various symptoms. Nevertheless, significant alterations still exist among currently employed workers.

Several systematic studies of the various clinical effects from styrene exposure have been conducted in recent years. Many of these focused on possible neurological alterations and clearly demonstrated adverse findings at moderate styrene exposures (less than most national standards). The most extensive of these is that by investigators from the Institute of Occupational Health of Finland (Seppäläinen and Härkönen, 1976; Lindström et al, 1976; Härkönen, 1977; Härkönen et al, 1978). They examined 98 workers employed at 24 plants manufacturing FRP products. Air concentrations were sampled and ranged from 5 to 300 ppm. The means of five post-workday urine mandalic acid (MA) concentrations ranged from 7 to 4,700 mg/l with a linear relationship existing between the log of the MA concentration and the log of the styrene concentration. The mean value of MA in the population was 808 mg/l and corresponded to an exposure of about 40 ppm of styrene. Abnormal EEG's were found in 30% of those with MA in excess of 700 mg/l, compared to about 10% for those with lower MA values and normal controls (Seppäläinen and Härkönen, 1976). Lindström et al (1976) noted increased visuomotor inaccuracy (symmetry drawing and Bourdon-Wiersma tests) and lowered psychomotor performance (Mira test) for various MA concentrations ranging from 800-2,000 mg/l (See also: Härkönen et al, 1978). Fatigue, irritation, difficulty in concentration, nausea, dizziness, and lightheadedness were more frequently reported by the styrene-exposed workers than by unexposed controls (Härkönen, 1977).

A deteriorating EEG among styrene workers has also been described by Klimkova-Deutschova et al (1973). Alterations of nerve conduction have been documented by Rosén et al (1978) who observed an increased duration and decreased amplitude of sensory action potentials. Lilis et al (1978) suggested the possibility of a decrease in peroneal nerve conduction velocity. The decrease, however, was only associated with length of employment and not intensity of exposure. Subjective symptoms of physical and mental tiredness after styrene exposure have been reported by Klimkova-Deutschova et al (1973) and Cherry et al (1980), among others. The reported neurological findings among styrene-exposed workers are less serious than those reported in some other groups occupationally exposed to solvents, such as as painters (Lindström, 1980; Hane et al, 1977). However, the continued exposure to concentrations causing the observed abnormalities may lead to more serious central nervous system impairment.

In addition to neurological disturbances, alterations of pulmonary function have been noted among styrene-exposed workers. Lorimer et al (1977), in an examination of 494 polymerization workers, found a reduction in FEV_1 to be associated with intensity of exposure and with urine MA concentration. Further, the workers in the higher exposed group reported a greater percentage of acute and recurrent lower respiratory symptoms. Härkönen (1977) reported an increase in incidence of chronic bronchitis which correlated with intensity of exposure. On the other hand, Axelson et al (1978) did not find any abnormalities of pulmonary function among 27 FRP boat manufacturing workers.

Laboratory test results have been rather unremarkable. Lorimer et al (1977) suggested the possibility of increased lymphocytosis among polymerization workers and Checkoway (1982) found a correlation between decreased red blood cell count and increased basophil count with styrene/ butadiene exposure. Increased blood enzyme concentrations, characteristic of abnormal liver function, occasionally have been reported. Lorimer et al (1977) found a significant increase in concentrations of GGTP (gamma glutamyl transpeptidase) and Hotz (1980), noted increased concentrations of OCT (orinthine carbamoyl transferase) and ALAT (alanine amino transferase) among polymerization workers.

SUMMARY OF HUMAN MORTALITY AND MORBIDITY

The limited epidemiological studies of mortality associated with styrene exposure do not demonstrate any excess cancer risk that can be attributed to styrene, although elevated risks for leukemia have been noted in exposed groups. However, these excesses may be related to exposures

to other chemicals, such as benzene. The studies are severely limited because of the relatively low styrene exposure of the groups studied. They provide no guidance on
carcinogenic risk in populations much more heavily exposed
as in the FRP industry. The significant clinical findings
among styrene-exposed workers are largely limited to abnormalities of the central nervous system, where a variety of
objective and subjective symptoms have been reported in the
FRP industry. Concern for long-term degenerative neurological disease exists for continued long-term exposure in this
industry. The possibility of pulmonary effects from styrene
exposure has also been noted.

REFERENCES

Axelson O, Gustavson J (1978). Some hygienic and clinical observations on styrene exposure. Scand J Work
Environ Health 4:215-219 (suppl 2).

Checkoway H, Williams TM (1982). A hematology survey of
workers at a styrene-butadiene synthetic rubber manufacturing plant. Am Ind Hyg Assoc 43:164-169.

Cherry N, Waldron HA, Wells GG, Wilkinson RT, Wilson HK,
Jones S (1980). An investigation of the acute behavioral
effects of styrene on factory workers. Brit J Ind Med
37:234-240.

Delzell E, Monson RR (1982). Mortality among rubber workers:
VI. Men with potential exposure to acrylonitrile. J Occ
Med 24:767-769.

Frentzel-Beyme R, Thiess AM, Wieland R (1978). Survey
of mortality among employees engaged in the manufacture of
styrene and polystyrene at the BASF Ludwigshafen works.
Scand J Work Environ Health 4:231-239 (suppl 2).

Hane M, Axelson O, Blume J, Hogstedt C, Sundell L, Ydreborg
B (1977). Psychological function changes among house
painters. Scand J Work Environ Health 3:91-99.

Härkönen H (1977). Relationship of symptoms to occupational styrene exposure and to the findings of electro-
encephalographic and psychological examinations. Int Arch
Occ Environ Health 40:231-239.

Härkönen H, Lindström K, Seppäläinen AM, Asp S, Hernberg
S (1978). Exposure-response relationship between styrene
exposure and central nervous functions. Scand J Work
Environ Health 4:53-59.

Hotz P, Guillemin MP, Lob M (1980). Study of some hepatic effects (induction and toxicity) caused by occupational exposure to styrene in the polyester industry.
Scand J Work Environ Health 6:206-215.

Huff JE (1983). National toxicology program and carcinogen bioassays of the United States. This volume.

IARC (1979). Evaluation of the carcinogenic risk of chemicals to humans; some monomers, plastics and synthetic elastomers, and acrolein. Vol 19.

Klimkova-Deutschova E, Jandova D, Salamanova Z, Schwartzova K, Titman O (1973). Recent advances concerning the clinical picture of professional styrene exposure. Cs Neurol 36:20-25.

Lilis R, Lorimer WV, Diamond S, Selikoff IJ (1978). Neurotoxicity of styrene in production and polymerization workers. Environ Resh 15:133-138.

Lindström K, Härkönen H, Hernberg S (1976). Disturbances in psychological functions of workers occupationally exposed to styrene. Scand J Work Environ Health 3:129-139.

Lindström K (1980). Changes in psychological performances of solvent-poisoned and solvent-exposed workers. Am J Ind Med 1:69-84.

Lorimer WV, Lilis R, Nicholson WJ, Anderson H, Fischbein A, Daum S, Rom W, Rice C, Selikoff IJ (1976). Clinical studies of styrene workers: Initial findings. Environ Health Persp 17:171-181.

McMichael AJ, Spirtas R, Gamble JF, Tousey PM (1976). Mortality among rubber workers: Relationship to specific jobs. J Occ Med 18:178-185.

Meinhardt TJ, Young RJ, Hartle RW (1978). Epidemiologic investigations of styrene-butadiene rubber production and reinforced plastics production. Scand J Work Environ Health 4:240-246 (suppl 2).

Monson RR, Fine LJ (1978). Cancer mortality and morbidity among rubber workers. J Natl Cancer Inst 61:1047-1053.

Nicholson WJ, Selikoff IJ, Seidman H (1978). Mortality experience of styrene-polystyrene polymerization workers. Scand J Work Environ Health 4:247-252 (suppl 2).

O'Berg MT (1980). Epidemiologic study of workers exposed to acrylonitrile. J Occ Med 22:245-252.

Ott MG, Kolesar RC, Scharnweber HC, Schneider EJ, Venable JR (1980). A mortality survey of employees engaged in the development or manufacture of styrene-based products. J Occ Med 22:445-460.

Rosen I, Haeger-Aronsen B, Rehnström S, Welinder H (1978). Neurophysiological observations after chronic styrene exposure. Scand J Work Environ Health 4:184-194 (suppl 2).

Seppäläinen AM, Härkönen H (1976). Neurophysiological findings among workers occupationally exposed to styrene. Scand J Work Environ Health 3:140-146.

Spirtas R, Van Ert M, Gamble J, Wolf P, McMichael AJ (1976). Toxicologic, industrial hygiene and epidemiologic considerations in the possible association between SBR manufacturing and neoplasms of lymphatic and hematopoietic tissues. In: Proceedings of NIOSH Styrene-Butadiene Briefing. US Department of Health, Education and Welfare Publication (NIOSH) 77-129.

Tossavainen A (1978). Styrene use and occupational exposure in the plastics industry. Scand J Work Environ Health 4:7-13 (suppl 2).

Table 1

Occupational exposures to styrene

Process	Percent of styrene production	Number of workers involved	Typical exposure (ppm)
Monomer production	100	10,000	1 - 20
Polymer production (PS,ABS,SBR)	60 - 70	50,000	1 - 20
Reinforced plastics production (FRP)	20 - 30	200,000	20 - 300
Polymer processing (PS,ABS,SBR)	5 - 10	1,000,000	0.01 - 1

From Tossavainen, 1978

Table 2

Population and follow-up characteristics of
three studies of styrene exposed workers

Study	Country	Analysis cohort size	Percent traced	Number of deaths analyzed	Percent of total
Ott et al. 1980	USA	2904	97.4	303	10.4
Nicholson et al. 1978	USA	560	100.0	83	14.8
Frentzel-Beyme et al. 1978	GER	1960	93.0Ger 29.0For	73	3.7

	Minimum exposure (years)	Minimum latency (years)	Years follow-up	Exposures
Ott et al. 1980	1	1	1940-1975	<10 ppm (3)[a]
Nicholson et al. 1978	5	10	1960-1975	<20 ppm (5)
Frentzel-Beyme et al. 1978	1 mo.	1 mo.	1956-1976	< 1 ppm[b]

----

[a] ( ) = Estimated average exposure
[b] Current measurement after installation of controls

Table 3

Expected and observed mortality experiences of
560 individuals employed in styrene production
and polymerization prior to 1 May 1955, followed
ten years after onset of exposure
(1 May 1960 -- 31 December 1975)

Cause of Death	Expected	Observed	SMR
All causes	106.41	83	78
Cancer	21.01	17	81
Cancer of the lung	6.99	6	117
Leukemia	0.79	1	126
Lymphomas	1.25	1	80
Other cancer	11.98	9	75
Heart and circulatory dis.	56.35	52	92
Respiratory diseases	6.64	1	15
Other causes of death	22.41	13	58

From Nicholson et al. 1978

Table 4

Observed and expected deaths by cause for total production
and non-professional research employees (2310 men), 1940-1976

Causes	Observed deaths	Expected deaths US white males	SMR	Expected deaths, Company comparison	SMR
All causes	282	357.8	79	287.6	98
Malignant neoplasms	55	64.2	87	65.0	85
Respiratory system	14	20.8	67	23.9	59
Digestive system	16	18.0	89	21.2	67
Lymphatic and hematopoietic system except leukemia	6	4.5	133	2.6	230
Leukemia	6	2.9	207	1.6	375
Other sites	13	18.0	72	15.7	83
Cardiovascular disease	143	172.4	83	141.5	101
Nonmalignant respiratory dis.	12	14.3	84	10.0	120
All other causes	72	106.9	67	71.1	101

From Ott et al. 1980

Industrial Hazards of Plastics and Synthetic Elastomers, pages 279–286
© 1984 Alan R. Liss, Inc., 150 Fifth Ave., New York, NY 10011

PREVENTION OF STYRENE HAZARDS - HYGIENIC APPROACHES

Kalliokoski, P[1], Koistinen, T[1], Jääskeläinen, M[2]

1. University of Kuopio, Finland
2. Technical University of Helsinki, Finland

Exposure to styrene is much higher in reinforced plastic plants than in other industrial plants using styrene. A comprehensive review of the industrial hygiene surveys conducted in plants manufacturing styrenic polymers is given by Kalliokoski elsewhere in this book. Exposure to styrene is most problematic during manual lamination with open molds. Therefore, our study of methods for the reduction of exposure to styrene has been limited to the manual production of fiberglass laminates by either the hand lay-up or the spray-up technique.

MODIFICATION OF THE RESIN

It is possible to substitute other monomers for styrene. Vinyl toluene has been suggested as a suitable substitute. According to Hebert (1975), the quality of the laminates manufactured with vinyl toluene based resins is comparable to the quality of products manufactured with styrene based resins. He also found that vinyl toluene reduces monomer loss by 36-41 % when substituted for styrene on the basis of weight, and the concentration of monomer in workplace air was 21-44 % lower when vinyl toluene was used. These reductions occurred because the vapor pressure of vinyl toluene (at 20°C) is 145 Pa, whereas the vapor pressure of styrene (at 20°C) is 595 Pa. The benefit of a lower airborne concentration of monomer, however, may be nil, because vinyl toluene is probably more dangerous than styrene (Heinonen and Vainio, 1980). Nor does the substitution of vinyl toluene for styrene appear to be economically feasible.

Most commercial unsaturated polyester resins contain some volatilization inhibitors for styrene. Paraffin and thixotropic agents are generally used. Paraffin builds a film on the surface of the laminate which, after lamination, rapidly reduces the evaporation of styrene from the product. The use of high concentrations of film forming additives has previously caused poor interlaminar adhesion during multistage lamination. Several resin manufacturers, however, are now marketing so-called low styrene emission (LSE) resins. They claim that the delamination problems have been overcome by properly adjusted resin formulations (Voskamp and Studenberg, 1981). Tests for the evaporation of styrene have yielded significant decreases in the emission of styrene. Styrene evaporation rates of 20 g/m^2 or less have been reported for LSE resins, whereas the corresponding value for the conventional resins is about 100 g/m^2 (Anon., 1980; Nylander, 1979; Voskamp and Studenberg, 1981). These results do indeed appear convincing, but it should be noted that the tests were conducted and published by the resin companies.

Kalliokoski and Jantunen (1981) performed a series of laboratory tests with several styrene suppressed resins. When they investigated the exposure to styrene during the lamination of small experimental plates, they found that an increased paraffin content in the resin did not clearly reduce the level of exposure to styrene. The first unbiased field study on the effect of LSE resins, by Schumacher et al. (1981), gave similar results.

We performed a set of full-scale experiments this year in order to get a satisfactory answer about the effect of LSE resins. The concentrations of styrene in the breathing zone were determined by charcoal tube sampling and gas chromatography. In addition, an infrared analyzer was used to monitor fluctuations in the concentration of styrene at a stationary sampling site near the worker.

Again, the results of the studies conducted in the first two reinforced plastics plants were not very encouraging (see Tables 1 and 2). However, the experimental conditions were not identical enough when the different types of resins were tested; workers' behaviour and background activity varied from test to test.

Table 1. Airborne concentrations of styrene during the
lamination of containers (surface area 1.3 m²).

Resin	Type	Concentration of styrene (ppm)	
		Breathing zone	Stationary sampling site near the worker
A	conventional	109	79
B	LSE	72	91
C	fast conventional	94	221

Table 2. Airborne concentrations of styrene during the
lamination of wall elements (surface area 8.2 m²).

Resin	Type	Concentration of styrene (ppm)	
		Breathing zone	Stationary sampling site near the worker
D	conventional	123	113
B	LSE	35	46
E	conventional	46	82

We then continued the study by establishing identical working
conditions for each test. These experiments were conducted
in a small reinforced plastic boat plant. The volume of the
workplace was 2025 m³. Its rate of ventilation was 1.7 m³/s.
The products were manufactured in a spray booth of 72 m³.
We compared four resins (the properties of which are
presented in Table 3) during both hand lay-up and spray-up
work. Each resin was used on a separate day, and no other
lamination work was performed simultaneously during the
study period. Table 3 also includes the results of the
customary tests for the evaporation of styrene. These tests
indicated that the name of LSE resins is justified; the
emission of styrene from the best LSE resin was less than
1 % of the emission from the conventional resin when the
test was conducted in the usual manner with non-reinforced
polyester resins. Interestingly enough, the difference was
much smaller when reinforced resins were compared.

Table 3. Properties of the resins tested.

Property	Resin			
	F	G	H	I
Type	conventional	LSE	LSE	LSE
styrene content (%)	41.8	43.0	41.1	40.3
peak exotherm (°C)	107	96	103	102
gel time (min)	53	44	50	56
evaporation of styrene (g/m²)				
- resin only	365	8.8	24.4	2.8
- with fiberglass	74	8.9	20.7	7.5

The working schedule for each day was similar. A rowboat
was laminated with a spray gun in the morning, and a small
motorboat was laminated by hand lay-up technique in the afternoon. The consumption of resin was about 50 kg for a rowboat
and 30 kg for a motorboat.

The benefit of LSE resins became obvious during the
hand lay-up work. Table 4 indicates that LSE resins reduced
the concentration of styrene by 50 % or even more. The
differences between various LSE resins were minor. In both
the evaporation test and during use at the workplace resin
F was the worst and resin I the best from the point of view
of industrial hygiene. However, resins G and H were not
notably different in the field experiment, although resin
G was clearly better in the evaporation test. This is not
surprising, because the surface of the resin is not touched
in the evaporation test, when the film forming additives can
efficiently shut off the resin from the air, whereas the
major occupational exposure occurs when the surface is rolled
out.

Table 4. Airborne concentrations of styrene, during the
lamination of small motorboats.

		Concentration of styrene (ppm)						
		Breathing zone			Stationary sampling site near the worker			General workroom
Resin	Type	Deck	Hull	Total	Deck	Hull	Total	atmosphere
F	conventional	92	103	100	42	129	105	22
G	LSE	72	86	82	26	55	45	10
H	LSE	40	75	62	35	63	55	10
I	LSE	26	55	44	22	47	40	5

Table 5 shows that similarly high atmospheric levels of styrene were measured for all the resins during spray-up application.

Table 5. Airborne styrene concentrations during the manufacture of rowboats by the spray-up technique.

		Concentration of styrene (ppm)			
		Breathing zone		Stationary sampling site near the worker	
Resin	Type	Spraying	Rolling	Spraying	Rolling
F	conventional	120	86	116	34
G	LSE	83	56	87	33
H	LSE	102	54	133	30
I	LSE	103	40	135	31

Thus, the effect of LSE resins was minimal when the spray-up technique was used or small products were laminated by rolling, when almost the whole surface of the product is simultaneously being worked on. But LSE resins are effective when large products are made and only a part of the surface is being laminated at any given time.

VENTILATION

Isolation is a solution to the problem of occupational exposure to styrene in the automated and the closed mold production of reinforced plastic. Open mold work does not permit complete enclosure, and the extent of success in reducing the exposure to styrene depends on the capture velocity and the prevention of eddy formation at the working site.

Kalliokoski (1976) found a reduction of 80-90 % in the concentration of styrene in the breathing zone when a laminar airflow of 0.5 m/s was directed over small laminates in a laboratory test. A Norwegian research group (Stinessen, 1978) designed cabins for lamination (157 m^3) with a laminar flow type of ventilation. The 8-hour time-weighted average exposures of styrene were 20-41 ppm during a full-scale test period when boats (5.5m) long were being laminated. During spraying, however, the airborne concentrations of styrene exceeded 100ppm. The lamination of small (3 m) boats in a spray boothwith an adjustable working door and an airflow was investigation in a Swedish study (Isaksson, 1976). A control velocity over 0.5 m/s at the opening of the booth was found sufficient to reduce the concentration of styrene in the breathing zone from over 200 ppm to about 1 ppm during the spray-up application of the resin. The worker stayed outside the booth while spraying. In the hand lay-up work, adequate control of the exposure could not be achieved by this method because the worker used a roller with a short shaft, and he had to enter the booth. Even though the main airflow came from behind the worker, he was exposed to relatively high concentrations of styrene (over 150 ppm) due to strong eddy formation over the mold. Better results were obtained with local exhaust ducts. When large volumetric exhaust air flows (0.6 m³/s/m²) were used, concentrations of styrene as low as 3-10 ppm were measured.

We are now conducting a comprehensive series of ventilation experiments in order to solve the problem of exposure to styrene during hand lay-up work. The full-scale experimental arrangement, in principle, is similar to that used in Isaksson's study. Our aims are to investigate how the exposure to styrene can be minimized by the proper location of the mold in the booth and to continue the local exhaust experiments. We are also studying the use of air curtains and jets. The results of the most interesting tests conducted so far are summarized in Table 6. LSE resin (type H in Table 3) was used with the exception of test 3.1, where a conventional resin (type F) was used. The average consumption of resin per boat was 47 kg.

Table 6. Airborne concentrations of styrene in the breathing zone during the lamination of motoroat hulls (4.1 m x 1.5 m x 0.5 m) in various conditions of ventilation.

Experiment	Special ventilation arrangement	Air velocity at working site (m/s)	Length of roller shaft (m)	Mold position	Conc. of styrene (ppm)
1. General ventilation of 3.3 m³/s					
1.1	-	0.3	0.14	upright	48
1.2	air curtain of 0.09 m³/s over the boat edge	0.4	1.14	tilted	17
1.3	blowing 0.07 m³/s of air longitudinally	0.6-1.3	1.14	upright	58

Table 6. cont.

Experiment	Special ventilation arrangement	Air velocity at working site (m/s)	Length of roller shaft (m)	Mold position	Conc. of styrene (ppm)
2.	Exhaust ducts of 2.8 m³/s				
2.1	ducts in the center of the boat	0.4	1.14	upright	5
2.2	ducts near the upper edge of the boat	0.5	1.14	tilted	3
2.3	as 2.2 and air curtain of 0.06 m³/s over the boat edge	0.4	1.14	tilted	8
3.	Booth with ventilation of 3.3 m³/s				
3.1	(conventional resin)	1.0	0.14	upright	135
3.2	as 3.1 but with LSE resin	1.0	0.14	upright	81
3.3	as 3.2	1.0	0.14	tilted	110
3.4	as 3.2	1.0	1.14	tilted	5
3.5	air curtain of 0.09 m³/s over the boat edge	1.0	1.14	tilted	66

Although the use of a LSE resin reduced the level of exposure again, it was not the final solution. Other engineering control methods also turned out to be necessary. Concentrations below 10 ppm were achieved by the use of either local exhaust ducts or a booth for lamination. In the experiments with local exhaust, the capture velocity exceeded 0.3 m/s. The booth provided good control only if the work was done in the door opening. We also got relatively good results by the use of air curtains and good general ventilation.

All the experiments in Table 6 were carried out so that the clean air flowed from behind the worker towards the mold. We also took samples when the worker stood on the other side of the boat; these invariably showed high concentrations. We will continue our series of experiments. However, it may well turn out that the use of respiratory protective devices during lamination and the isolation of

lamination work stations to prevent the passive exposure of
the other employees constitute the only economically feasible
solution that provides the necessary reduction in exposure
to styrene during the manufacture of large open mold
products.

LIST OF REFERENCES

Anon. (1980). LSE capability with good interlaminar adhesion.
  Reinfoeced Plastics (3): 72.
Herbert NT (1975). Comparison of the vapor pressure
  relationship vinyl toluene and styrene in corrosion resistant
  resins. Midland: The Dow Chemical Co., 25 pp.
Heinonen T, Vainio H (1980). Vinyltoluene induced changes in
  xenobiotic-metabolizing enzyme activities and tissue
  glutathione content in various rodent species. Biochem
  Pharmac 29:2675.
Isaksson G (1976). [Methods to reduce the exposure to styrene
  in the reinforced plastic industry.] Stockholm: Sveriges
  Plastförbund, 79 pp (in Swedish).
Kalliokoski P (1976). The reinforced plastics industry - a
  problematic work environment. Työterveyslaitoksen tutki-
  muksia 122, 130 pp (in Finnish).
Kalliokoski P, Jantunen M (1981). Control of airborne styrene
  in reinforced plastics industry. 2nd World Congress of
  Chemical Engineering. Montreal Oct. 4-9, 1981.
Nylander P (1979). LSE polyesters has effect on the environment.
  Plastforum Scandinavia (9):130 (in Swedish).
Schumacher RL, Breysse PA, Carlyon WR, Hibbard RP, Klainman
  GD (1981). Styrene exposure in the fiberglass fabrication
  industry in Washington State. Amer Ind Hyg Assn J 42:143.
Stinessen KO (1978). Ventilation for glass reinforced plastic
  industry. Sandefjord: Norges Teknisk-Naturvitenskapelige
  Forskningsråd, 85 pp (in Norwegian).
Voskamp AJ, Studenberg JE (1981). A new low styrene emission
  laminating resin. 36th Annual Conference, Reinforced Plastics
  Composites Institute. The Society of the Plastics Industry,
  Feb. 16-20, 1981.

Industrial Hazards of Plastics and Synthetic Elastomers, pages 287–296
© 1984 Alan R. Liss, Inc., 150 Fifth Ave., New York, NY 10011

BIOLOGICAL MONITORING AND HEALTH SURVEILLANCE OF WORKERS
EXPOSED TO STYRENE

Jorma Järvisalo, M.D. and Antero Aitio, M.D.

Institute of Occupational Health

Haartmaninkatu 1, SF-00290 Helsinki 29, Finland

Styrene or material which contains styrene is used in
many industrial branches, and the workers exposed to
styrene are often exposed to numerous other (toxic)
compounds. This communication, however, deals with the
biological monitoring and health surveillance of styrene
exposure per se, and the points discussed are best applicable to the fabrication or application of plastics that
contain styrene, especially in the reinforced plastics
industry.

Both biological monitoring and health surveillance
contribute to the protection of workers from the damage
that may follow occupational exposure to toxic chemicals.
Biological monitoring is an activity that in occupational
settings is primarily directed to the assessment of
workers' exposure or the effects of workers' exposure at
the individual level; its final goal is to minimize
workers' risk of impaired health due to the exposure.
Health surveillance often involves measures of biological
monitoring, but as an approach it is more general and
deals with the examination of workers' health and changes
in their state of health. Health surveillance very often
has aspects that are not merely preventive. Almost synony-
mous to the term health surveillance are the terms medical
surveillance and health or medical examinations (Aitio et
al., 1983; Zielhuis, 1980).

The health risks that may be caused by exposure to
styrene can be classified into the following categories:
1. mutagenicity and clastogenicity, 2. carcinogenicity;

3. teratogenicity; 4. neurotoxicity; and 5. others
(including, e.g., irritative and metabolic effects). The
present communication touches only the issues of
categories 1,2, and 4.

MONITORING OF EXPOSURE TO STYRENE

The monitoring of exposure to styrene by the use of
biochemical specimens may be done by analyses of styrene
or its metabolites in human body fluids or other specimens
(e.g., WHO, 1983).

Of the various metabolites of styrene (for the metabolic diagram, see Vainio et al. in these proceedings),
measurements of the urinary excretion of mandelic acid
(being the main metabolite) or phenyl glyoxylic acid, or
the sum of the two acids have been used (WHO, 1983).
Recently, the analysis of 4-vinylphenol, a product of ring
hydroxylation has been suggested as one additional aspect
of the monitoring of exposure to styrene (Pfäffli et al.,
1981). The compound is excreted, both in conjugated and
unconjugated forms, in the urine of styrene exposed
workers. Some of the styrene 7,8-oxide is also conjugated
with glutathione, and after further metabolism in the
liver and the kidneys, the product(s) are excreted in the
urine as mercapturic acids.

The excretion of these metabolites cannot be used to
indicate the health risk that may follow exposure to
styrene; rather they indicate exposure to styrene.
Approximately 50-85 % of the styrene in the body is
excreted as mandelic acid, 15-40 % as phenyl glyoxylic
acid (for ref. see WHO, 1983). Measurement of mandelic
acid in the urine is commonly used to assess exposure to
styrene. Spot specimens are collected directly after the
daily work shift on the last workday of the week. If the
specimen is collected on the next morning instead, the
concentration of mandelic acid in the urine will have
decreased to a third of the concentration on the previous
afternoon.

In studies where the concentrations of the acid found
in urine specimens after the daily exposure period have
been correlated with concentrations of styrene in the
ambient air, the urinary mandelic acid/styrene in air
relation has shown some sigmoidality (Engström et al.,

1976; Fields & Horstman, 1979; Elia et al., 1980). The concentration of mandelic acid in the urine that corresponds with the level of exposure to styrene at 100 cm^3/m^3 (ppm), on average, ranges approximately from 13 mmol/l (or gram creatinine) to 20 mmol/l. The wide variation in the results may to a great extent be explained by variation in the biological and hygienic samplings, differences in the conditions of exposure at the sites where the measurements have been taken, and partly also by the workers' personal hygienic and biological factors.

Some authors regard the sum of mandelic acid and phenylglyoxylic acid as a better indicator of exposure to styrene than the concentration of styrene alone (Guillemin & Bauer, 1978). The sum concentration of the acids in the urine at exposure to a level of 100 $cm^3 m^3$ (ppm) is about 17 mmol/l.

The excretion of 4-vinylphenol in the urine is parallel with that of mandelic acid (Pfäffli et al., 1981). Arene oxidation is a minor metabolic route of styrene in man, and the concentration of phenol may be determined in conditions where the study of this route is important. Exposure to styrene has also been measured through the analysis of the concentration of styrene in exhaled air, in venous blood, or in subcutaneous fat specimens (Engström et al., 1978; Fields & Horstman, 1979; Wolff et al., 1977; Wolff et al., 1978). Of these, the measurement of styrene in the blood or expiratory air is of lesser practical importance: the first due to kinetics and the latter due to the difficulty of standardizing sampling of the alveolar air in practical occupational monitoring. The kinetic behavior of styrene in the blood or exhaled air after exposure has ceased is such that most of the styrene in the body is excreted very rapidly; consequently the concentration of styrene in the two specimens very much depends on the conditions of exposure prior to the collection of specimens. For the analysis of the exposure pattern of a full workday, several successive specimens would be needed (e.g., Fernandez & Caperos, 1977; Fields & Horstman, 1979; Ramsay et al., 1980). The use of air specimens is also obscured by the further polymerization of styrene in the bags used to collect the specimens.

An additional difficulty is the methodologic detection limit of the analysis of styrene in the blood. In most cases of occupational exposure, the analysis of styrene in blood specimens taken on the morning after the workday under study is no more possible (e.g., Wolff et al., 1978).

Due to all these factors, the measurement of styrene in the blood or expiratory air is of minor practical importance compared to the analysis of, e.g., mandelic acid in the urine. Styrene analyses may be important when one aims at the assessment of exposure in specific conditions: peak level exposure; the analysis of the exposure pattern over a full day; or the collection of detailed information in connection with research projects.

Compared with the measurement of styrene in the blood or exhaled air, the measurement of styrene in adipose tissue taken from the buttocks area may be of special value (Wolff et al., 1977, 1978; Engström, 1978). Although the sampling and the specimen pretreatment methods have not been standardized yet, the concentration of styrene in body fat may in the future be used to indicate body burden of the compound. The half time of styrene in body fat ranges from two to four days (Engström et al., 1978). The measurement may also be used in conditions where the metabolic disposition of styrene becomes overloaded, a situation which leads to the decreased uptake of styrene in the body but to the continued uptake of styrene in adipose tissue. However, to date only limited basic data is available.

CLASTOGENIC AND MUTAGENIC EFFECTS OF STYRENE

The bulk of the published evidence suggests that occupational exposure to styrene in the reinforced plastics industry enhances the frequency of chromosomal aberrations in peripheral blood lymphocytes. Despite the time consuming analyses and the poorly developed methods to assure the quality of such analyses, the analysis of chromosomal aberrations seems to be the method of choice for the study of the clastogenic effects of exposure to styrene in workers. A clear limitation is caused by the lack of resources for such analyses, but automation, at least in part, may improve the possibilities for such analyses in the near future (Norppa et al., 1981; WHO,

1983).

It would be much easier to analyze sister chromatid exchanges (SCEs) in the same cells. However, there are few published studies on the effects of exposure to styrene on workers' SCE rates, and the results are so far inconclusive (WHO, 1983; Norppa et al., 1981; Camurri et al., 1983). Consequently, the value of this measurement in the occupational biological monitoring of the effects of styrene remains uncertain.

There is also some evidence that exposure to styrene may cause micronuclei detectable in peripheral blood cells (Norppa et al., 1981). Their detection would be easier to automate than the analysis of chromosomal aberrations in blood lymphocytes. However, further research is needed to establish the value of the micronuclei analysis as one component of occupational health surveillance.

Currently there is no evidence to support the use of bacterial mutagenicity testing to assessment the mutagenic risk of exposure to styrene. Nor is there any data on the value of the use of various tumor markers in the occupational monitoring of the effects of styrene. Some very limited evidence indicates that measurements of either the adducts of styrene with DNA or other blood tissue macromolecules or the metabolites of styrene adducts may in the future become methods monitoring the extent to which styrene exposure (through its electrophilic metabolites) is attacking body macromolecules (see Vainio et al., 1981).

BIOLOGICAL MONITORING IN THE ASSESSMENT OF NEUROTOXICITY

There is no biochemical or physiological method presently available which would be simple enough for practical occupational health care and could also be used to predict the exposed workers' risk of damage to the central or peripheral nervous system. There is, however, a real need for such methods. Neurophysiological methods (e.g., electroencephalography and electroneuromyography) are too complex for purposes of health surveillance. The current complex neuropsychological test batteries are also both time-consuming and need special expertise; consequently, they cannot be added to health surveillance programs carried out due to exposure to neurotoxic agents (e.g. Seppäläinen, 1982; Lindström, 1982). In recent years, studies where symptom questionnaires were used have produced valuable information on the nature of the damage to the central nervous system caused by exposure to neuro-toxic agents (Hänninen, 1983). Such questionnaires could probably be simplified for practical occupational health care.

When interpreting the results of biological monitoring of exposure to styrene, some factors must be considered when the results are being interpreted. First of all, at best the data provide information on the exposure on the workday when the specimens were collected. Secondly, the reference values used for measurements of the two acids have been derived from the hygienic limit values. Consequently, such reference limit values do not reflect solely medical or toxicological evidence for safe work conditions (e.g., Rantanen et al., 1982; Järvisalo & Tossavainen, 1982).

APPLICATION OF MEASUREMENTS OF BIOLOGICAL MONITORING IN HEALTH SURVEILLANCE DUE TO STYRENE EXPOSURE

The application of measurements of biological moni-toring in occupational health surveillance must take into account several issues (Table 1). The listed factors, at least, should be carefully considered before measurements are taken.

The measurement of urinary mandelic (+ phenyl glyoxylic) acid is applicable for the assessment of workers' individual exposure. The methods have been standardized, and recent interest in quality control issues has led to improved analytical quality. Due to the nature of such measurements, the results are specific to the exposure conditions of a very limited time period (Rantanen et al., 1982).

Although applicable to certain specific conditions where the pattern of exposure in given work surroundings should be clarified in detail, measurements of the urinary excretion of 4-vinylphenol (Pfäffli et al., 1981) or of the levels of styrene in exhaled air, the blood, or body fat are not yet methods of practical biological moni-toring. Special attention should be paid to the methodologic development of the analysis of styrene in adipose tissue, because this measure could be used to monitor the exposure of a rather long time period.

Table 1. Factors to be considered before biological monitoring is applied in occupational health surveillance.

---

    1. When and where to apply?
      - exposure at work
      - other occupational and
        non-occupational exposures

    2. Validity questions
      - Is the measurement the right one
        for these particular conditions?
      - sensitivity and specificity of
        the method
      - quality of the measurement (including
        the collection, transport, and analysis
        of specimens)

    3. Data interpretation
      - reference values or groups
      - inferences (What to do with
        the results?)

Of the methods currently available for the detection
of mutagenic damage in workers exposed to styrene, the
analysis of chromosomal aberrations in lymphocytes of the
peripheral blood is the method of choice. The scientific
evidence for the applicability of sister chromatid
exchange or micronuclei analyses is presently either
lacking or controversial (Norppa et al., 1981; WHO, 1983).
However, the wider use of aberration analyses is pro-
hibited by the minute resources available to occupational
biological monitoring. In addition, because aberrations
are thought to be correlated with chemical carcinogenesis
and because almost nothing is known of the value of
increased aberration rates in the prediction of malignant
outcome among exposed persons, one should give special
consideration to the various issues listed in Table 1.
Especially important is that decisions about the implications or actions to which the results of the measurements will lead be made before the measurements are
carried out. Thus also the workers monitored by such

measurements must be provided with thorough information
about the use of the measurements and about their results.

As to methodologic aspects of aberration analyses,
automation and both the statistical and the analytical
quality of the measurements must be given special attention. Automation will probably make it possible to
increase the number of cells counted and also to standard-
ize the analysis of various types of aberrations.

In conclusion, there are many methods of biological
monitoring available for use in connection with occupational health surveillance because of exposure to
styrene. Many of the methods, however, lack extensive
background information; in some cases information is
lacking to such an extent that the methods must be considered methods for research or for use in special circumstances. One must keep in mind that the exposure of
workers to styrene does not occur as a sole phenomenon but
in combination with other occupational exposures, and in
some cases workers' non-occupational exposures must also
be taken into account decisions are made about the rel-
evance or contents of health surveillance because of
exposure to styrene.

Aitio A, Riihimäki V, Vainio H (1983). "Biological monitoring and health surveillance of workers exposed to
chemicals". Washington, D.C., Hemisphere Publ. Co., in
press.
Camurri L, Codeluppi S, Pedroni C, Scarduelli L (1983).
Chromosomal aberrations and sister chromatid exchanges in
styrene exposed workers. Mut Res, in press.
Elia VJ, Anderson LA, MacDonald TJ, Carson A, Buncher CR,
Brooks SM (1980). Determination of urinary mandelic and
phenylglyoxylic acids in styrene exposed workers and a
control population. Am Ind Hyg Assoc J 41:922.
Engström J, Bjorström R, Åstrand I, Ovrum P (1978).
Exposure to styrene in a polymerization plant. Uptake in
the organism and concentration in subcutaneous adipose
tissue. Scand J Work Environ Health 4:324.
Engström K, Härkönen H, Kalliokoski P, Rantanen J (1976).
Urinary mandelic acid concentration after occupational
exposure to styrene and its use as a biological test.
Scand J Work Environ Health 2:21.
Fernandez J, Caperos J (1977). Exposition au styrène. I.
Etude experimentale de l'absorption et l'excretion

pulmonaires sur des sujets humains. Int Arch Occup Environ Health 40:1.

Fields RL, Horstman SW (1979). Biomonitoring of industrial styrene exposures. Am Ind Hyg Assoc J 40:451.

Guillemin MP, Bauer D (1978). Biological monitoring of exposure to styrene by analysis of combined urinary mandelic and phenylglyoxylic acids. Am Ind Hyg Assoc J 39:873.

Hänninen H (1983). Behavioral methods in the assessment of early impairments in central nervous functioning. In Aitio A, Riihimäki V, Vainio H (eds): "Biological Monitoring and Health Surveillance of Workers Exposed to Chemicals", Washington, D.C., Hemisphere Publ. Co., in press.

Järvisalo J, Tossavainen A (1982). Exposure to neurotoxic agents in Finnish working environments. Trends and assessment of exposure. Acta Neurol Scand 66, suppl. 92:37.

Lindström K (1982). Behavioral effects of long-term exposure to organic solvents. Acta Neurol Scand 66, suppl. 92:131.

Norppa H, Vainio H, Sorsa M (1981). Chromosome aberrations in lymphocytes of workers exposed to styrene. Am J Ind Med 2:299.

Pfäffli P, Hesso A, Vainio H, Hyvönen M (1981). 4-Vinyl-phenol excretion suggestive of arene oxide formation in workers occupationally exposed to styrene. Toxicol Appl Pharmacol 60:85.

Ramsay JC, Young JD, Karbowski RJ, Chenoweth MB, McCarthy LP, Braun WH (1980). Pharmacokinetics of inhaled styrene in human volunteers. Toxicol Appl Pharmacol 53:54.

Rantanen J, Aitio A, Hemminki K, Järvisalo J, Lindström K, Tossavainen A, Vainio H (1982). Exposure limits and medical surveillance in occupational health. Am J Ind Med 3:363.

Seppäläinen A-M (1982). Neurophysiological findings among workers exposed to organic solvents. Acta Neurol Scand 66, suppl. 92:109.

Vainio H, Sorsa M, Rantanen J, Hemminki K, Aitio A (1981). Biological monitoring in the identification of the cancer risk of individuals exposed to chemical carcinogens. Scand J Work Environ Health 7:241.

Wolff MS, Daum SM, Lorimer WV, Selikoff IJ (1977). Styrene and related hydrocarbons in subcutaneous fat from polymerization workers. J Toxicol Environ Health 2:997.

Wolff MS, Lilis R, Lorimer WV, Selikoff IJ (1978). Biological indicators of exposure in styrene polymerization workers. Styrene in blood and adipose tissue and mandelic and phenylglyoxylic acids in urine. Scand J Work Environ Health 4:114.

World Health Organization, International Programme on Chemical Safety (1983). "Environmental health criteria document on styrene", Geneve: WHO, in press.

Zielhuis RL (1980). Recent and potential advances applicable to the protection of worker's health. Biological monitoring. In Logan D, Berlin A, Yodaiken R (eds): "Assessment of toxic agents at the workplace - roles of ambient and biological monitoring", Luxembourg, Commission of the European Communities, in press.

CHAPTER IV. OCCUPATIONAL HAZARDS OF POLYETHYLENE AND POLYPROPYLENE

Industrial Hazards of Plastics and Synthetic Elastomers, pages 299-307
© 1984 Alan R. Liss, Inc., 150 Fifth Ave., New York, NY 10011

PRODUCTION,PROCESSING AND DEGRADATION PRODUCTS OF POLYETHY-
ETHYLENE AND POLYPROPYLENE

Ariel Hoff

Department of Polymer Technology,
The Royal Institute of Technology,
S-100 44 Stockholm,Sweden

PRODUCTION METHODS

Polyethylene is the most widely produced thermoplastic
in the world. Polypropylene is the fourth main bulk plastic
in the world,which has achieved the highest increase in the
rate of growth. There is three main routes used for production of polyethylene: High pressure method, Philipps method,
and Ziegler process. Polypropylene is produced according to
the Ziegler method.

High Pressure Method

The bulk of commercial polyethylene is produced by the
high pressure process. In this route ethylene is polymerised
at high pressures and high temperatures. A free radical initiator such as bensoyl peroxide or azo-bis-isobutyronitril
is commonly used. The pressure varies between 1000 and 3000
atm, and the temperature between 80 and 300°C. This process
may be operated continuously by passing the reactants through
narrow bore tubes or through steered reactors or by a batch
process. Because of the high heat of polymerisation care
must be taken to avoid runaway reaction. This can be done by
heaving a high cooling surface-volume ratio in the appropriate
part of a continuous reactor,and in addition by running water
or a somewhat inert liquid as benzen through the tubes to dilute the exotherme. In a typical process 10-30% of the monomer
is converted to polymer. After a polymer-gas separation the
polymer is extruded into a ribbon and then granulated. The
reaction has two particular characteristics: the high exother-

mic reaction,and the critical dependence on the monomer concentration. At high temperatures (e.g. 200°C) much higher
pressures are required to obtain a given concentration or a
density of monomer than at 25°C. By varying temperature ,
pressure,initiator type and composition,and incorporating
chain transfer agents, and by injecting the initiator into
the reaction mixture at various points in the reactor it is
possible to vary independently of each other polymer characteristics such as branching,molecular weight, molecular weight
distribution. The polymer obtained by this process has a
density of about 0.915-0.94 g/cm^{3}, a molecular weight of
about 50.000 and contains about 20-40 branches per 1000 carbon atoms,some exceeding 10 carbon atoms in length.

Ziegler Process for Polyethylene

   This type of polymerisation is sometimes referred to   coordination polymerisation,since the mechanism involves a
catalyst-monomer coordination complex or some other driving
force that controls the way in which the monomer approaches
the growing chain. In a typical process ethylene is fed under
low pressure into the reactor which contains liquid hydrocarbon to act as a diluent. A typical catalyst is a blend of aluminium thriethyl and titanum tetrachloride. The reaction is
carried out at temperatures below 100°C in the absence of
oxygen and water. When the reaction is completed the catalyst
is destroyed by the action of ethanol,water or caustic alkali.
Molecular weights of 200.000 and higher are reached,the density
of the polymer is about 0.945 g/cm^{3}.

Ziegler Process for Polypropylene

   There are many points of resemblance between the production
of polypropylene and polyethylene using Ziegler-type cataly-
sis. A typical catalyst system may be prepared by reacting
titanum chloride and aluminium tributyl or aluminium diethyl
monochloride in naphtha under nitrogen to form a slurry consisting of about 10% of catalyst and 90% naphtha. The properties of the polymer are strongly dependent on the catalyst
consumption and its particle shape and size. Propylene is
charged into the polymerisation vessel under pressure,whilst
the catalyst solution and the reaction diluent(usually naphtha)
are metered separately. Reaction is carried out at temperatures
of about 60°C for approximately 8 hours. In a typical process

a 80-85% conversion to polymer is obtained. The molecular weight of the polymer can be controlled in a variety of ways, e.g. by use of hydrogen as a chain transfer agent or by variations in the molar ratio of catalyst components,the polymerisation temperature,the monomer pressure or the catalyst concentration. The separation of the reaction ingredients involves the transfer of the reaction mixture to a flush drum to remove the unreacted monomer. The residual slurry is centrifugated to remove the bulk of the solvent together with most of the atactic material which is soluble in naphtha. The remaining material is then treated with an agent which decomposes the catalyst and dissolves the residue (methanol with hydrochloric acid). The solution of residues in methanol is removed by a centrifuging operation and the polymer is washed and dried at about 80°C. At this stage the polymer may be blanded with antioxidants,extruded and cut into pellets.Polypropylene obtained in this way is an isotactic polymer with degree of crystallinity of 60-70%,and with density of about 0.91 g/cm^3.

Philipps Process

In this process ethylene dissolved in a liquid hydrocarbon is polymerised on a supported metal oxide catalyst at about $130-160^{\circ}$C and at pressure of about 30-40 atm. The preferred catalyst is one which contains 5% of chromium oxides, mainly CrO_3, on a finely divided silica-alumina catalyst (75-90% silica) and which has been activated by heating to about 250°C. After reaction the mixture is passed to a gas-liquid separator where the ethylene is flashed off,catalyst is then removed from the liquid product of the separator and the polymer separated from the solvent by either flashing off the solvent or by precipitating the polymer by cooling. The commercial products have a molecular weight of about 40.000 and the highest density of any commercial polyethylenes (ca. 0.96 g/cm^3).

PROCESSING METHODS

The main processing metods and the processing temperatures for polyethylene and polypropylene are summarized in Table 1.

TABLE 1

The Main Processing Methods and the Processing
Temperatures for Polyethylene and Polypropylene
(W.J.Roff and J.R.Scott,1971)

Processing Method	Temperature (^{o}C)	
	Polyethylene	Polypropylene
Extrusion:		
generally		210–270
pipes	140–170	
film	200–340	270–300
Injection Moulding	150–370	230–260
Compression Moulding	130–230	

Extrusion

Approximately three quarters of polyethylene and one
third of polypropylene produced is formed into products by
means of extrussion process. This process consists of metering
polymer (usually in granular form) into a heated barrel where
they are compacted. The resultant melt is then forced under
pressure through an orifice to give a product of constant
cross section.The screws usually have a length–diameter ratio
in excess of 16:1, and a compression ratio of between 2.5:1,
and 4:1. Well over the half the polyethylene extruded is
converted into film.

Injection Moulding

In this process the polymer is melted and injected in a
mould which is at a temperature below the melting point of the
polymer,so that the latter can harden.

Blow Moulding

Many articles,bottles and containers,in particular, are
made by blow moulding. In one typical process a hollow tube
is extruded vertically downwards on to a spigot. Two mould
halves close on to theextrudate and air is blown through the
spigot to inflate the extrudate so that it takes up the shape
of the mould

Compression Moulding

In this process the polymer is heated in a mould, compressed to shape and cooled. This process is slow since
heating and cooling of the mould must be carried out in each
cycle. This method is used for production of large blocks
and sheets.

DEGRADATION PRODUCTS

Polyethylene and polypropylene are rather stable against
heating in an inert atmosphere and in vacuum ,whereas in air
they are much more vulnerable. In this respect high density
polyethylene   is somewhat more stable than low density polyethylene   , and polyethylene is more thermostable than
polypropylene   ,Figure 1.

Figure 1. Thermal Gravimetric Analysis of low density-(1)
and high density (2) polyethylenes and polypropylene (3)
in air, heating rate 10°/min. (E.A.Boettner,G.L. Ball,
and B.Weiss, 1976).

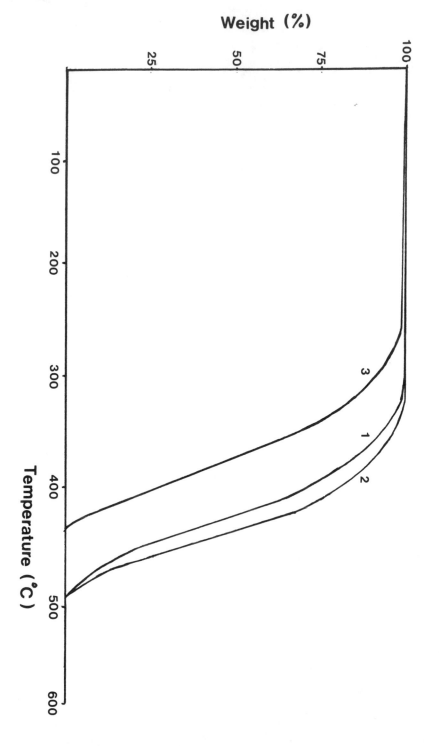

The difference between polyethylene and polypropylene depends
on the fact that every monomer unit of polypropylene contains a tertiary carbon atom which is more apt
to rupture than a secondary carbon atom. The difference between LDPE and HDPE depends on the fact that the former is
more branched than the latter, and the branch points contain
tertiary carbon atoms. The volatile products formed on thermal degradation of these polymers are different hydrocarbons.
Heating of the polymers in an oxygen-containing atmosphere,
e.g. in air dramatically increases the aptitude of the polymers to degradation. The observed weight loss of polyethylene
in the air commences of about 50-80°C lower than in nitrogen.
In the case of polypropylene the effect is much more pronounced.
Heating of the polymers in air leads as a first step to oxygen
absorbtion which is eventually followed by the formation of
hydroperoxide groups (reaction 1). The hydroperoxides are
very unstable at temperatures of about 200°C. Decomposition
of these labile groups results in formation of free
radicals (rection 2). The process of thermal oxidation of
polyethylene and polypropylene can be schematically expressed
in a following way:

$$RH + O_2 \longrightarrow ROOH \tag{1}$$

$$ROOH \longrightarrow RO\cdot + \cdot OH \tag{2}$$

$$RO\cdot + RH \longrightarrow ROH + \cdot R \tag{3}$$

$$\cdot OH + RH \longrightarrow H_2O + \cdot R \tag{4}$$

$$R\cdot + O_2 \longrightarrow ROO\cdot \tag{5}$$

$$ROO\cdot + RH \longrightarrow ROOH + \cdot R \tag{6}$$

$$RO\cdot \longrightarrow R\cdot + products \tag{7}$$

$$2ROO\cdot \longrightarrow O_2 + products \tag{8}$$

Although there is differences in kinetics of oxidation of
polyetylene and polypropylene as well as differences in the
volatile products formed during oxidation of these polymers
the main features of the oxidation process are rather close.
The oxidation process results in chain scission, formation of
functional groups within the oxidised polymer fragments,

and in the formation of volatile products. For example,44 and 51 volatile products were identified by gas chromatographic-mass spectrometric analysis in the course of oxidation of polyethylene and polypropylene,respectively. These volatile products include hydrocarbons,alcoholes,ethers,epoxides, aldehydes,lactones,ketones,acids and hydroxy acids (A.Hoff,S.Ja — cobsson,1981 and A.Hoff,S.Jacobsson,1982) . The oxygen-containing compounds constitute an absolute majority and the hydrocarbons a minority of products detected during oxidation of these polymers. The results of the quantitative measurments of some volatile oxygen-containing products obtained on oxidation of these polymers at $280^{\circ}C$ are given in Table 2.

TABLE 2

Amounts of some volatile products formed on oxidation of polyethylene (4min) and polypropylene (2min) at $280^{\circ}C$, (g/g) (A.Hoff,S.Jacobsson,1981 and 1982).

Compound	Polyethylene	Polypropylene
Formaldehyde	1780	12100
Acetaldehyde	1950	23000
α-methylacrolein		8800
Propanal	370	
Acrolein	280	
Butanal	700	
Pentanal	540	
Acetone	250	12350
2-butanone	320	600
2-pentanone	290	2800
Acetylacetone		5500
Formic acid	3000	2600
Acetic acid	1200	12800
Propionic acid	700	

REFERENCES

1. Roff WJ, Scott JR (1971), Fibres, Films, Plastics and Rubbers, London: Butterworks, p
2. Boettner EA, Ball GL, Veiss B (1970), Combustion products from the incineration of plastics, Report N EPA-670/2-73-049,

University of Michigan, Ann Arbour

3. Hoff A, Jacobsson S (1981), Thermo-oxidative Degradation of Low-density Polyethylene Close to Industrial Processing Conditions, J Appl Polym Sci 26:3409

4. Hoff A, Jacobsson S (1982), Thermal Oxidation of Polypropylene Close to Industrial Processing Conditions, J Appl Polym Csi 27:2539

Industrial Hazards of Plastics and Synthetic Elastomers, pages 309–312
© 1984 Alan R. Liss, Inc., 150 Fifth Ave., New York, NY 10011

OCCUPATIONAL HAZARDS OF POLYETHYLENE AND POLYPROPYLENE
MANUFACTURE.

A.J.M. Slovak

Divisional Medical Officer

Fisons Occupational Health Service, Derby Road,
Loughborough, Leicestershire. England.

Both polyethylene (PE) and polypropylene (PP) are
important bulk products in the world plastics industry. They
are manufactured in large plants with high volume throughput
and high on-line occupancy. A number of standard grades of
each polymer are produced, although the grade range and
technical performance varies between manufacturers.

MANUFACTURE OF POLYETHYLENE

The manufacture of PE is a relatively simple process,
the seeming complexity of the plant coming from the
recycling and stripping operations which are part of any
continuous operation. The health and safety hazards of the
process are summarised in Table 1.

```
                          ETHYLENE
                             |
                        ┌─────────────┐
                        │ COMPRESSION │ _ _ _ _ NOISE
                        │   STAGES    │            95-105 dB.
                        └─────────────┘  Cooling
                             |
                        ┌─────────────┐
                        │ SEPARATION  │
                        │   STAGES    │
ADDITIVES───────────────└─────────────┘
                             >|
  - Antioxidants        ┌─────────────┐
  - Slip-additives      │  EXTRUSION  │ _ _ _ _ Falls
  - Antistatics         └─────────────┘
  - Silica                   |
                        ┌─────────────┐
                        │ CUTTING AND │ _ _ _ _ Abrasions
                        │   DRYING    │            Lacerations
                        └─────────────┘
                             |
                        ┌─────────────┐
                        │ STORAGE AND │
                        │   BAGGING   │
                        └─────────────┘
```

Table 1. Occupational hazards of PE manufacture.

The main hazard of the PE process is noise in the compression room. Operators take part in a hearing preservation and audiometry programme but the risk is greatly minimised by the provision of a sound-proofed control-room. Entry into the compression room is unnecessary except for brief spells. The addition of small amounts of a wide range of additives presents little risk at the extrusion stage other than from falls resulting from the slippery floor. At the cutting and drying stage minor lacerations are common even when stout protective gloves and clothing are worn. Whilst these injuries may appear mundane, cuts and falls account for the vast majority of lost time accidents in the PE process.

MANUFACTURE OF POLYPROPYLENE

The health and safety hazards are summarised in Table 2:-

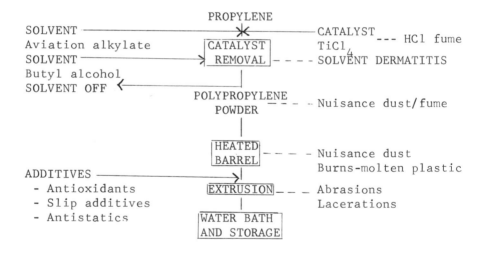

Table 2. Occupational hazards of PP manufacture.

As with PE, the main acute causes of lost-time accidents are simple; lacerations and adherence burns. Solvent dermatitis is not uncommon but is usually trivial unless neglected. Simple masks are effective for the dusty operations in this process and respirators are available in any case of breakdown in the catalyst recovery procedure leading to HCl release.

Health Surveillance

Biological monitoring of PE and PP manufacturing workers has proved unrewarding other than for the obvious risks. Epidemiological data on morbidity and mortality is not available in print and given the short manufacturing history, low exposure and small worker numbers, any data presently generated would tend to be inconclusive.

In the Soviet Union, Akhmetova [1] reported visual effects in PE workers suggestive of tunnel vision but these findings have not been confirmed elsewhere.

PE and PP have been evaluated for carcinogenic risk with negative findings. [2,3]

THERMAL DEGRADATION and PYROLYSIS of PE and PP

The emissions resulting from the thermal degradation and pyrolysis of PE and PP have been extensively studied. [4-10] The predominant combustion products are CO, CO_2, H_2O and carbon (as soot), the most toxic being CO. Irritant combustion products are also obtained in small amounts mainly low M.W. aldehydes. For PE the main irritants are acrylic acid, acrolein and formaldehyde; for PP they are formaldehyde and crotonaldehyde.

Under most test conditions PE and PP are less toxic and irritant than chlorine and nitrogen containing plastics and most natural materials. However like other plastics they are prone to rapid surface transmission of flame and release burning droplets which can cause rapid spread of fire and serious burn injuries.

A well-authenticated case of meat-wrappers asthma was described by Skerfving et al [11] in 1980 which was ascribed by the authors to thermal degradation products of PE coming off at 200-220°C. It is unclear whether the patient was also exposed to P.V.C. fumes or other substances capable of inducing asthma in the workplace or elsewhere.

It should be noted that PE and PP may be bonded or otherwise combined with different plastics and many other substances and health effects may result as a consequence of pyrolysis of such combined substances which may be due not to PE or PP but to those other substances or which are unique to the combination. Some caution and rigour of analysis needs to be applied in these circumstances before ascribing a health effect to a particular cause.

References

1. Akhmetova ET (1977). State of the eyes in synthetic ethanol and high-pressure polyethylene production workers. Vestn. Oftalmol. (4):58-9.
2. IARC Monographs (1979). Ethylene and polyethylene. 19:157-186.
3. IARC Monographs (1979). Propylene and polypropylene. 19:213-230.
4. Potts WJ, Lederer TS, Quast JF (1978). A study of the inhalation toxicity of smoke produced upon pyrolysis and combustion of polyethylene foams. Part I. Laboratory Studies. J. Combust.Toxicol.5(4):408-33.
5. Kuhn RL, Potts WJ, Waterman TE (1978) Part II. Full scale fire studies. J. Combust.Toxicol.5(4):434-64.
6. Hilado CJ, Kosola KL (1978). Effect of pyrolysis temperature on toxicity of gases from a polyethylene polymer. J. Combust Toxicol.5(3):339-43.
7. Hilado CJ, Huttlinger NV, Brauer DP (1978). Effect of pyrolysis temperature and airflow on toxicity of gases from a polyethylene polymer. J. Combust.Toxicol.5(4):465-75.
8. Hilado CJ, Brauer DP (1979). Polyethylene: effect of pyrolysis temperature and airflow on toxicity of pyrolysis gases. J. Combust.Toxicol.6:63-8.
9. Hilado CJ, Schneider JE, Brauer DP (1979). Toxicity of pyrolysis gases from polyproylene. J. Combust.Toxicol. 6:109-16.
10. Hilado CJ, Huttlinger NV (1980). Toxicity of pyrolysis gases from polypropylene fibers containing a flame retardant J. Combust.Toxicol.7:134-40.
11. Skerfving S, Akesson B, Simonsson B (1980). "Meat wrappers' asthma" caused by thermal degradation products of polyethylene. Lancet 1 (8161):211.

Industrial Hazards of Plastics and Synthetic Elastomers, pages 313–317
© 1984 Alan R. Liss, Inc., 150 Fifth Ave., New York, NY 10011

OCCUPATIONAL HAZARDS OF POLYETHYLENE AND POLYPROPYLENE PROCESSING.

A.J.M. Slovak

Divisional Medical Officer

Fisons Occupational Health Service, Derby Road, Loughborough, Leicestershire. England.

The main volume usages of polyethylene (PE) are in the manufacture of films, sheets and layered packaging materials. A small amount of PE is used for moulded items but polypropylene (PP) is more often used in this application. PP is also widely used for fibre in the manufacture of woven goods such as sacking and carpets.

Almost all processing applications of PE and PP technology conform to a fundamental pattern whereby the raw material is melted and then extruded, spun, moulded or otherwise manufactured according to the physical form of the polymer required. Thus the basic hazards of processing are much the same as those of the final stages of polymer production namely adherence burns and lacerations. The latter are especially common in the trimming and general finishing of rigid moulded items. Another common hazard is from pyrolysis products if melt temperature is poorly controlled although this seldom causes more than minor irritant symptoms unless there is a serious malfunction of processing machinery.

The main specific hazards of PE and PP processing come from substances added to polymer base mixtures especially for moulding operations and to a lesser extent for fibres manufacture. These substances are blowing agents, pigments and talc.

The function of blowing agents is to introduce a cellular structure into plastics systems which, unlike polyurethanes, do not have their own inherent foaming characteristics. This has the effect of increasing bulk and altering rigidity characteristics in a polymer product whilst reducing weight. Typical blow-moulded products are automobile dashboards and bumpers (fenders). The simplest blowing agent is nitrogen gas but its applications are limited and more often blowing agents are relatively simple, solid chemicals incorporated at a low concentration into pre-moulding mixes. During the thermoplasticising process they decompose spontaneously in a controlled way to give gas bubbles which produce the foaming effect.

One of the most common of these blowing agents is azodicarbonamide (AC) which has now been shown to be a cause of occupational asthma. It is manufactured predominantly as a fine yellow powder and is milled to particle sizes in the 2-10 micron range. This technically most useful particle size distribution is almost exclusively in the respirable range. In 1980, the author carried out an epidemiological survey of a plant manufacturing AC [1]. This prevalence study revealed a high level of occupational asthma (18.5%) among people who had been working on the plant. Although all those who had developed asthma had been relocated away from the plant a proportion of those (7 out of 13) who had stayed working with AC for more than 3 months after getting their first symptoms had developed general airways hyperactivity to common environmental irritants such as SO_2 and tobacco smoke. This hyperactivity persisted for up to 7 years.

It is obvious that exposure in the manufacture of such a product is likely to be much higher than when it is used in a mixture in plastics production. However the first published report by Ferris et al [2] came from the latter source. Additionally the author is aware of a number of outbreaks of asthma ascribed to AC in the plastics industry at least one of which carried a higher incidence than that reported in the prevalence study referred to above.

The range of pigments which may be incorporated into moulded and woven products is immense. Most modern pigments are of low toxicity and seldom pose a hazard to workpeople. However some caution needs to be exercised with at least two classes of colouring agents. These are benzidine based dyestuffs and reactive dyestuffs. The former now contain very little or no benzidine when manufactured but certain processes may cause reversion to benzidine in appreciable quantity. The latter are a cause of occupational asthma although the incidence appears to be much lower than AC.

Talc filling is added to certain moulding grades of PP to increase rigidity and sometimes to aid finish quality. It is usually present in considerable quantities in these grades, the quantity most often being in the 20-40% range. Talc pneumoconiosis [3] has been a well established clinical entity for many years although it is still a relatively rare finding and has a low incidence in comparison with coalminers pneumoconiosis or silicosis. There is no published evidence that talc constitutes a real hazard in the plastics industry.

Polypropylene pipework and other mouldings are now used extensively in domestic and industrial applications. PP is easy to cut and has also proved very easy to weld so that a large industry has evolved at small workshop level producing welded PP items including vessels and complex pipework. The ease of working with PP has encouraged the unskilled amateur to try his hand with occasionally disastrous results. It is particularly in this application that pyrolysis effects may be most hazardous since many workshops undertaking this type of work are small, poorly ventilated and poorly supervised.

The storage of PE and PP as raw materials is not totally without hazard. Under certain circumstances, particularly when stored in the cold for long periods, some additives may creep to the surface of the plastic. This can cause an irritant dermatitis whose aetiology often proves difficult to track down.

Protection and Prevention

To prevent minor skin abrasions and irritation it is sensible to wear stout protective gloves when handling PE and PP. Workrooms should be well ventilated to remove heat and pyrolysis products. Protective clothing should be able to withstand plastic burns.

Exposure to talc, pigments and blowing agents may be reduced by the purchase of premixed master batches from major manufacturers. Alternatively careful attention to enclosure and extract ventilation at loading, mixing and moulding stages can greatly reduce the escape of hazardous substances. In particular the substitution of AC paste for powder is normally associated with a greatly reduced or

total absence of new allergic symptoms. In occupational
asthma, the protective efficacy of masks and ventilated
helmets has yet to be properly evaluated.

Biological Monitoring

     If occupational hygiene surveys indicate a risk of
exposure after all practicable protective and preventative
measures have been introduced, it may be appropriate to
monitor the health of the workforce for the specific risks
of their exposure.

     In the case of occupational asthma there are no hard
and fast rules for medical surveillance, nor should there
be since some asthmas come on more quickly than others and
have different clinical characteristics. With AC, the
author found[1] that 50% of cases occurred within 3 months of
exposure and 75% within the first year. Thus it would seem
useful to monitor at 3 months and 1 year and annually thereafter and to encourage the reporting of symptoms between
these times. The methods that the author has used are to
carry out preliminary screening with a specially designed
questionnaire and to follow this up in suspect cases using
serial peak expiratory flow rate (PEFR) records according to
the techniques developed by Burge[4]. These techniques
require workpeople to measure and record their own PEFR
every one or two hours for several weeks at work and away
from work. This is surprisingly easy to organise and
80-90% compliance can be obtained in most studies. The use
of pre and post shift spirometry to detect occupational
asthma is now discredited and should be discarded. In some
cases where serial PEFR results are negative or give
equivocal results in the face of a positive clinical history,
it may be necessary to have recourse to serial studies of
airways reactivity by histamine or methacholine challenge
tests or by bronchial provocation tests with suspect causal
agents. These latter studies should only be performed in
specialist centres with adequate resuscitation facilities.

     Where talc-filled grades of polymer constitute a
predominant part of the manufacturing output of an
industrial undertaking, it may be prudent to monitor the
health of the workforce for this risk. The standard MRC
respiratory questionnaire is ideal for this purpose and
should be used in combination with periodic clinical

examination and X-ray of the chest.

Policy and Management

There are certain occupational illnesses, among them
occupational asthma, which can be effectively concealed by
those who have developed them. This is particularly the case
if there is a risk of job loss if the disease is declared or
if the compensation available is inadequate. For these
reasons occupational disease generally, and occupational
asthma in particular, tend to be underdiagnosed. It is
essential therefore for industries and sovereign national
states to have policies which do not penalise those
unfortunate enough to develop occupational diseases since
early discovery of disease coupled with proper relocation is
highly effective in preventing any further illness.

References

1. Slovak AJM (1981). Occupational asthma caused by a
plastics blowing agent, azodicarbonamide. Thorax 36:906-909.
2. Ferris BG, Peter JM, Burgess WA, Cherry RV (1977).
Apparent effect of an azodicarbonamide on the lung: a
preliminary report. J. Occup. Med. 19:424-5.
3. Hunter D (1976). "The Diseases of Occupations".
London: Hodder and Stoughton, p. 987-8.
4. Burge PS (1982). Advances in the diagnosis of
occupational asthma. In Proceedings of a Society of
Occupational Medicine Research Panel Symposium. Slovak AJM
(ed.): "Occupational Asthma", London: S.O.M.

Industrial Hazards of Plastics and Synthetic Elastomers, pages 319–334
© 1984 Alan R. Liss, Inc., 150 Fifth Ave., New York, NY 10011

THE PRODUCTION AND USE OF SOME THERMOPLASTICS AND THEIR
CHEMICAL OCCUPATIONAL HAZARDS

Professor Bo Holmberg

National Board of Occupational Safety and Health
Unit of Occupational Toxicology, Research Dept.

S-17184 Solna, Sweden

The plastics can be classified into two main groups
(Miles, Briston 1979), thermosets and thermoplastics.
Thermosets are irreversibly hardening upon heat application and thermoplastics may be repeatedly softened by
heating and hardened by lowering temperature.

The thermosets are phenoplasts, aminoplasts, polyesters,
epoxy resins, and polyurethane. To the thermoplastics
belong polyolefins, polyvinyls, polystyrene and copolymers,
polyamides, acrylic polymers, fluorocarbon polymers,
cellulose esters and ethers, high-performance, and heat-
resistant thermoplastics (Brydson 1975) - such as silicones.

Acrylic Polymers.

The acrylic polymers (Horn 1960) can be divided into two
main groups, polyacrylates and polymethacrylates (Miles,
Briston 1979). Other acrylics are acrylic rubbers,
acrylic fibers and copolymers between styrene/methyl
methacrylate and ethylene/methyl acrylate.

All monomers for the two main acrylic polymer groups are
esters of either acrylic acid or methacrylic acid, most
often the methyl ester.

Methyl acrylate is a liquid (boiling point 80°C) and is
synthezised according to one of four methods. One synthesis involves a reaction between ethylene oxide and
hydrogen cyanide to give ethylene cyanohydrin. This
intermediary product reacts with methyl acrylate and
sulfuric acid to give methyl acrylate and ammonium
bisulphate. A second possibility, the Reppe process,
involves a carbonylation of acetylene to give ethylene
carbonyl which is treated with methyl alcohol to give
the end product. Lactic acid (may also be acetylated
and esterified via acetoxy methyl propionate) can also
be heated (475-550°C) to give methyl methacrylate
(Miles, Briston 1979).

Methyl methacrylate is a liquid boiling at 101°C. In
one process it may (Sittig 1975) be synthesized by a
reaction between acetone and a cyanide (HCN; NaCN)
giving a cyanohydrin, which gives methacrylamide
sulphate with warm sulphuric acid. Treatment with warm
methyl alcohol results in methyl methacrylate and
ammonium bisulphate.

The main polymerization processes for acrylates and
methacrylates are bulk, emulsion, suspension, polymerization (Sittig 1975; Miles, Briston 1979).

When the monomer is transported or handled, a small
amount (5-15 ppm) of an inhibitor (for instance hydroquinone) is added to prevent spontaneous polymerization.
The polymerization is catalysed by adding e g benzoyl
peroxide or azoisobutyronitrile to the monomer, from
which the inhibitor has been removed. A polymethyl
methacrylate sheet is molded by allowing the polymerization to proceed between glass plates at temperatures
between 40° and 75° for 12-40 hours. Process generated
heat has to be carefully controlled. Complete curing is
done at 90°C and annealed at 140°C.

Emulsion polymerization proceeds in water. Water soluble
catalysts are hydrogen peroxide, ammonium persulfate,
potassium persulfate, sodium metabisulfite and tert.-
butyl hydroperoxide. Nonionic surface active agents are
used as emulsifiers. The temperature rises to 70°C and
the polymerization takes about 15 mins. The acrylic
emulsions obtained are used for leather-coatings and in
the manufacture of water-based paints.

Polymethyl methacrylate can be injection molded, extruded, or compression molded. The final goods can be processed by conventional wood or metal working machinery provided the generated heat is regulated. This plastic can also be glued (by polymer solutions in chloroform or ethylene chloride) and be dissolved in solvents or emulsified/suspended in water for paints. Polymethyl methacrylate is used for glazing, signs, display materials, sanitary ware, telephones, instrument panels, protective goggles and shields. Acrylic polymers are used for contact lenses, artificial dentures, for textile and paper treatment and for acrylic paints.

Acrylic fibers are important man made fibers. The fibers consist of high molecular weight acrylonitrile polymers. The melted or dissolved viscous polymer is extruded through a die. For dyeing the acrylonitrile have to be copolymerized to improve colouring. Polyacrylonitriles are used among other things for making blankets, ropes, and "furs". Other polyacrylonitriles are copolymers with styrene and butadiene.

Acrylic acid has a high acute toxicity towards laboratory animals by i p injection (LD_{50} rats 24 mg/kg bw) and skin application (LD_{50} rabbits 245-950 mg/kg bw). The vapors cause nasal and eye irritation (Documentation of TLVs 1981) to rats and in absence of human data to guide, a TLV has been set at 10 ppm.

No other data on chronic effects or special chronic effects are available than from one study on teratogenicity and fetotoxicity (Singh et al 1972). Acrylic acid injected 3 times i p in rats in four doses up to 8 mg/kg bw caused skeletal abnormalities and embryotoxic effects at highest doses.

Methyl acrylate is a little more acutely toxic than its parent acid. Respiration difficulties, salivation, body weight decrease is described at moderate inhalation levels (IARC 1979).

Liquid methacrylic acid causes skin injuries in rabbits (Documentation of TLVs 1981) and vapors cause eye irritation in rats at around 1000 ppm. Slight kidney injury occurs in rats inhaling 300 ppm, 6h/d during 20 d.

No other information seem to be available; the TLV is set to 20 ppm.

For its methyl ester, methyl methacrylate (MMA), a little more toxicological information is available. Inhalation toxicity data on rats show a low acute toxicity (Tansy, Kendall 1979) as well as a reversible reduction of neuronal discharge rates of hypothalamus and hippocampus of rats (Innes, Tansy 1981). Damage to tracheal mucosa (Tansy et al 1980) was observed in rats. MMA is mainly exhaled (IARC 1979) as CO_2 in rats (88 % within 10 d). MMA is clastogenic to rats (IARC 1979) but two carcinogenicity studies do not permit an evaluation of its possible carcinogenicity.

Acrylamide (propenamide) is made from acrylonitrile and is the monomer for production of polyacrylamides. Polyacrylamides are effective flocculants for waste water treatment, but are also used as paper strengtheners, surface coatings, and textile treatment.

Monomeric acrylamide is known for its neurotoxic effects (NIOSH 1976) with symptoms such as tingling of fingers, of lower limbs, coldness, sweeting of hands, skin erythema, muscle weakness, dizziness, ataxia. The substance can penetrate the skin. Several studies have been published showing a spectrum of symptoms indicating injuries on the peripheral nervous system and the CNS. Symptoms are consistent with experimental data which show similar type of effects and also show myelin and axon degeneration in rats, monkeys and cats fed with acrylamide (NIOSH 1976) as well as changes in EEG of cats (Kuperman 1958).

The literature contains many theories on the mechanism of action of acrylamide suggesting local effects on axon metabolism (Cavanagh 1964), intracellular transport of substances (Pleasure et al 1969), or simply a selective toxicity on nerve cells (Spencer, Schaumburg 1974).

The urinary excretion of radioactivity after a single i v dose of acrylamide in the rat was 40% within the 1d and 65 % within 4 d (Hashimoto, Aldridge 1970). Meta-

bolites were still present in tissues after 14 d. This
suggests an accumulation. The distribution of acrylamide
between soft tissues, including nervous tissue, showed
high values in blood, but with radioactivity bound to
proteins, e g in brain two weeks after exposure.

Acrylonitrile (AN; vinyl cyanide) has a high acute
toxicity for animals (IARC 1979; NIOSH 1980) with LC_{50}
between 140 and 455 ppm by inhalation for mice, rats,
and guinea pigs. It damages liver, kidney, CNS, and
causes haemorrhage. Human exposures seem not to have
such dramatic consequences. However, irritation, skin
erythema and symptoms of effects on CNS, respiratory
fatigue, headache, irritability occur by inhalation
and/or skin exposure to AN (US Dept Labor 1978).

AN is metabolized to cyanide, which is then transformed
to thiocyanate and excreted in the urine in a greater
proportion (>10 %) after oral administration than when
administrated by other routes (Gut et al 1975). AN
decreases GSH in liver, lung, kidney, and adrenals
(Szabo et al 1977; Vainio, Mäkinen 1977). The toxic
effects of AN can be counteracted by cysteine (Hashimoto,
Kanai 1965). MFO inducers can also prevent AN induced
effects on serum and liver enzymes (Duverger-van Bogaert
et al 1978).

AN is mutagenic to Salmonella and Escherichia (IARC
1979) after metabolic activation. Urine samples from i
p injected cats were directly mutagenic to Salmonella
(Lambotte-Vandepaer et al 1980).

In a 12 m p o study on rats with up to 300 mg AN/l,
stomach papillomas, Zymbal gland, and CNS tumors appeared
and in parallel inhalation experiments tumors of the
CNS and the Zymbal gland occurred (US Dept Labor 1978).
In another study on rats, mammary gland tumors and
forestomach tumors were observed (Maltoni et al 1977)
in p o administered animals and also CNS tumors in the
inhalation series.

Alarmed by preliminary information on elevated cancer
risks for lung and large intestine among AN exposed

workers (US Dept Labor 1978), US (O'Berg 1980), German
(Thiess et al 1980), Japanese (Sakurai et al 1978), and
British (Werner, Carter 1981) studies have thrown some
light on the possible elevated cancer risk among AN
workmen. Of these studies, the Japanese was not planned
to reveal carcinogenic effects; it did also in fact not
detect any thing else than possibly slight liver damage
(Sakurai et al 1978). Thiess et al (1980) found a tendency to an elevation of the mortality risk in lung
cancer among 1345 workers, coinciding with similar
findings in a US company (O'Berg 1980). Lung cancer and
possibly stomach cancer mortalities were elevated in
the British study (Werner, Carter 1981). All studies
together point to the same direction and warrants, together with animal data, AN to be regarded as a carcinogen.

Fluorocarbon Polymer

The fluorocarbon polymer family (Rudner, 1958) consists
of two major members, namely polytetrafluoroethylene
(PTFE, teflon), and polymonochlorotrifluoroethylene
(PCTFE, fluorothene). Some smaller members of copolymers
with e. g. vinylidene fluoride, acrylate, and silicones
(Miles, Briston 1979) also belong to this group.

The commercially most important of the two types, PTFE
is produced from tetrafluoroethylene, C_2F_4, which is
obtained by pyrolysis of chlorodifluoromethane at
$800^\circ C$. C_2F_4 is a gas and polymerizes after addition of
initiators, such as sodium oxydisulfate (Sittig 1975).
The polymerization takes place under pressure in presence
of excess water.

Although being a thermoplast, PTFE cannot be extruded
or formed by injection molding. Its high melt viscosity
makes it necessary to sinter the PTFE particles by
pressure and heat in a molding process. PTFE is resistant
to acids, alkalis, and solvents. It has outstanding
mechanical, electrical and chemical characteristics.

The monomer for PCTFE production is produced from hexa-
chloroethane. The processing is similar to the processing
of PTFE. PCTFE is used for package film, pharmaceutical
tablets and capsules, valves and aerospace hardware.

The PTFE monomer, tetrafluoroethylene, has a low toxicity by inhalation to rats and mice (IARC 1979).

The dominating effect in terms of industrial toxicology of PTFE production and handling is the polymer fume fever obtained by exposure to thermodegradation products of solid or liquid (Wegman, Peters 1974; Åkesson et al 1980) fluorocarbon polymer contaminating tobacco. It is characterized by influenza-like symptoms (Documentation of TLVs 1981), fever, chills, respiratory discomfort. The component(s) responsible for the toxicity in the fume has been under dispute. Some workers argue that particulate products (Birnbaum et al 1968), others that gaseous compounds, like carbonyl fluoride (Coleman et al 1968) or other compounds formed between 300 and 500^{0}C, are responsible. It has also been forwarded that pyrolysis products injure leucocytes (cited in Anon. 1972), which release endogenous pyrogens.

Clorotrifluoroethylene (CTFE) is the monomer for PCTFE. Hexafluoropropene (HFP) can be formed by PCTFE pyrolysis (Potter et al 1981). Both CTFE and HFP show a dose-dependent nefrotoxicity by inhalation in rats. CTFE is the most potent nefrotoxin showing signs of necrosis in proximal tubuli 24 h after a 4 h 220 ppm exposure (Potter et al 1981).

Polyamides

Polyamides are also called nylon polymers (Kohan, 1973) and are named after the number of carbon atoms in the parent amino acid (Miles, Briston 1979), as for instance Nylon 6 from ω-amino caproic acid, Nylon 11 from 11-amino-undecanoic acid and Nylon 12 from dodecyl lactam. They were originally developed as fiber polymers but have since also gained importance as extruded and molded materials.

Nylon 6,6 is obtained by the reaction of adipic acid and hexamethylene diamine. The raw materials are dissolved in boiling methanol (Miles, Briston 1979), from which the "nylon salt" is precipitated. After reprecipitation from water, the "nylon salt" is heated to 280^{0}C under nitrogen and the reaction is stopped at any desired time by addition of a monofunctional acid or amine. The water is removed and the hot nylon can be spinned or transformed to pellets for later molding or extrusion.

Nylon 6 is made by condensation of ξ-caprolactam (2-ketohexamethyleneimine) in water under nitrogen and at 250°C during 12 hours. The remaining low molecular polymer material is forced out by high pressure nitrogen and the high molecular polymer is cooled and broken up to chips. The synthesis can be done in a bulk or a continuous process.

Nylon 11 is the polymerization product of 11-amino-undecanoic acid. The monomer is simply heated to 200°C with addition of phosphoric and hypophosphoric acid (Kohan 1973). The molten polymer can be spun into fibers or extruded.

Nylon 12 is made by a similar process as Nylon 11 although the starting material is dodecyl lactam. The polymerization process is slower and acid catalysts as well as $300-350^\circ$C are necessary conditions.

The nylon family members can be copolymerized among themselves giving copolymers with lower softening points than the homopolymers. For some polymers added softeners (n-ethyl sulfonamide, butyl benzene sulfonamide, o-p-toluene sulfonamide, or urea) further lowers the softening points.

Nylon films are used for packaging of fats, greases, oils and foods. Fibers are used for textiles, ropes, cords, safety belts, racket filaments, fishing lines, tooth brushes and many other common day goods. Poly-amides are also used for production of adhesives for can linings, paper treatment, shoe cements, and painting inks. Molded nylons are used for gears, meter wheels, conveyor chains and similar technological uses. Glass reinforced nylon can be molded into valves, filters, and junction boxes in the electrical industry to mention only a few examples of their use. Casted nylon is used for self lubricating cog-wheels, gears, slide and rota-tional bearings.

The monomer for Nylon 6 production, caprolactam, has a low acute toxicity for laboratory animals (IARC 1979). The toxic picture dominates by CNS effects: convulsions,

mydriasis, and temperature depression. It is metabolized
completely in rabbits and partly excreted in the urine
as the amino acid in the rat (Goldblatt et al 1954).
Also workers exposed to nylon 6 monomer show symptoms of
CNS influence: from irritability and confusion (Hohensee
1951) to disorders in the nervous system, genito-
urinary tract etc (Pekov 1975 in IARC 1979).

Cellulose Polymers

Being a natural substance, cellulose has found use early
in the history of the polymer technology (Yarsley et al
1964). In fact, cellulose nitrate was the first polymer
synthesized. Cellulose nitrate (celluloid) is prepared
from fibrous cellulose, formerly cotton linters, by
treatment with nitric acid in sulfuric acid and water
for 30-60 mins. at 35-40^{0}C. A small batch size is
needed to minimize risk of explosions. The cellulose
nitrate is washed and dried. The celluloid can be extruded at 85^{0}C to rods, tubes or shaped profiles or can
be casted by a continuous process to films. The solvent
used is a mixture of ether and alcohol. Celluloid
sheets can be molded after heating to 110^{0}C. Celluloids
are used for making photographic film, combs, knife
handles, spectacles, tooth brushes, frames, table tennis
balls, etc. Cellulose nitrate was formerly used for
production of rayon (chardonnet silk)(Yarsley et al
1964).

Cellulose acetate is obtained from wood pulp by treatment with acetic anhydride in glacial acetic acid in
diluted sulfuric acid and methylene chloride as a solvent at 50^{0}C (Sittig 1975). Cellulose acetate can be
extruded with added plastisizers to sheets and films, or
by casting in an organic solvent. It can also be injection molded, blow molded (for bottles), or made into
fibers (acetate silk, acetate rayon) after dissolving in
acetone and forced through spinnerets.

Rayon became in 1924 (Miles, Briston 1979) the established term for artificial silk made by cellulose
esters. The two main types are viscose (xanthate) and
saponified acetate (Yarsley et al, 1964). The first step
is the production of alkali-cellulose by treating pure
sulphite-pulp with 17-21 % NaOH at room temperature

for 20-60 mins. The material is aged for a couple of
days with atmospheric oxygen. The aging gives shorter
cellulose chains. 30-40 % by weight of CS_2 is then
sprayed to the alkali-cellulose in rotating xanthating
churns. The process must be cooled. The resulting sodium
cellulose xanthate is stirred with alkali during 2-6 h
to obtain the viscose. The viscose has to be filtered
and ripened, after which time it can be pressed through
spinnerets into coagulation baths with 10-15 % sulphuric
acid and sodium sulfate at 40-55°C. Further treatment,
such as washing, drying, softening, and dyeing is necessary to get the wanted raw material for textile production.

A recognized occupational chemical hazard in rayon
production is exposure to carbon disulphide (WHO 1979).
For every kg of viscose 20-30 g carbon disulphide and
4-6 g hydrogen sulphide will be emitted. About 0,6-1,0
kg/h of viscose is used in a textile yarn factory. Even
more carbon disulphide may be emitted in production of
cellophane, where viscose usage is 1800-2000 kg/h.

The toxicity of carbon disulphide has been reviewed (WHO
1979). Retention of inhaled carbon disulphide is 40-50 %
at equilibrium 1-2 h after start of exposure. 10-30 %
of absorbed substance is exhaled and 70-90 % is
excreted as metabolites in urine. Carbon disulphide
poisoning is characterized by a variety of symptoms,
where effects on the central and peripheral nervous
system as well as cardiovascular effects are predomina-
ting in chronic industrial exposures. At about 3 ppm
(10 mg/m^3) sensory polyneuritis may be detected in long
term exposed workers and at about 10 ppm the coronary
mortality is 5 times higher than expected. The mechanism
underlying the carbon disulphide effects is under debate.
A disturbance of the metabolism of metals, vitamins,
lipids or of metabolic enzymes themselves has been
proposed. When considering the carbon disulphide associated elevated risk for deaths in myocardial infarctions
it may be of interest to note that heart muscle degeneration has been observed in rats (Ekvärn et al 1977)
exposed to disulfiram, which is metabolized to carbon
disulphide.

High-Performance Thermoplastics

To this group of polymers the following materials
belong (Miles, Briston 1979), namely the chlorinated

polyethers, polyacetals, polycarbonates, polysulfones
and the polyphenylene oxides. Of these, the polyacetates
and polycarbonates are the most important.

The polyacetates are made up of formaldehyde (e.g.
polyformaldehyde) treated with an initiator (organic
acids, amines, alkyl phosphines, arsines, stibines or
metal carbonyls) with an organic solvent (such as
hexane) as a medium. Formaldehyde copolymerizing with a
cyclic ether, such as ethylene oxide, may give copolymers
with unique properties as bridges between plastics and
metals. They may even be substitutes for metals in the
plumbing (ball-cocks, valves, pumps) and car industries.

Polycarbonates may be produced in a process (Miles,
Briston 1979) from bisfenol A, at 40^0C during 1 to 3
hours in a solvent (chlorobenzene, methylene chloride,
o-dichlorobenzene, pyridine). Catalysts used are tertiary
or quaternary amines or aluminium chloride, stannic
chloride, titanium tetrachloride, or benzyl triethyl
ammonium chloride (Sittig 1976).

The polycarbonates can be injection molded or extruded
at temperatures between 240-300^{0}C. These thermoplasts
are glass transparent and are used for household wares
(e.g. measuring cups, bowls), laboratory animal cages,
car lights, etc.

If one polymerizes bisfenol A instead with a sulfone,
for instance 4,4'-dichlorodiphenyl sulfone, in dimethyl
sulfoxide in alkali at 150^{0}C (Sittig 1976) it gives a
polysulfone. The polysulfones are processed by molding
(95^{0}C) or extrusion (300-360^{0}C). Electrical applications
(circuit carriers, terminal blocks), the car and aircraft industries, as well as household consumption goods
(camera parts, coffee makers) are technological areas
for polysulfones.

Heat-Resistant Thermoplastics

Silicones (Ranney 1977) are semiorganic polymers of
alternating silicon and oxygen atoms (siloxane lin-
kages). The polysiloxanes can be linear, branched or
cyclic in structure. The silicones are prepared from

substituted chlorosilanes, $R_n SiCl_{(4-n)}$. The methyl chlorosilane is produced by a direct process from powdered silicon and methyl chloride at $250-280^oC$ (with Cu as catalyst). Other substituted chlorosilanes can be made by a variety of processes (Miles, Briston 1979). The polymerized product may be a fluid, resin or elastomer depending on the chemical structure.

Silicone fluids are used in the paint industry, as lubricants and as water repellants for leather, textile and paper. Silicone resins have an important use for high temperature insulation in the electric industry, for water proof treatment and as laminates (often with reinforcers like asbestos, glass fiber or mica) in the electric and aircraft industries. Elastomers have found use as cable insulation, gaskets, and for medical tubes, valves, stoppers, etc.

Polysiloxanes used as surface coatings and in cosmetics have been investigated particularly for their possible skin penetrating properties and their abilities to affect the reproductive system and its function.

References

Anonymous (1972). Polymer-fume fever. Lancet ii:27-28

Birnbaum HA, Scheel LD, Coleman WE (1968). Toxicology of the pyrolysis products of polychlorotrifluoroethylene. Am Ind Hyg Assoc J 29:61-65

Brydson JA (1975). Plastic materials. London: Newness-Butterworths

Cavanagh JB (1964). The significance of the "dying-back" process in experimental and human neurological disease. Int Rev Exp Pathol 3:219-267

Coleman WE, Scheel LD, Kupel RE, Larkin RL (1968). Identification of toxic compounds in the pyrolysis products of poly(tetrafluoroethylene). Am Ind Hyg Assoc J 29:33-40

Documentation of TLVs (1981). Supplemental documentation. Cincinnati: American Conference of Governmental Industrial Hygienists, Inc

Duverger-van Bogaert M, Lambotte-Vandepaer M, Noel G, Roberfroid M, Mercier M (1978). Biochemical effects of acrylonitrile on the rat liver, as influenced by various pretreatments of the animals. Biochem Biophys Res Commun 83:1117-1124

Ekvärn S, Jönsson M, Lindqvist NG, Holmberg B, Kronevi T (1977). Disulfiram-induced myocardial and skeletal-muscle degeneration in rats. Lancet ii:770-771

Goldblatt MW, Farquharson ME, Bennett G, Askew BM (1954). ε-Caprolactam. Br J Ind Med 11:1-10

Gut I, Nerudova J; Kopecky J, Holecek V (1975). Acrylonitrile biotransformation in rats, mice, and Chinese hamsters as influenced by the route of administration and by phenobarbital, SKF 525-A, cysteine, dimercaprol, or thiosulphate. Arch Toxicol 33:151-161

Hashimoto K, Kanai R (1965). Studies on the toxicology of acrylonitrile. Metabolism mode of action and therapy. Ind Health 3:30-46

Hashimoto K, Aldridge WN (1970). Biochemical studies on acrylamide, a neurotoxic agent. Biochem Pharmacol 19:2591-2604

Hohensee F (1951). Uber die pharmakologische und physiologische Wirkung des ε-caprolactams. Faserforsch Textiltech 8:299-303

Horn MB (1960). Acrylic resins. New York: Reinhold Publ Corp.

IARC (1979). Some monomers, plastics and synthetic elastomers, and acrolein. Monographs on the evaluation of the carcinogenic risk of chemicals to humans. Vol 19 Lyon: International Agency for Research on Cancer.

Innes DL, Tansy MF (1981). Central nervous system effects of methyl methacrylate vapor. Neurotoxicol 2:515-522

Kohan MI (Ed.)(1973). Nylon Plastics. New York: J Wiley & Sons.

Kuperman AS (1958). Effects of acrylamide on the central nervous system of the cat. J Pharm 123:180-192

Lambotte-Vandepaer M, Duverger-van Bogaert M, de Meester C, Poncelet F, Mercier M (1980). Mutagenicity of urine from rats and mice treated with acrylonitrile. Toxicology 16:67-71

Maltoni C, Ciliberti A, Di Maio V (1977). Carcinogenicity bioassays on rats of acrylonitrile administered by inhalation and ingestion. Med Lav 68:401-411

Miles DC, Briston JH (1979). Polymer Technology. New York: Chemical Publ Corp Inc.

NIOSH (1976). Criteria for a recommended standard... Occupational exposure to acrylamide. Washington: NIOSH, Dept of Health, Education and Welfare. Publ No 77-112

NIOSH (1980). A recommended standard for occupational exposure to acrylonitrile. Washington: Dept of Health, Education and Welfare. Publ No 78-116

O'Berg MT (1980). Epidemiologic study of workers exposed to acrylonitrile. JOM 22:245-252

Pleasure DE, Mishler KC, Engel WK (1969). Axonal transport of proteins in experimental neuropathies. Science 166:524-525

Potter CL, Gandolfi AJ, Nagle R, Clayton JW (1981). Effects of inhaled chlorotrifluoroethylene and hexafluoropropene on the rat kidney. Toxicol Appl Pharmacol 59:431-440

Ranney MW (1977). Silicones. Vol 1. Rubber, Electrical Molding Resins and Functional Fluids. New Jersey: Noyes Data Corp.

Rudner MA (1958). Fluorocarbons. New York: Reinhold Publ Corp.

Sakurai H, Onodera M, Utsunomiya T, Minakuchi H, Iwai H, Matsumura H (1978). Health effects of acrylonitrile in acrylic fibre factories. Brit J Ind Med 35:219-222

Singh AR, Lawrence WH, Autian J (1972). Embryonic-fetal toxicity and teratogenic effects of a group of metha-crylate esters in rats. J Dent Res 51:1632-1638

Sittig M (1975). Pollution Control in the Plastics and Rubber Industry. Park Ridge: Noyes Data Corp

Spencer PS, Schaumburg HH (1974). A review of acrylamide neurotoxicity. - Part I. Properties, uses and human ex-posure. Can J Neurol Sci 1:143-150

Szabo S, Bailey KA, Boor PJ, Jaeger RJ (1977). Acrylo-nitrile and tissue glutathione: Differential effect of acute and chronic interactions. Biochem Biophys Res Commun 79:32-37

Tansy MF, Kendall FM (1979). Update on the toxicity of inhaled methyl methacrylate vapor. Drug Chem Toxicol 2:315-330

Tansy MF, Hohenleitner FJ, White DK, Oberly R, Landin WE, Kendall FM (1980). Chronic biological effects of methyl methacrylate vapor. III Histopathology, blood chemistries, and hepatic and ciliary function in the rat. Env Res 21:117-125

Thiess AM, Frentzel-Beyme R, Link R, Wild H (1980). Mor-talitätsstudie bei Chemiefacharbeitern verschiedener Produktionsbetriebe mit Exposition auch gegenüber Acryl-nitril. Zbl Arbeitsmed 30:259-267

US Department of Labor (1978). Occupational exposure to acrylonitrile (vinyl cyanide). Proposed standard and notice of hearing. Fed Reg 43:2586-2621

Vainio H, Mäkinen A (1977). Styrene and acrylonitrile induced depression of hepatic nonprotein sulphydryl content in various rodent species. Res Commun Chem Pathol Pharmacol 17:115-124

Wegman DH, Peters JM (1974). Polymer fume fever and cigarette smoking. Ann Internal Med 81:55-57

Werner JB, Carter JT (1981). Mortality of United Kingdom acrylonitrile polymerization workers. Brit J Ind Med 38: 247-253

WHO (1979). Carbon disulphide. Env. Health Criteria 10. Geneva: World Health Organization.

Yarsley VE, Flavell W, Adamson PS, Perkins NG (1964). Cellulosic Plastics. London: Iliffe Books Ltd

Åkesson B, Högstedt B, Skerfving S (1980). Fever induced by fluorine-containing lubricant on stainless steel tubes. Brit J Ind Med:307-309

CHAPTER V. OCCUPATIONAL HAZARDS OF POLYURETHANE

Industrial Hazards of Plastics and Synthetic Elastomers, pages 337–346
© **1984 Alan R. Liss, Inc., 150 Fifth Ave., New York, NY 10011**

PRODUCTION AND USES OF POLYURETHANES
(INCL. THERMODEGRADATION)

Christina Rosenberg
Institute of Occupational Health
Haartmaninkatu 1
SF-00290 Helsinki 29, Finland

INTRODUCTION

The first polyurethane product was synthesized in
Germany in 1937 when Dr. Otto Bayer (IG Farben Industrien)
made elastomeric products by reacting toluene diisocyanate
with various polyfunctional alcohols (polyols). In 1945,
further research led to the discovery of both flexible and
rigid polyurethane foam. At the same time, similar research was conducted by ICI in UK and by DuPont in USA in
the field of elastomers, adhesives, and coatings (Buist,
1978). In 1981, polyurethane foams constituted about 5 %
of the total sales of plastic in the USA. The consumption
of flexible and rigid polyurethane foam totalled about
0.79 million tons in 1981; two thirds of this was flexible
foam. The principal applications of flexible foam were as
cushioning or padding material in furniture and in vehicles, whereas rigid foam was mainly used in insulation
and construction (Mod Plast Int 1982). In 1981 the consumption of polyurethanes, including, foams, elastomers,
adhesives, and paints, was approx. 0.98 million tons in
Western Europe and 0.28 million tons in Japan.

POLYURETHANE PRODUCTION AND USES

Polyurethanes are generally considered to cover all
products of reaction between isocyanates and polyhydroxy
compounds. Isocyanates react with compounds which have
active hydrogen atoms such as alcohols, amines, water,
carboxylic acids, and phenols thus forming, e.g., urethane

or urea groups (Saunders and Frisch 1965). For the formation of macromolecules, both components should be at least bifunctional. Three dimensional networks can be made by use of polyisocyanates or triols or by using diisocyanate in excess (allophanate formation) (Buist 1978). Isocyanates are generally produced for commercial use by phosgenation of the corresponding amine. The most common isocyanates used in industry are (Brydson 1979): toluene diisocyanate (TDI), 4,4'-methylenediphenyl diisocyanate (MDI), hexamethylene diisocyanate (HDI) and its derivatives, 1,5-naphtalene diisocyanate (NDI), and polymethylene polyphenyl isocyanate (PAPI).

Flexible and Rigid Foam

Flexible polyurethane foam production generally utilizes polyols, isocyanates, catalysts, blowing agents, and surfactants (Buist 1978). The polyols are polyhydroxy compounds, polyethers or polyesters, which react with isocyanates to yield polyurethanes. The first flexible foams were based on the reaction between polyesters and TDI 65 (65:35 ratio of the 2,4- and 2,6-isomers). The polyesters are usually produced from adipic acid and a glycol (polydiethylene glycol adipates). Polyester foams possess high load-bearing properties and they are widely used in clothing, textile, lamination and footwear applications. The production of polyester foam represents less than 10 % of all flexible foam production. More resilient foams are obtained with polyether polyols. These are widely used as upholstery foams in, for example, the furniture and automobile industries. The most common polyethers are produced through polymerization of propylene and ethyleneoxides with trimethylol propane, 1,2,6-hexane triol, or glycerol as initiators.

The most important isocyanate used in flexible foam production is toluene diisocyanate, nowadays mostly the 80:20 mixture of the 2,4- and 2,6-isomers. Other isocyanates used are modified TDI, modified MDI, and mixtures of MDI and TDI. Tertiary amines are widely used as catalysts in both polyester and polyether foam production; these amines include N,N-dimethylethanolamine, triethylamine, triethylenediamine, and alkylmorpholines. Various organotin compounds (stannous octoate, dibutyl tin dilaurate) have also been used as catalysts, usually in combination with amines.

Water serves as blowing agent because of the formation of carbon dioxide gas in the reaction with the isocyanate group. The urea bridges thus obtained form the active points for cross-linking. The use of chlorofluoromethanes known as physical blowing agents, is based on the fact that the total gas present increases whereas the degree of cross linking remains unchanged. The concern about the risk the chlorofluorocarbons constitute for the earth's stratospheric ozone layer has led to the development of alternative blowing agents; one of which is dichloromethane. Surface active agents modify the reaction profile and the properties of the polymer. The agents are usually copolymers of polysiloxanes and polyalkylen oxide. Additional compounds used include anti-aging additives, fillers, colorants, and cell regulators.

The two chief methods of manufacture of foam are known as the one-stage process and the two-stage process. In the former, the isocyanate and the polyol compound react in the presence of the activation system. In the latter, a prepolymer, composed of the isocyanate and the polyol, is allowed to react with water, the catalysts and the other ingredients.

Flexible foams have polymer structure with a low degree of crosslinking. Polyols with high functionality yield tougher products, i.e. rigid foam. The rigid foams display a very low thermal conductivity and hence the major interest of rigid foams has been in the field of thermal insulation (Buist 1978). Typical applications are insulation of refrigerators, deep freezers, hot water storage cylinders, and pipelines in urban heating systems. Further applications are, e.g., in the manufacture of prefabricated houses and in the flotation material of small boats or barges. Rigid polyurethane foams are usually manufactured in an one-stage process. The isocyanate used is generally a polymeric form of MDI. The polyol is usually a polyether based on a tri, tetra, or hexafunctional compound such as glycerol, pentaerythritol or sorbitol (Brydson 1979). As with flexible foam, a variety of catalysts, surfactants and blowing agents are used in the manufacture of rigid foam. To improve the fire resistance of the foam, flame retardants are used. They are halogen and phosphorus-based compounds such as tri-

chloroethyl phosphate or oxypropylated phosphoric acid. In
addition fire retarding properties can be improved with
specific structures like isocyanurate rings or carbodi-
imide groups and by glass reinforcement.

Elastomers

Polyurethane elastomers have in general very good
mechanical properties such as high tensile strength and
tear and abrasion resistance. They also show good resistance to ozone and oxygen and to aliphatic hydrocarbons.
Polyurethane elastomers find use in machine construction
and in the automobile and shoe industries (oil seals, shoe
soles and heels, tires). Nearly all polyurethane elastomers are block copolymers which consist of alternating
blocks of flexible chains and highly polar relatively
rigid segments (Buist 1978). The flexible chains or soft
blocks usually comprise aliphatic polyesters or polyethers
such as poly(tetramethylene adipate) and poly(oxytetra-
methylene)glycol. The hard blocks are formed in the reaction of a diisocyanate, e.g., NDI, MDI or TDI, with a low
molecular-weight glycol or a diamine such as butane 1,4-
diol and 3,3-dichloro-4,4-diaminodiphenylmethane (MOCA).
The rubbery nature of the elastomers is due to the micro-
phase separation of the two blocks. Polyurethane elastomers can also be modified to give them thermoplastic
properties: these elastomers can be extruded, injection
molded, calandered, or compression molded like the typical
thermoplastics.

Surface Coatings

Polyurethane paints or lacquers vary considerably in
hardness and flexibility. Because of properties like
toughness, abrasion, chemical, and weather resistance,
they find use as coatings of metal surfaces in automobiles
and aircraft, and as finishing in boats and sports equipment and rubber goods. Many types of products are available. They can be, e.g., blocked isocyanate coatings or
one or two-pack coatings (Brydson 1979). In the blocked
system the isocyanate has been allowed to react, e.g.,
with a phenol. Upon heating of the blocked isocyanate, the
original isocyanate is liberated and reacts with the
polyol present in the formulation. In the one-pack system,

an adduct, composed of excess polyisocyanate and polyol, is dissolved in waterfree solvent. When applied to a sur- face, the film hardens as moisture in the air reacts with the free isocyanate groups. In the two-pack system, the isocyanate and the polyol are separate components, and curing occurs when these are mixed. A widely used polyiso- cyanate is the biuret adduct of HDI.

Miscellaneous Applications

There are numerous other applications of poly- urethanes, e.g., the manufacture of polyurethane fibers or crystalline molding compounds. Isocyanates, like MDI, are used as constituents in a binding system for making cores in foundries or as binders in the manufacture of particle board. Polyurethanes are also used as printing inks for plastics, as adhesives or in decoration as wood imitation (Brydson 1979, Buist 1978).

THERMODEGRADATION

Owing to the wide use of polyurethanes, there is great concern about the combustion products formed when finished products are destroyed by fire. Thermal degra- dation may also occur in certain industrial processes with or without direct application of open flame; these pro- cesses include hot wire cutting, textile lamination, or soldering of polyurethane-varnished wire in electronic assembly work (Hardy and Devine 1979). The generation of heat, smoke, and toxic gases during combustion has been extensively studied for some years. Woolley and Fardell (1977) reported in their work on flexible polyester and polyether foams that initially the decomposition appeared to proceed via a low temperature (200-300 °C) elimination of nitrogenous material. The particles in the smoke were propably a polymeric form of TDI. The smoke decomposed at high temperature (800 °C) to produce carbon monoxide and a range of nitrogen containing products (nitriles) es- pecially hydrogen cyanide. In an other study, the pyrol- ysis of flexible foam was investigated in an inert atmos- phere at temperatures ranging from 300 °C to 1000 °C (Hileman et al. 1975). The compounds detected included low-molecular weight alkanes and alkenes, aldehydes, ke- tones, nitriles, and TDI. The formation and release of

isocyanate vapours during thermodegradation of polyurethane varnishes was reported by Seemann and Wölcke
(1976). The mechanism of degradation of polyurethane foams
has been studied both with flexible and rigid foams (Foti
et al. 1981, Gaboriaud and Vantelon 1981, Grassie and
Zulfiqar 1978) as well as with model compounds (Chambers
et al. 1981). These studies confirm the theory that the
principal mechanism of decomposition of polyurethanes is
the reverse of the reaction in which they are formed,
there is thus generation of free isocyanate and polyol.
This appears to be followed by a polymerisation of the
isocyanates to yield polyureas in the case of flexible
foams and polycarbodiimides in the case of rigid foams.

AIR-BORNE POLLUTANTS IN MDI-BASED POLYURETHANE WORK

Industrial hygiene measurements were performed by our
laboratory both during manufacture and processing of MDI-
based polyurethanes. The main purpose of the study was to
survey the concentrations of isocyanates during polyurethane work. Possible degradation products as well as
other harmfull compounds such as amines used as catalysts
were measured.

Air sampling was carried out in two polyurethane production plants and in two foundries; the latter used MDI-
based resin in the molding sand. Samples were also collected during installation of polyurethane-isolated pipes
used in an urban heating system. Sampling was conducted in
the breathing zone of the workers both with stationary and
personal devices. The sampling time varied from 10 to 60
minutes. A joint mean of the results of personal and
stationary sampling was calculated for each sampling area.
In the first production plant, rigid foam was produced for
the insulation of refrigerators and deep freezers and in
the second plant for the insulation of hot water cylinders. In both plants the foam was MDI-based and produced
by a mixing-gun technique, i.e., the isocyanate and the
polyol components were mixed in the mixing head just before spraying. The MDI concentrations measured in both
plants were all very low and ranged from < 0.0005
mg/m^3 to 0.003 mg/m^3 (25 samples). In the second
plant, the concentrations of trietylamine and dimethylcyclohexylamine, used as catalysts, were also measured.
The concentrations were 0.03 mg/m^3 and 0.3 mg/m^3 re-

spectively. The current threshold limit value is according to ACGIH[1] 0.2 mg/m^3 for MDI and 100 mg/m^3 for tri-ethylamine. The measured air concentrations in both production plants were well below these values. No threshold limit value has been given for dimethylcyclohexylamine.

In the foundries, a phenol-formaldehyde-MDI-based resin was used as a binder in the molding sand. The air concentrations of the following substances were measured; MDI, phenol, aldehydes, benzene; in addition triethylamine was measured in the second foundry. The results are summarized in Table 1. The MDI concentrations in the first foundry ranged from 0.004 to 0.015 mg/m^3 while in the second foundry no air-borne MDI was detected. The phenol and formaldehyde concentrations were of the same order of magnitude in the two foundries the highest value for phenol being about 4 % and for formaldehyde about 10 % of the respective threshold limit value (see Table 2). No benzene was detected, and only trace amounts of acetaldehyde and triethylamine were found.

The insulated pipes used in urban heating are based on a mechanically strong sandwich of polyethylene, rigid MDI-based polyurethane and glass-reinforced polyester. For joining the culverts, shrinking sleeves of crosslinked polyethylene are used. During installation work, which includes welding, the polyurethane foam may be exposed to heat and may even occasionally burn. Air samples were collected during welding operations. In addition to MDI, the concentrations of aldehydes and organic acids were measured. The results are presented in Table 2. The MDI concentrations were relatively high, approaching the recommended threshold limit value. The samples were, however, collected close to the welding point and thus they do not, in this case, represent true breathing zone samples. The results nevertheless indicate that considerable MDI exposure is possible during this type of work. Only trace concentrations of the other pollutants were detected.

[1] American Conference of Governmental Industrial Hygienists

Table 1. Air-borne contaminants in two foundries using a polyurethane binder.

Location	No. of samples n	Concentration + SD[a] (mg/m^3)					
		MDI	Phenol	Formalde-hyde	Acetalde-hyde	Benzene	Triethyl-amine
1. Foundry[b]							
Sand making	4	0.007+0.002	0.23 (n=1)	0.05 (n=1)	-[c]	-	-
Core making	4	0.004+0.003	0.6 +0.4	0.3 +0.3	-	-	-
Casting	4	0.015+0.015	0.75+0.4	0.05+0.05	-	nd[d] <0.1	-
2. Foundry							
Core making	3	nd[d]<0.001	0.2 +0.01	0.1 +0.1	0.06+0.02	-	0.6 +0.4
Casting	2	nd[d]<0.001	0.06+0.05	0.04+0.02	0.07+0	-	0.03+0
After casting	2	nd[d]<0.001	0.1 +0.1	0.04+0.04	0.1 (1x<0.02)	-	0.06+0.01
Threshold limit value[e]		0.2	19	3	180	30	100

a) SD = standard deviation
b) MDI results from Rosenberg and Pfäffli 1982.
c) - = not measured
d) nd = not detected
e) According to American Conference of Governmental Industrial Hygienists (1982).

Table 2.   Air-borne contaminants during installation of urban heating

Compound	No. of samples n	Concentration $\pm SD^{a)}$ mg/m^3	Threshold limit value[b] mg/m^3
MDI	5	0.1+0.08 (1x<0.0005)[c]	0.2
Formaldehyde	3	0.1+0	3
Total aldehydes (calc. as form- aldehyde)	3	0.7+0.3	
Formic acid	3	0.2+0 (1x<0.01)[c]	9
Acetic acid	3	0.25 (2x<0.01)[c]	25

a) SD = Standard deviation
b) According to American Conference of Governmental Industrial Hygienists (1982)
c) not included in the mean

The results of this small scale survey show that the concentrations of air-borne contaminants, the isocyanate MDI in particular, are generally low during the manufacture and processing of MDI-based polyurethane. Isocyanates are, however, powerful respiratory irritants and sensitizers and can even at low concentrations, cause permanent impairment of pulmonary functions (IARC 1979). Therefore all means available, including process technology, ventilation, and respiratory protection, should be used to minimize exposures and the associated risk to health.

ACKNOWLEDGEMENTS

I wish to thank Ms. Pirkko Pfäffli, M.Sc., for her critical reading of my manuscript and the Swedish Work Environment Fund for financial support.

REFERENCES

BRYDSON JA (1979). "Plastics Materials." London: Butterworths, p 631.

BUIST JM (1978). "Developments in Polyurethane-1." London: Applied Science Pulsishers Ltd, p 1-280.

CHAMBERS J, Jiricny J, Reese CB (1981). The thermal decomposition of polyurethanes and polyisocyanurates. Fire and Materials 5:133.

FOTI S, Maravigna P, Montaudo G (1981). Mechanisms of thermal decomposition in totally aromatic polyurethanes. J Polym Sci 19:1679.

GABORIAUD F, Vantelon JP (1981). Thermal degradation of polyurethane based on MDI and propoxylated trimethylol propane. J Polym Sci 19:139.

GRASSIE N, Zulfiqar M (1978). Thermal degradation of the polyurethane from 1,4-butanediol and methylene bis(4-phenyl-isocyanate). J Polym Sci 16:1563.

HARDY HL, Devine JM (1979). Use of organic isocyanates in industry-some industrial hygiene aspects. Ann Occup Hyg 22:421.

HILEMAN FD, Voorheers KJ, Wojcik LH, Birky MM, Ryan PW, Einhorn IN (1975). Pyrolysis of a flexible urethane foam. J Polym Sci 13:571

IARC Monographs (1979). Some Monomers, Plastics and Synthetic Elastomers, and Acrolein 19.

MOD Plast Int (1982). Materials 1982, Special Report. Jan:39.

ROSENBERG C, Pfäffli P (1982). A comparison of methods for the determination of diphenylmethane diisocyanate (MDI) in air samples. Am Ind Hyg Assoc 43:160.

SAUNDERS JH, FRISCH KC (1965). "Polyurethanes chemistry and technology." New York: Interscience Publishers. p 65.

SEEMANN J, Wölcke U (1976). Über der Bildung der toxisher Isocyanatdämpfe bei der thermischen Zersetzung von Polyurethanlacken und ihren polyfunctionellen Härtern. Zbl Arbeitsmed 1:2.

WOOLLEY WD, Fardell PJ (1977). The prediction of combustion products. Fire Res 1:11.

Industrial Hazards of Plastics and Synthetic Elastomers, pages 347–363
© 1983 Alan R. Liss, Inc., 150 Fifth Ave., New York, NY 10011

Carcinogenesis Results on Seven Amines, Two Phenols, and One Diisocyanate Used in Plastics and Synthetic Elastomers

J. E. Huff

National Institute of Environmental Health Sciences
National Toxicology Program
Research Triangle Park, NC 27709   U.S.A.

Welcome, O life!
I go to encounter for the millionth time
the reality of experience
and to forge in the smithy of my soul
the uncreated conscience of my race.

James Joyce (1882-1941)
A Portrait of the Artist as a Young Man (1916)

Since the U.S. National Toxicology Program (NTP) was given responsibility for the U.S. National Cancer Institute (NCI) Carcinogenesis Bioassay Program in July 1981, nearly 75 Technical Reports have been published or prepared in draft form (Huff, Moore, and Rall, 1983; Huff and Moore, 1983). This paper gives summary information from those ten NCI or NTP Technical Reports containing carcinogenesis results from 40 studies on chemicals used in making plastics and synthetic elastomers. Carcinogenesis data on phthalate esters and on styrenes are given in separate papers (Huff and Kluwe, 1983; Huff, 1983).

The seven amines, two phenols, and one diisocyanate tested by the NTP and the NCI in carcinogenesis studies represent a diverse group of chemicals (Table). As seen in the Table, seven were administered via diet, two by drinking water, and one by gastric intubation. Other information contained in the Table includes chemical purity, the laboratories contracted to do the studies, the doses or dose ranges used, a single word conclusion of results, and the reference.

TABLE. Carcinogenesis Studies Information

Test Chemical	Purity (%)	Testing Laboratory	Exposure Route	Doses	Interpretation Results	Reference
11-Amino-undecanoic acid	>99	Litton Bionetics	Feed	0; 7,500; 15,000 ppm	Positive	NTP, 1982a
Bisphenol A	>99	Litton Bionetics	Feed	0; 1,000 to 10,000 ppm	Equivocal	NTP, 1982b
Caprolactam	100	Litton Bionetics	Feed	0; 3,750 to 15,000 ppm	None	NTP, 1982c
2,6-Dichloro-p-phenylenediamine	>99	Litton Bionetics	Feed	0; 1,000 to 6,000 ppm	Positive	NTP, 1982d
Melamine	>95	Litton Bionetics	Feed	0; 2,250 to 9,000 ppm	Positive	NTP, 1983a
4,4'-Methylene-dianiline dihydro-chloride	>98	EG&G Mason Res. Inst.	Water	0; 150; 300 ppm	Positive	NTP, 1983b
4,4'-Oxy-dianiline	>98	EG&G Mason Res. Inst.	Feed	0; 150 to 800 ppm	Positive	NCI, 1980a
Phenol	>98	Hazleton Labs, America	Water	0; 2,500; 5,000 ppm	None	NCI, 1980b
2,6-Toluene-diamine dihydrochloride	>99	EG&G Mason Res. Inst.	Feed	0; 50 to 500 ppm	None	NCI, 1980c
2,4-/2,6-Toluene diisocyanate	77 to 90	Litton Bionetics	Gavage	0; 30 to 240 mg/kg	Positive	NTP, 1983c

Single copies of the complete Technical Reports on each chemical are available without cost, while supplies last, from the National Toxicology Program, Public Information Office, P.O. Box 12233, Research Triangle Park NC, 27709, USA.

METHODS -- Male and female inbred Fisher 344/N rats and male and female hybrid B6C3F$_1$ mice were used in these carcinogenesis studies. Concurrent control and treated groups contained 50 animals of each species and sex. Except for toluene diisocyanate, the test chemicals were incorporated into the feed or into the water and supplied to the animals for 103 to 106 consecutive weeks beginning at 4 to 6 weeks of age. 2,4-/2,6-Toluene diisocyanate was given by gavage in corn oil 5 days per week for 105 to 106 weeks. Estimated maximally tolerable doses (EMTD) were determined by preceding 13-week studies. In general, each species/sex was divided into three groups: 0 (control), a low dose (one-half EMTD), or a high dose (EMTD). The study design conformed to the NCI Guidelines for Carcinogen Bioassay in Small Rodents (Sontag, Page, and Saffiotti, 1976).

All animals that died during the study (and were not excessively autolyzed or cannibalized) or that were killed at the end of the exposure period were subjected to a gross necropsy and a complete histopathological examination. Statistical analyses of survival differences among groups were done by life table methods (Cox, 1972; Tarone, 1975). For tumor incidence data pairwise comparisons were made by Fischer's exact tests (Gart, Chu and Tarone 1979), and the significance of dose response trends was assessed by Cochran-Armitage tests (Armitage, 1971). Later Technical Reports (NTP, 1982a, 1983a,b,c) adopted additional statistical methods. These tests assume either all tumors of a particular site are fatal to the animals or are incidental in animals dying of another cause (Peto et al., 1980). Background incidences of neoplasms in untreated controls have been reported by Haseman, Huff, and Roorman (1983) and have been summarized by Huff and Haseman (1983).

RESULTS -- Abbreviated findings from each of the studies are reported herein, particularly emphasizing chemically related increased incidences in neoplasms. Complete technical details and comprehensive results for each study are given and available in the appropriate Technical Report. Arbitrarily the chemicals are presented in alphabetical order.

● 11-Aminoundecanoic Acid (CAS No. 2432-99-7) -- The monomer
used in the manufacture of the polyamide, nylon-11, is
synthesized from ricinoleic acid isolated from castor oil
bean. Carcinogenesis studies were conducted by giving diets
containing 0; 7,500; or 15,000 ppm to rats (104 weeks) and
mice (103 weeks) (Dunnick et al., 1982; NTP, 1982a).

Nonneoplastic effects included decreases in mean body weight
for high-dose male rats and for male and high-dose female
mice; compared to controls survivals were decreased for high
dose-male rats and for high-dose male and female mice; a
dose-related increased incidence of hyperplasia of the transitional epithelium of the kidney and urinary bladder in
rats of each sex; and mineralization of the kidney in dosed
mice of each sex. Neoplastic nodules of the liver in dosed
male rats (1/50, 9/50: P<0.01, 8/50:   P<0.01) and transitional-cell carcinomas of the urinary bladder in high-dose
male rats (0/48, 0/48, 7/49:   P<0.01) were observed at
increased incidences compared with controls.   Malignant
lymphomas occurred at an increased rate in low-dose male
mice (2/50, 9/50: P<0.05, 4/50).

● Bisphenol A (CAS No. 80-05-7) --   4,4'-Isopropylidene-
diphenol is a monomer used in the manufacture of epoxy,
polycarbonate, and corrosion-resistant unsaturated   polyester-styrene resins found in reinforced pipe, adhesives,
flooring, water main filters, artificial teeth, nail polish,
food packaging materials, and interior coatings for cans and
drums.   Carcinogenesis studies were conducted by feeding
diets containing 0; 1,000; or 2,000 ppm to rats and 0;
1,000; or 5,000 ppm to mice for 103 weeks (NTP, 1982b).

Mean body weights of rats of either sex and of low- and
high-dose female mice and high-dose male mice were lower
than those of the controls throughout the study. Food consumption by dosed female rats was 70% to 80% that of the
controls throughout most of this study, and reduced body
weight gain was probably due to reduced food consumption.
Food consumption by dosed male rats was 90% that of
controls and appear to be similar among all groups of mice.
Survivals among groups were comparable. A compound-related
increased incidence of multinucleated giant hepatocytes was
observed in male mice (1/49, 41/49, 41/50); no increase in
liver neoplasms was observed.

Leukemias occurred at increased incidences in dosed rats: In males, the dose-related trend (P<0.05: 13/50, 12/50, 23/50: P<0.05) was statistically significant and the incidence in the high-dose group was significantly more than in controls. The numerical increases seen in female rats were not statistically significant (7/50, 13/50, 12/50). In low-dose male mice, an increased incidence of lymphomas/leukemia was observed (2/49, 9/50: P<0.05; 5/50).

Interstitial-cell tumors of the testes occurred at statistically significant incidences in dosed male rats; since this lesion normally occurs at a high incidence in ageing F344 male rats, the increase was not considered compound related (P<0.005: 35/49, 48/50: P<0.005, 46/49: P<0.005). The trend statistic was positive for fibroadenomas of the mammary gland in male rats (P<0.05: 0/50, 0/50, 4/50).

● Caprolactam (CAS No. 105-60-2) -- Aminocaproic lactam is the monomer used to produce polycaprolactam, known as nylon 6, a fiber or resin used to make carpets, knit fabrics, hosiery, thread, hair brushes, flotation devices, and food contact films. Carcinogenesis studies were conducted by giving diets containing 0; 3,750; or 7,500 ppm to rats and 0; 7,500; or 15,000 ppm to mice for 103 weeks (Huff, 1982; NTP, 1982c).

Mean body weights were decreased and dose related for rats and mice throughout most of the study. Feed consumption was reduced in high dose male and female rats to about 70-80% of controls. Survivals were comparable among all groups. The incidences of animals with neoplasms at a specific anatomical site did not differ significantly between treated and control groups.

● 2,6-Dichloro-p-phenylenediamine (CAS No. 609-20-1) -- 1,4-Diamino-2,6-dichlorobenzene is a possible intermediate for use as a polyurethane curative and as a monomer in the manufacture of polyamide fiber. Carcinogenesis studies were conducted by administering diets containing 0; 1,000; or 2,000 ppm to male rats, 0; 2,000; or 6,000 ppm to female rats, and 0; 1,000; or 3,000 ppm to mice for 103 weeks (NTP, 1982d).

Mean body weights were lower for rats (in particular for 6,000 ppm females) and for 3,000 ppm mice throughout the studies. Survivals were comparable among groups, except for

2,000 ppm male rats, with 21/50 (42%) surviving until the end of the study.

Ectopic hepatocytes were observed at an increased incidence in the pancreas, and nephrosis was observed in increased severity in dosed rats of either sex. No increase in any tumor type was observed in treated male or female rats when compared to controls.

Increased incidences of liver tumors were observed in mice of both sexes. In male mice, the incidence of hepatocellular adenomas was related to dose and the incidence was greater in the high-dose group (P<0.005: 4/50, 7/50, 15/50: P<0.01); the incidence of hepatocellular carcinomas was 12/50, 13/50, 17/50. In female mice, hepatocellular carcinomas exhibited a dose-related trend, but no single dose group had a statistically significant increased incidence of either adenomas (4/50, 4/50, 9/50) or carcinomas (P<0.05; 2/50, 2/50, 7/50) alone. Combining hepatocellular adenomas or carcinomas gave a positive dose-related trend, and the incidence in the high-dose group was significantly increased (P<0.005: 6/50, 6/50, 16/50: P<0.05).

● Melamine (CAS No. 108-78-1) -- 2,4,6-Triamino-s-triazine is a starting material in the manufacture of polymeric amino resins and thermosetting plastics used in electrical equipment and housewares (buttons, table tops, wall coverings, and dinnerware). Carcinogenesis studies were conducted by feeding diets containing 0; 2,250; or 4,500 ppm to male rats and to male and female mice; female rats had diets with 0; 4,500; or 9,000 ppm for 103 weeks (Melnick et al., 1983); NTP, 1983a).

Mean body weights of high-dose male rats were lower than those of the controls after about week 60; high-dose female rats had lowered mean body weights most of the second year, but final weights were comparable. Food consumption for dosed rats was >97% that of the controls. Survival of high-dose male rats was significantly lower (P<0.05) than that of the controls (30/49 versus 19/50). Survivals of all other rat groups were comparable.

Mean body weights of high-dose male mice were lower than that of controls after week 50. Mean body weights of low-dose male and dosed female mice were comparable to controls. Survival of high-dose male mice was significantly less

(P<0.02) than that of the controls (39/49 versus 28/50). Survivals of other groups of mice were similar.

Chronic inflammation, distinguishable from the nephropathy observed in ageing F344/N rats, was significantly increased in the kidneys of dosed female rats (4/50, 17/50: P<0.01, 41/50: P<0.01) and is attributed to the administration of melamine. For male rats the rates were 2/49, 3/50, 6/49.

Transitional-cell carcinomas in the urinary bladder of male rats occurred with a positive trend and the incidence in the high-dose group was significantly higher (P<0.016) than that in the controls (P<0.005: 0/45, 0/50, 8/49: P<0.05). A transitional-cell papilloma was observed in the urinary bladder of an additional high-dose male rat. Seven of the eight high-dose male rats with the transitional-cell carcinomas also had bladder stones. An association (P<0.001) was found between bladder stones and bladder tumors in male rats. One transitional-cell papilloma occurred in a single female rat in each dosed group: 1/49 (papilloma, NOS), 1/49, 1/47; none had bladder stones.

Acute and chronic inflammation and epithelial hyperplasia of the urinary bladder were found in increased incidence in dosed male mice (1/45, 11/47, 13/44). The incidence of bladder stones in dosed male mice was increased relative to controls (2/45, 40/47, 41/44); no evidence of bladder tumor development was seen in this species. Also, four high-dose female mice had bladder stones without any tumors.

● 4,4'-Methylenedianiline Dihydrochloride (CAS No. 13552-44-8) -- MDA is used primarily as a chemical intermediate in the closed system production of isocyanates and polyisocyanates. Other uses are as a curing agent for epoxy resins and urethane elastomers, a dye intermediate, and a corrosion inhibitor. Carcinogenesis studies were conducted using drinking water containing 0, 150, or 300 ppm administered to rats and mice for 103 weeks (NTP, 1983b).

Survivals were comparable among groups except for male mice receiving the high dose; in that group survival was lower than that in controls (P<0.01: 40/50 versus 32/50). Mean body weights were reduced in high-dose female rats and in high-dose male and female mice. Water consumption was reduced in a dose-related manner in both sexes of rats (g/day: male -- 28, 24, 21; female -- 19, 18, 16). No compound-related clinical effects were observed.

The thyroid glands and the livers both of rats and mice were the major target organs for 4,4'-methylenedianiline dihydrochloride. Chemical-related nonneoplastic lesions of the thyroid glands in female rats included follicular cysts (0/47, 3/47, 7/48) and hyperplasia (1/47, 3/47, 8/48). The incidence of thyroid follicular cell hyperplasia was elevated in high-dose male (0/47, 3/49, 18/49) and female mice (0/47, 0/47, 23/48). The incidences of thyroid neoplasms in the high-dose groups were elevated compared with those of control groups for both sexes of both species. Thyroid follicular cell carcinoma was increased in male rats (0/49, 0/47, 7/48: P<0.05). Follicular cell adenoma was increased in high-dose female rats (0/47, 2/47, 17/48: P<0.001), in high-dose male mice (0/47, 3/49, 16/49: P<0.001), and in high-dose female mice as compared with controls. In female rats, thyroid C-cell adenoma was also elevated in a dose-related manner (0/47, 3/47, 6/48: P<0.05).

Dose-related increases in nonneoplastic lesions were observed for male rats (nonspecific liver dilatation) and for male and female rats (fatty metamorphosis and focal cellular change). Liver degeneration was present in 80% of the low-dose and 60% of the high-dose male mice but was not found in controls. Neoplastic nodules of the liver were observed at greater incidences for low- and high-dose male rats as compared with controls (1/50, 12/50: P<0.005, 25/50: P<0.001). Hepatocellular adenoma was increased in a dose-related manner in dosed female mice (3/50, 9/50, 12/50: P<0.05). Hepatocellular carcinoma was observed in greater incidence in dosed male mice (10/49, 33/50: P<0.001, 29/50: P<0.001) and in high-dose female mice (1/50, 6/50, 11/50: P<0.005).

Male rats had a dose-related increase in kidney mineralization. Nephropathy was increased in dosed mice of both sexes; renal papillary mineralization was greater in high-dose male and female mice than in the controls.

Other tumors that were elevated in dosed animals included adrenal pheochromocytomas in male mice (2/48, 12/49: P<0.01, 14/49: P<0.005), alveolar/bronchiolar adenoma in female mice (1/50, 2/50, 6/49: P<0.05), and malignant lymphomas in female mice (13/50, 28/50: P<0.005, 29/50: P<0.005).

Uncommon tumors were observed in dosed animals at low incidences but may be important because the historical control incidences are very low: bile duct adenoma in 1/50 high-dose male rats, transitional-cell papillomas of the urinary bladder in female rats (low dose, 2/50; high dose, 1/50), and granulosa cell tumors of the ovary in female rats (low dose, 3/50; high dose, 2/50).

• 4,4'-Oxydianiline (CAS No. 101-80-4) -- This chemical is an intermediate in the manufacture of high-temperature-resistant straight polyimide and poly(esterimide) resins. Carcinogenesis studies were conducted by administering diets containing 0, 200, 400, or 500 ppm to rats and 0, 150, 300, or 800 ppm to mice for 104 weeks (NCI, 1980a).

A dose-related decrement in mean body weights was observed for all groups of dosed rats and mice. Survivals were significantly ($P < 0.05$) shortened in the high-dose female rats (40/50 versus 13/50) and in the low- and mid-dose female mice (42/50 versus 33/50 and 33/50).

The thyroid glands and the livers were target organs for 4,4'-oxydianiline in rats and mice. In rats, hepatocellular carcinomas (males: 0/50, 4/50, 23/50: $P < 0.001$, 22/50: $P < 0.001$; females: 0/50, 0/49, 4/50, 6/50: $P < 0.05$) and neoplastic nodules (males: 1/50, 9/50: $P < 0.01$, 18/50: $P < 0.001$, 17/50: $P < 0.001$; females: 3/50, 0/49, 20/50: $P < 0.001$, 11/50: $P < 0.05$) occurred at incidences that were dose-related, and the incidences in all dosed groups (except low-dose females) were higher than those in the controls. In high-dose female mice hepatocellular carcinomas and hepatocellular adenomas occurred at incidences significantly higher than those in the controls (carcinomas: 4/50, 7/49, 6/48, 15/50: $P < 0.01$; adenomas: 4/50, 6/49, 9/48, 14/50: $P < 0.01$). In male mice the combined incidence of liver cell neoplasms (adenomas or carcinomas) were increased in the low-dose group: 29/50, 40/50: $P < 0.05$, 34/49, 36/50.

The occurrence of follicular-cell neoplasms of the thyroid gland was dose-related among rats: carcinoma--males: 0/46, 5/47: $P < 0.05$, 9/46: $P < 0.005$, 15/50: $P < 0.001$; females: 0/49, 2/48, 12/48: $P < 0.001$, 7/50, $P < 0.01$; adenoma--males: 1/46, 1/47, 8/46: $P < 0.05$, 13/50: $P < 0.001$; females: 0/49, 2/48, 17/48: $P < 0.001$, 16/50: $P < 0.001$.

In female mice, follicular cell adenomas of the thyroid gland were more frequent in the high-dose group than in the controls (0/46, 0/43, 0/42, 7/48: P<0.01).

In mice, adenomas in the harderian glands occurred in all dosed groups at incidences that were significantly (P<0.005) higher than the incidence in the controls (males: 1/50, 17/50, 13/49, 17/50; females: 2/50, 15/50, 14/50, 12/50).

Other tumors occurring among male mice at increased incidences were adenomas in the pituitary (1/37, 0/44, 0/34, 7/35: P<0.05) and hemangiomas of the circulatory system (0/50, 0/50, 5/49: P<0.05, 5/50: P<0.05).

● Phenol (CAS No. 108-95-2) -- About 90% of the 1,000,000,000 kg produced in the U.S. is used to manufacture phenolic (phenol formaldehyde) resins, caprolactam, bisphenol A, alkyl phenols, and adipic acid. The remainder is used to produce salicylic acid, phenacetin, disinfectants, antiseptics, wood preservatives, and so on. Carcinogenesis studies were conducted by providing drinking water containing 0; 2,500; or 5,000 ppm to rats and mice for 103 weeks (NCI, 1980b).

Dose-related depressions in mean body weight occurred in rats and mice of each sex. Rats (80% and 90% of controls) and mice (75% and 50-60% of controls) given water containing phenol drank less than did the corresponding controls. Survivals were comparable among all groups.

An increased incidence of neoplasms was detected in low-dose male rats (leukemia--18/50, 30/50: P<0.05, 25/50; pheochromocytoma of the adrenal glands--13/50, 22/50: P<0.05, 9/50; C-cell carcinoma of the thyroid--0/50, 5/49: P<0.05, 1/50). No chemical-related neoplasms were observed in female rats or in mice.

● 2,6-Toluenediamine Dihydrochloride (CAS No. 15481-70-6) -- This chemical is used as an intermediate in the production of dyes for furs and textiles and of flexible polyurethane foams and elastomers. Carcinogenesis studies were conducted by giving diets containing 0, 250, or 500 ppm to rats and 0, 50, or 100 ppm to mice for 103 weeks (NCI, 1980c).

Mean body weights were comparable among groups of rats and mice, except for female rats who showed a dose-related decrease (83% and 73% of controls). Survivals were not different among groups.

In male rats, dose-related trends were significant for islet cell adenomas of the pancreas ($P<0.05$: 0/45, 1/46, 4/45) and for neoplastic nodules/carcinomas of the liver ($P<0.05$: 0/50, 2/50, 4/50). In female mice, hepatocellular carcinomas showed a dose-related trend ($P<0.05$: 0/50, 0/49, 3/49) but combined with adenomas of the liver removed any significance (4/50, 3/49, 7/49). In male mice, lymphomas in the low-dose group were significantly increased (2/50, 8/50: $P<0.05$, 2/50).

● 2,4-/2,6-Toluene Diisocyanate (CAS No. 26471-62-5) --Most of the 282,000,000 kg TDI produced in the U.S. and the 2,300,000 kg TDI imported is used in the manufacture of flexible polyurethane foams and to produce polyurethane coatings for lacquers and wood finishes. Carcinogenesis studies were conducted by administering via gavage in corn oil 0, 30, or 60 mg/kg to male rats; 0, 60, or 120 mg/kg to female rats and female mice; and 0, 120, or 240 mg/kg to male mice for 105-106 weeks (NTP, 1983c).

Mean body weights decreased relative to controls in all dosed rat groups throughout most of the study: at the end of experiments, -14% and -22% for males and -26% and -25% for females. During the second year of the study, mean body weights of high-dose male mice were less (-18%) than those of the controls, but were -8% at the end of the study. Survivals in all groups of dosed rats in the 2-year studies were shorter ($P<0.01$) than that of the controls. A dose-dependent pattern of cumulative toxicity in rats began at 70 weeks and culminated in excessive mortality (males: 14/50, 36/50, 42/50; females: 14/50, 31/50, 44/50). Survival of high-dose male mice in the 2-year study was significantly shorter than that of the controls (males: 46/50, 40/50, 26/50: $P<0.001$; females: 34/50, 43/50, 33/50).

Acute bronchopneumonia occurred at increased incidences in groups of dosed rats (males: 2/50, 6/50, 14/50; females: 1/50, 10/50, 25/49). Cytomegaly of kidney tubular epithelium was found in 45/48 (94%) low-dose male mice and 41/50 (82%) high-dose male mice and in none of the controls.

Because of the reduced survival in rats, the statistical procedures that adjust for intercurrent mortality (life table and incidental tumor tests) were regarded as more meaningful than the "unadjusted" analyses in the evaluation of tumor incidence data.

Subcutaneous tissue fibromas or fibrosarcomas (combined) in male rats occurred with a positive trend and the incidence in the high-dose group was higher (P<0.005: 3/50, 6/50, 12/50: P<0.001). The same tumor comparisons in female rats were 2/50, 1/50, 5/50: P<0.001. Mammary gland fibroadenomas in female rats occurred with a positive trend, and the incidences in low- and high-dose groups were higher than that in controls (P<0.001: 15/50, 21/50: P<0.01, 18/50: P<0.05; terminal rates: 13/36, 36%; 15/19, 79%; 4/6, 67%).

Pancreatic acinar cell adenomas in male rats occurred with a positive trend and the incidence in the high-dose group was higher than that in the controls (P<0.05: 1/47, 3/47, 7/49: P<0.05). The incidences of pancreatic islet cell adenomas in female rats were higher in low- and high-dose groups than in controls (0/50, 6/49, 2/47), the terminal rates being 0/36, 3/19, 2/6. An islet cell carcinoma was also observed in a low-dose female rat.

The incidences of female rats with neoplastic nodules in the liver occurred with a positive trend and the incidence in the high-dose group was higher than that in the controls (P<0.05: 3/50, 8/50, 8/48: P<0.05). Hepatocellular adenomas in female mice occurred with a positive trend and the incidence in the high-dose group was higher than that in the controls (P<0.005: 2/50, 3/50, 12/50: P<0.005).

Hemangiomas or hemangiosarcomas (combined) of the circulatory system in female mice occurred with a positive trend; the incidence in the high-dose group was significantly higher than that in the controls (P<0.05: 0/50, 1/50, 5/50: P<0.05).

CARCINOGENESIS SUMMARY -- Of the ten chemicals studied in rodents, six gave evidence of causing carcinogenic responses, three gave no such evidence, and one was considered equivocal. Transitional-cell carcinomas of the urinary bladder were induced in male rats by 11-aminoundecanoic acid (Dunnick et al., 1983; NTP 1982a) and by melamine (Melnick et al., 1983; NTP 1983a). Epithelial hyperplasia of the urinary bladder was increased in male and female rats by 11-aminoundecanoic acid and in male mice by melamine; (high-dose male rats (2/49) and female mice (4/50) had marginal occurrences). In the 11-aminoundecanoic acid studies, an increased incidence of calculi of the urinary bladder was seen in males in the high-dose group (1/48, 1/48, 5/49). The rats with calculi were not the ones that had tumors of the urinary bladder. Hyperplasia of the transitional epithelium of the kidney and bladder was associated with the administration of 11-aminoundecanoic acid in male and female rats. In particular, hyperplasia was associated with calculi of the urinary bladder in 2/5 high-dose males. Of the 20 high-dose male rats with hyperplasia of the urinary bladder, 7 had transitional-cell carcinomas and 1 other had a papilloma; the remaining 12 had hyperplasia in the apparent absence of urinary bladder tumors. Hyperplasia of the renal pelvis was also observed in the 13-week study in 2/9 females that received 18,000 ppm and in 1/10 males and 6/10 females that received 21,000 ppm.

In the melamine studies an increased incidence of urinary bladder stones occurred in high-dose male rats: of the 49 urinary bladders of this group examined histologically, 7 had transitional-cell carcinomas with stones, 1 had transitional-cell carcinoma without stones, 3 had stones without evidence of carcinoma, and 38 had neither stones nor transitional-cell carcinomas. A significant association was found to exist between bladder stones and bladder tumors. Of the three rats with bladder stones and no transitional-cell carcinomas, one had a transitional-cell papilloma and one other had epithelial hyperplasia in the urinary bladder.

In the 13-week studies, bladder stones were observed in male rats at doses as low as 750 ppm melamine, while in female rats no bladder stones were seen in animals receiving less than 15,000 ppm. The circumferential relationship of the prostate gland to the urethra in males may increase the tendency for obstruction of the urethra and thereby account for the apparent difference in susceptibility between male and female rats in developing bladder stones. The incidences of urinary bladder stones in male mice fed diets containing melamine increased in the 13-week and chronic studies in comparison with those of the controls. In the chronic study, acute and chronic inflammation and epithelial hyperplasia were observed in the urinary bladders; however, there was no apparent evidence of bladder tumor development.

Species variations may account for the lack of such tumors in male mice. Although there was a significant association between urinary bladder stones and urinary bladder tumors in male rats fed diets containing 4,500 ppm melamine, the data from the carcinogenesis studies are not sufficient to determine whether the tumors developed as a consequence of the bladder stones. Additional experimental data on urinary bladder tumorigenesis induced by melamine or melamine stones might resolve this issue.

Liver cell neoplasms were increased by six of the ten chemicals tested in 13 of the 40 studies: 11-aminoundecanoic acid (male rat); 2,6-dichloro-p-phenylenediamine (male and female mice); 4,4'-methylenedianiline dihydrochloride (male rats and both sexes of mice); 4,4'-oxydianiline (male and female rats and mice); 2,6-toluenediamine dihydrochloride (marginal in male rats); and 2,4-/2,6-toluene diisocyanate (female rats and mice).

Neoplasms of the thyroid gland were seen in rodents receiving the structurally related chemicals 4,4'-methylenedianiline dihydrochloride (NH_2-$\emptyset$-CH_2-$\emptyset$-NH_2) and 4,4'-oxydianiline (NH_2-$\emptyset$-O-$\emptyset$-NH_2). Both induced follicular-cell neoplasms of the thyroid glands, whereas 4,4'-methylenedianiline increased C-cell neoplasms in female mice as well.

The two phenol chemicals (bisphenol A and phenol) -- although judged not carcinogenic -- were associated with neoplasms of the hematopoietic system. The marginally significant increase in leukemias in male rats, along with an increase (not statistically significant) in leukemias in female rats and a marginally significant increase in the incidence of lymphomas in male mice, suggest that exposure to bisphenol A may correspond with increased cancers of the hematopoietic system. Phenol increased the numerical occurrence of leukemia in both dosed groups of male rats (36% versus 60% and 50%); only the low-dose group was statistically different from controls. Other chemicals related to increases in hematopoietic lesions include 11-aminoundecanoic acid (lymphoma in male mice), 4,4'-methylenedianiline dihydrochloride (lymphoma in female mice), and 2,6-toluenediamine dihydrochloride (lymphoma in male mice).

Uncommon neoplasms were caused by 4,4'-oxydianiline (harderian gland lesions in male and female mice) and by 2,4-/2,6-toluene diisocyanate (hemangioma/hemangiosarcoma in female mice).

For caprolactam the numbers of chemically treated animals with neoplasms at a specific anatomical site did not differ significantly from those in the controls. To compare neoplastic incidence rates for the control and treated rodents in these ten studies with the background rates in more than 2000 rodents of each species/sex, see the summary compilations reported by Huff and Haseman (1983).

REFERENCES

1. Armitage, P. (1971). Statistical Methods in Medical Research. John Wiley & Sons, Inc., New York, pp. 362-365.

2. Cox, D.R. (1972). Regression Models and Life Tables. J. R. Stat. Soc. B34: 187-220.

3. Dunnick, J., Huff, J.E., Haseman, J.K., and Boorman, G.A. (1983). Lesions of the Urinary Tract in Fisher 344 Rats and B6C3F$_1$ Mice After Chronic Administration of 11-Aminoundecanoic Acid. Fund. Appl. Toxicol. (in press).

4. Gart, J., Chu, K., and Tarone, R. (1979). Statistical Issues in Interpretation of Chronic Bioassay Tests for Carcinogenicity. J. Nat. Cancer Inst. 62(4): 957-974.

5. Haseman, J.K., Huff, J.E., and Boorman, G.A. (1983). Use of Historical Control Data in Carcinogenicity Studies in Rodents (Draft NTP Document).

6. Huff, J.E. (1982). Carcinogenesis Bioassay Results from the National Toxicology Program. Environ. Health Perspect. 45: 185-198.

7. Huff, J.E. (1983). Styrene, Styrene Oxide, Polystyrene, and β-Nitrostyrene/Styrene Carcinogenicity in Rodents. (These Proceedings).

8. Huff, J.E. and Haseman, J.K. (1983). Background Neoplasms in "Untreated" Fisher 344/N Rats and B6C3F$_1$ Mice. (These Proceedings).

9. Huff, J.E. and Kluwe, W.M. (1983). Phthalate Esters Carcinogenicity in F344/N Rats and B6C3F$_1$ Mice. (These Proceedings).

Carcinogenesis Results for Ten Plastics Chemicals

10. Huff, J.E. and Moore, J.A. (1983). Carcinogenesis Studies Design and Experimental Data Interpretation/ Evaluation at the National Toxicology Program. (These Proceedings).

11. Huff, J.E., Moore, J.A., and Rall, D.P. (1983). The National Toxicology Program: 1978-1983. (These Proceedings).

12. Melnick, R.L., Boorman, G.A., Haseman, J.K., Montali, R.J., and Huff, J.E. (1983). Urolithiasis and Bladder Carcinogenicity of Melamine in Rodents. Toxicol. Appl. Pharmacol. (in press).

13. NCI (1980a). NCI Bioassay of 4,4'-Oxydianiline for Possible Carcinogenicity, CAS No. 101-80-4, TR No. 205, 131 pages, National Cancer Institute, Bethesda, MD.

14. NCI (1980b). NCI Bioassay of Phenol for Possible Carcinogenicity, CAS No. 108-95-2, TR No. 203, 123 pages, National Cancer Institute, Bethesda, MD.

15. NCI (1980c). NCI Bioassay of 2,6-Toluenediamine Dihydrochloride for Possible Carcinogenicity, CAS No. 15481-70-6, TR No. 200, 160 pages, National Cancer Institute, Bethesda, MD.

16. NTP (1982a). NTP Carcinogenesis Bioassay of 11-Aminoundecanoic Acid (CAS No. 2432-99-7) in F344 Rats and B6C3F1 Mice (Feed Study), TR No. 216, 116 pages, National Toxicology Program, Research Triangle Park, NC.

17. NTP (1982b). NTP Carcinogenesis Bioassay of Bisphenol A (CAS No. 80-05-7) in F344 Rats and B6C3F1 Mice (Feed Study), TR No. 215, 116 pages, National Toxicology Program, Research Triangle Park, NC.

18. NTP (1982c). NTP Carcinogenesis Bioassay of Caprolactam (CAS No. 105-60-2) in F344 Rats and B6C3F1 Mice (Feed Study), TR No. 214, 129 pages, National Toxicology Program, Research Triangle Park, NC.

19. NTP (1982d). NTP Carcinogenesis Bioassay of 2,6-Dichloro-p-phenylenediamine (CAS No. 609-20-1) in F344 Rats and B6C3F1 Mice (Feed Study), TR No. 219, 121 pages, National Toxicology Program, Research Triangle Park, NC.

20. NTP (1983a). NTP Carcinogenesis Bioassay of Melamine (CAS No. 108-78-1) in F344/N Rats and B6C3F1 Mice (Feed Study), TR No. 245, 171 pages, National Toxicology Program, Research Triangle Park, NC.

21. NTP (1983b). NTP Carcinogenesis Studies of 4,4'-Methylenedianiline Dihydrochloride (CAS No. 13552-44-8) in F344/N Rats and B6C3F1 Mice (Drinking Water Studies), TR No. 248, 182 pages, National Toxicology Program, Research Triangle Park, NC.

22. NTP (1983c). NTP Carcinogenesis Studies on Commercial Grade 2,4(86%)- and 2,6(14%)-Toluene Diisocyanate (CAS No. 26471-62-5) in F344/N Rats and B6C3F1 Mice (Gavage Studies), TR No. 251, 187 pages, National Toxicology Program, Research Triangle Park, NC.

23. Peto, R., Pike, M., Day, N., Gray, R., Lee, P., Parish, S., Peto, J., Richard, S., Wahrendorf, J. (1980). Guidelines for Simple, Sensitive, Significant Tests for Carcinogenic Effects in Long-Term Animal Experiments, In: Long-Term and Short-Term Screening Assays for Carcinogens: A Critical Appraisal, Supplement 2: 311-426, IARC Monographs on the Evaluation of the Carcinogenic Risk of Chemicals to Humans, International Agency for Research on Cancer, Lyon, France, 426 pages.

24. Sontag, J.M., Page, N.P., and Saffiotti, U. (1976). Guidelines for Carcinogen Bioassay in Small Rodents. NCI Carcinogenesis Technical Report (TR 1), DHEW, Washington, DC, 65 pages.

25. Tarone, R.E. (1975). Tests for Trend in Life Table Analysis. Biometrika 62: 679-682.

Industrial Hazards of Plastics and Synthetic Elastomers, pages 365–372
© **1984 Alan R. Liss, Inc., 150 Fifth Ave., New York, NY 10011**

PRODUCTION AND MANUFACTURE OF EPOXY THERMOSETS

Jukka M. Martinmaa

Department of Wood and Polymer Chemistry

University of Helsinki, Meritullinkatu 1 A
SF-00170 Helsinki 17, Finland

INTRODUCTION

After the first patent by Dr. P. Castan (Switzerland) in 1938 and the commercial introduction in 1947 epoxy resins have conquered more areas of application in a shorter period than any synthetic resin. Today epoxy resins are used where product design calls for a high-performance plastic material, especially in small items in which a slightly higher cost is acceptable. The great popularity of epoxy thermosets in these applications is attributed to the resin's many outstanding physical and chemical properties. The relatively high price of the starting materials somewhat limits the wider use of epoxies in more common areas.

SYNTHESIS OF LINEAR EPOXY RESINS

The chemical basis which makes the existence of epoxy thermosets possible is the oxirane ring or epoxy group:

$$\overset{\diagdown}{-}C \overset{O}{\underset{\diagdown}{-}} C \overset{\diagup}{-}$$

The most common monomer with an epoxy group is epichlorhydrin:

$$CH_2 \overset{O}{-} CH-CH_2Cl$$

The comonomer to which epichlorhydrin usually reacts is bis-phenol A prepared from phenol and acetone:

$$HO-C_6H_4-C(CH_3)_2-C_6H_4-OH$$

As a result of their addition reaction, diglycidyl ethers are formed:

$$\underset{H_2C-CHCH_2}{\overset{O}{\diagup\!\!\!\!\diagdown}} -\!\!\left[O-C_6H_4-\underset{\underset{CH_3}{|}}{\overset{\overset{CH_3}{|}}{C}}--C_6H_4-OCH_2\underset{OH}{\overset{}{C}}HCH_2 \right]_n\!\!-O-C_6H_4-\underset{\underset{CH_3}{|}}{\overset{\overset{CH_3}{|}}{C}}--C_6H_4-OCH_2\underset{}{\overset{O}{C}}H-CH_2$$

Diglycidyl ethers are manufactured in special resin kettles. In a typical procedure the monomers are mixed in a desired proportion and heated up to about 50 °C. During 30 minutes an alkali hydroxide catalyst is slowly added and at the same time the temperature is rised to 115 °C. After further heating over a period of 30 minutes, the sodium hydroxide catalyst, salt, and unreacted monomer are washed off with hot water. After a final drying procedure the resin is ready for use. At this stage the resin molecules are linear chains with glycidyl (epoxy) groups in both ends. The inner part of a resin molecule contains ether and hydroxyl groups.

The average value of n in the structural formula usually does not exceed 20. When n is in the region from 10 to 20 the resin is a high-melting solid and comparable to a thermoplastic. If n approaches zero the molecular weight is low and, therefore, we have a liquid resin. The viscosity and molecular weight are controlled using epichlorhydrin in excess in relation to bisphenol A in the reaction mixture.

Other epichlorhydrin-derived resins are the epoxidized novolacs and o-cresol novolacs, p-aminophenol epoxies, and epoxidized glycidyl esters. Due to the absence of traces of chlorine, cycloaliphatic epoxy resins are also very popular. To improve the flame resistance of the resin tetrabrombisphenol A is used together with bisphenol A. The proportion of the bromine component is determined according to the flame-resistance requirements of the end-product.

Solid epoxies are considered nontoxic, but when heated they create vapors.

CURING

All epoxy resins discussed before have a long storage life. Actually they are monomers in the sense that they must be activated to get the final thermoset structure. This process is called curing. In curing the epoxide groups take part in a polymerization reaction that makes the plastic to a dense, closely knit chemical structure which is insoluble and infusible. The term epoxy resin is used to indicate both the uncured and cured forms of epoxy thermosets.

Methods to initiate the curing process are many and varied. Usually this is brought about with catalysts or specific chemicals called hardeners. Catalysts, such as tertiary amines are used when an epoxy-epoxy or an epoxy-hydroxyl reaction is desired. The catalyst to resin ratio depends on the particular epoxy resin, but varies usually between 5 and 20%.

Hardeners are chemical compounds, usually primary amines which undergo an addition-type reaction with the epoxide groups and, therefore, become a part of the resulting giant macromolecular structure. To get a product with desired properties the amount of hardener has to be calculated stoichiometrically. Amounts up to 125 parts of hardener per 100 parts of resin are possible. Greater deviations from the calculated stoichiometric proportions will result to an uncured or poor-quality structure. Liberation of by-products does not occur, but heat is evolved. In the fully cured state practically all of the epoxide groups have reacted with the active groups of the hardener. The properly cured resin is odorless, tasteless and nontoxic.

Selection of a proper curing agent depends upon the application requirements: temperature, viscosity, mass, electrical properties, resistance to chemicals, etc. Many types of materials are used as curing agents with or without accelerator: aliphatic amines (diethylene triamine, DETA; triethylene tetramine, TETA; triethylamine), amidopolyamines $(CH_3(CH_2)_n CONHCH_2CH_2NHCH_2CH_2NH_2)$, acid anhydrides (phthalic anhydride, PA; pyromellitic dianhydride, PMDA; hexahydrophthalic anhydride, HHPA), aromatic amines (m-phenylene diamine, MPDA; diaminodiphenyl sulfone, DDS or DADS), aromatic polyols, and sulfides, etc.

Different curing agents require different curing conditions and have different curing times. The various base resins in combination with a definite curing agent produce thermosets having somewhat different prperties. Thus, a great amount of thermosets with specific properties can be obtained.

Resins can be formulated to cure rapidly or slowly with or without applied heat. The time range from the moment of mixing the hardener and resin under which the blend is applicable before gelation, is called the pot life. After the pot life period the resin cannot be formed.

The pot life strongly depends on the type of hardener selected. Aliphatic amines are used when a short room temperature cure is desired. Patching compounds, sealants, and coatings are examples of rapid curing resins that harden at room temperature. Because a short pot life causes high exotherm, aliphatic amines cannot be used in large castings or thick laminates.

Most amines in this category are skin-irritating so that they are to be handled with care and according to the instructions. Secondary cyclic amines and primary amines are often used in conjunction. In this way the pot life, exotherm, and cure rate can be modified.

Anhydride hardeners have lower reactivity and are, therefore, less troublesome than aliphatic amines. Anhydrides, however, must be handled with the same care as organic acids. When used, they require heating, usually in the range between 130 and 200°C, often in conjunction with amines to speed up the curing reaction. Some types of adhesives, casting resins, and laminating resins are cured in this way.

A group of modifying hardeners are the polysulfide rubbers called thiokols. They are nontoxic, react with low rate and exotherm, thus making massive castings possible.

The hardener can also be selected to modify the end-structure of a cured resin. This is usually done to improve the flexibility of the base resin. Liquid amide hardeners derived from vegetable oils are often used for this purpose.

The viscosity of many uncured liquid epoxy resins can be so high that they cannot, for example, penetrate small cavities. To improve the fluidity in these cases a reactive diluent is added to the resin. Such compounds are usually monofunctional or difunctional glycidyl ethers. Reactive diluents behave as do the hardeners: they become a part of the final structure. Proper ratios of catalysts and accelerators are important to the speed of this reaction.

Epoxy resin-hardener systems manufactured by commercial formulators are commonly available in a wide range of degrees of reactivity. The resins are produced in forms ranging from low viscosity liquids to high melting solids. Because special safety hardeners have been developed by the epoxy formulators, these should be used whenever possible. By mentioning of this it is not intended to convey that epoxies are dangerous to handle, because most epoxy-formulations are non-hazardous. Safety instructions, however, should be given in written form and followed up periodically.

PHYSICAL PROPERTIES

The popularity of epoxy systems can be attributed to many of their useful properties. These include: durability, excellent electrical, thermal, and chemical resistance, ease of use, low shrinkage during cure, and an outstanding adhesion to a variety of metallic and nonmetallic surfaces. These desirable properties are not influenced by wet or high-humidity conditions, brine, or other aggressive environments. Some epoxy adhesives are stable up to $200^{\circ}C$ without any loss of strength.

APPLICATIONS

Owing to the adhesion properties, epoxy resins are widely used as adhesives, floorings, coatings, paints and primers, solders, and linings. Typical applications for epoxies are the reinforced and composite materials for various corrosion and abrasion resistant purposes. Solid resins are used primarily for coating, prepreg and molding. Heat-curing epoxy powders can be used as a decorative or durable product finish for many industrial and consumer articles as, for example, steel furniture, tubular and wirework articles, domestic, office, and hospital appliances.

Reinforced Epoxy Thermosets

Laminating methods are generally used to make reinforced articles. The good adhesion properties of epoxy resins to many substrates permit tight bonding with reinforcing ma-terials. Reinforcements for laminating purposes include: glass fibers and cloths, asbestos, cotton, metal foils, paper, and graphite, carbon, and synthetic fibers. These are effectively utilized to produce high strength epoxy composites. Other properties gained with reinforcing materials are: low weight, heat and corrosion resistance, and good electrical properties. In addition to pipe, tank, drum, and container linings, laminated glass fabrics are used for pressure vessels that operate at elevated temperatures. Wrappings made of a glass cloth saturated with catalyzed resin can be used to patch leakages in chemical storage and handling equipments. In moulding applications short chopped strands of reinforcement are widely used.

Encapsulation

The good electrical properties, high volume resistivity and low dissipation (power) factor together with low shrinkage during cure, make epoxy resins very important materials in hermetical encapsulations of electric and electronic components. The size of these components range from minute 0.1 mm semiconductors to high-voltage transformers by weight of about 25 kilograms. Additional examples are: transistors, switches, coils, insulators, and microchips. Due to its high mechanical strength the thermoset effectively protects the delicate part against shocks, dust, moisture, and looks.

Practically all of the encapsulations described above are carried out by solid resins. It has been estimated that the world production of these epoxy resins was about 30,000 metric tons in 1980. From this amount nearly 80% was used by the electronics industry for encapsulation of electronic components.

The most common processing method, transfer molding, is simple and permits use of inserts of any shape, material, or number. Other encapsulation methods are: injection molding, powder coating, and casting. In the dry-cote method the component is covered by the resin powder which is then cured by heat in an oven or the powder is sprayed directly onto the heated part. In this method the use of liquid materials is also possible. Potting and dipping are also useful.

Sensitive parts such as those on the data branch require a 100 per cent protection and isolation against electrical disturbances. Epichlorhydrin must be replaced in these applications with a non-chlorine monomer.

Adhesives and Paints

The excellent adhesion properties associated to low
curing shrinkage, heat and chemical resistance, and high
mechanical strength are the most important reasons for the
popularity of epoxy resins as adhesives. Epoxy adhesive kits
commonly available at hardware stores for household repar-
ations contain two tubes: the hardener and the viscous resin.
By mixing equal amounts from both tubes and applying the
mixture to the joint of two materials, good bonds with high
peel strength are obtained.

Because of their strong adhesion to various materials
epoxy paints are often selected to the most critical applications. The content of the jar with the hardener is
thoroughly mixed with the colored resin according to instructions, and then applied during the pot life period taking
care of a proper ventilation.

Coating

In coating applications epoxy resins are usually powdered. Because plastic powders are organic materials,
certain precautions must be undertaken to prevent the risk
of explosion. For this reason the concentration of the epoxy
powder within the booth and recovery unit is to be maintained
sufficiently low. Powders, depending on their specific type,
require a high ignition energy before an explosion can occur,
and then only within certain concentration limits. If solvents
are used the danger is much more obvious due to the lower
ignition energy.

Additives

Epoxy systems generally contain colorants, accelerators,
additives, and fillers, introduced for many practical reasons.
Colorants for epoxies are both inorganic and organic, for
example phthalocyanines.

Resins for moulding usually contain 65-70% of filler by
weight. Fillers are used to reduce the price (silica,
calcium carbonate), to reduce the density (hollow glass-
spheres), to improve impact and tensile strength (glass fibers,
polyamide fibers, carbon fibers), to improve workability

(aluminum powder, calcium carbonate), to improve heat resistance (carbon black, asbestos), for dimension stability (silica, steel powder), for low coefficient of thermal expansion (lithium aluminum silicate), for chemical resistivity (powdered carbon), for low moisture absorption (carbon powder, silica, mica), for good electric isolation (silica, mica), for good electric conductivity (aluminum powder, graphite), for low friction coefficient (graphite, molybdenum disulfide, mica), for abrasion stability (microscopic glass-spheres), and for high pressure resistance (microscopic glass-spheres).

All applications discussed above obviously require experience and experimentations based on sound knowledge of chemical reactions, and an appreciation of the toxicological aspects of many curing agents.

REFERENCES

Lee H, Neville K (1967). "Handbook of Epoxy Resins." New York: McGraw-Hill.

Milby RV (1973). "Plastics Technology." New York: McGraw-Hill.

Miles DC, Briston JH (1965). "Polymer Technology." London: Temple.

Schreiber B (1982). Epoxidformmassen. Kunststoffe 72:7

Industrial Hazards of Plastics and Synthetic Elastomers, pages 373–384
© 1984 Alan R. Liss, Inc., 150 Fifth Ave., New York, NY 10011

GENOTOXICITY OF EPOXIDES AND EPOXY COMPOUNDS

K. Hemminki and H. Vainio

Institute of Occupational Health

Haartmaninkatu 1, SF-00290 Helsinki 29, Finland

Introduction

Epoxides (oxiranes) are useful chemicals due to their spontaneous reactivity. Ethylene oxide ($CH_2 \cdot O \cdot CH_2$), propylene oxide ($CH_3 \cdot CH \cdot O \cdot CH_2$) and epichlorohydrin ($Cl \cdot CH_2 \cdot CH \cdot O \cdot CH_2$) are all synthesized in excess of 1 million tons per year. In addition to being used as precursors for a variety of important organic products, ethylene oxide and propylene oxide are used as chemical sterilising agents in hospitals as well as in pharmaceutical and food industry. The major use of epichlorohydrin is for the production of glycidyl ethers ($R \cdot O \cdot CH_2 \cdot CH \cdot O \cdot CH_2$) used in epoxy resins. In addition to being used as such, epoxides are generated metabolically from many substrates by monooxygenation. Many important industrial products such as ethene, propene, styrene, vinyl chloride and polycyclic aromatic hydrocarbon are known or suspected to be metabolized into epoxide intermediates in the cells.

The present review deals with genotoxic effects (e.g., reactions with nucleic acids, mutagenicity and carcinogenicity) of epoxides and glycidyl ethers with some industrial applications. A comprehensive overview on the genetic toxicity of epoxides has been published recently (Ehrenberg and Hussain 1981).

Mechanism of Action

In a neutral media epoxides react by a bimolecular

(S_N2) mechanism in a second order reaction, i.e., the rate depends on the concentration of the epoxide and the nucleophile. The alkylation of a nucleophile Y^- can be described as:

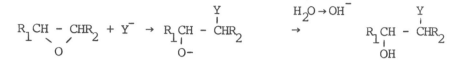

Hydrolysis of an epoxide ($Y^- = H_2O$) yields a glycol. The site of attack of Y^- is the less substituted, i.e., sterically more accessible carbon. Epoxides may also undergo an acid-catalyzed reaction in a low pH, e.g., in the stomach. An acid catalyzed nucleophilic substitution is a unimolecular (S_N1) reaction. It is likely that the alkylation of nucleic acids by the epoxides in a neutral medium (S_N2 mechanism) afford N-7 alkyl guanine as the main product, while a more heterogenous distribution of alkylation products may be expected in the acid reactions (see Lawley 1976).

Reaction Product with Nucleic Acids

In this chapter covalent nucleic acid adducts of epoxides and some compounds that are metabolized to alkyl epoxides will be surveyed (Table 1). Most of the binding studies with epoxides have been carried out in vitro using DNA, nucleosides or deoxyribonucleosides as nucleophiles. Table 1 only lists the presence (+) or absence (-) of an adduct. Brookes and Lawley (1961) studied the reaction products of butadiene oxide and ethylene oxide with guanosine and guanylic acid. For both compounds they identified a 7-alkyl derivative. For the bifunctional butadiene dioxide a di(guanin-7-yl)derivative was also detected. Windmueller and Kaplan (1961) have reported a reaction of ethylene oxide with N-1 of adenosine. Ehrenberg and coworkers have shown that ethylene oxide alkylates guanine at N-7 in vivo (Ehrenberg et al. 1974), and they have suggested that alkylation at DNA phosphate groups also takes place (Walles and Ehrenberg 1968). 2,3-Epoxybutane has been shown to react with N-7 of guanine in vivo (Paul and Pavelka 1971).

Reactions of propylene oxide with DNA, guanosine and adenine were studied by Lawley and Jarman (1972). They described 7-(2-hydroxypropyl)guanine and 3-(2-hydroxypropyl)-

Table 1. Adducts of epoxides and epoxide-forming compounds

Compound	Source	Guanine			Adenine				Cytosine		Reference
		7	1,7	$1-N^2$	1	3	N^6	$1-N^6$	3	$3-N^4$	
Bisphenol A glycidyl ethers (Epikotes 824, 1001, 1003)	In vitro, guanosine	+	-	-	-	-	-	-	-	-	Hemminki and Vainio 1980
Butadiene dioxide	In vitro, guanosine	+	-	-	-	-	-	-	-	-	Brookes and Lawley 1961
2,3-Epoxybutane	In vivo	+	-	-	-	-	-	-	-	-	Paul and Pavelka 1971
Epichlorohydrin	In vitro, deoxy-nucleosides	+	+	-	-	-	+	-	-	-	Hemminki et al. 1980c
Ethylene oxide	In vitro, guanosine, adenosine	+	-	-	+	-	-	-	-	-	Brookes and Lawley 1961; Windmueller and Kaplan 1961
Glycidaldehyde	In vitro, guanosine	-	-	+	-	-	-	-	-	-	Goldschmidt et al. 1968
Glycidol	In vitro, deoxy-nucleosides	+	+	-	-	-	+	-	+	-	Hemminki et al. 1980c
Glycidyl ethers, allyl, butyl, isopropyl, phenyl	In vitro, guanosine	+	-	-	-	-	-	-	-	-	Hemminki and Vainio 1980

Cont'd....

Table 1. Cont'd...

Compound	Source	Guanine			Adenine				Cytosine		Reference
		7	1,7	1-N^2	1	3	N^6	1-N^6	3	3-N^4	
p-Methylstyrene oxide	In vitro, deoxy-guanosine	+	–	–	+	–	–	–	–	–	Sugiura and Goto 1981; Hemminki et al. 1981b
Propylene oxide	In vitro, DNA	+	–	–	+	+	+	–	–	–	Lawley and Jarman 1972; Hemminki et al. 1980c
Styrene oxide	In vitro, deoxy-nucleosides	+	–	–	–	–	–	–	+	–	Hemminki et al. 1980c
Trichloropropylene oxide	In vitro, deoxy-nucleosides	+	+	–	–	–	+	–	+	–	Hemminki et al. 1980c
Vinyl chloride	In vivo	+	–	–	–	–	–	+	–	+	Osterman-Golkar et al. 1977; Green and Hathway 1978; Laib et al. 1981

+ = Adduct detected

adenine in DNA. Reactions with adenine in acetic acid also produced an N-1 and N-9 derivative in addition to dialky-lation products. Hemminki et al. (1980c) studied the alkylation products of deoxyribonucleosides by a series of epoxides with wide range of reactivity (Hemminki 1979) and mutagenicity (Hemminki and Falck 1979): propylene oxide, glycidol, epichlorohydrin, trichloropropylene oxide, and styrene oxide. The following products were identified: all the epoxides reacted with deoxyguanosine at N-7: the most reactive epoxides also formed 1,7-dialkylation products. It has been shown by Lyle et al. (1980) that nucleophilicity of N-1 atom greatly increases in guanosine upon substitution at N-7. Propylene oxide, glycidol and epichlorohydrin reacted with deoxyadenosine at N^6; this may, however, be due to rearrangement from N-1 to N^6, as S_N2 reactions normally involve ring nitrogens. Glycidol, trichloropropylene oxide and styrene oxide were observed to alkylate at N-3 of deoxycytidine. It has been shown later on using a fluorescence technique (Hemminki 1980) that styrene oxide also reacts with DNA and RNA at N-7 guanine (Hemminki and Vainio 1980). The studies with styrene oxide have recently been extended to p-methylstyrene oxide and, 3,5-dimethylstyrene oxide by us (Hemminki et al. 1981b) and to 3,4-dimethyl-, p-methyl-, m-methyl-, m-methoxy-, p-bromo-, m-chlorostyrene oxide by Sugiura and Goto (1981). For all the above compounds N-7 alkylation of guanosine or deoxyguanosine was observed. Sugiura and Goto observed substitution at C-7 as well as at C-8 (main site) of styrene oxide.

Bisphenol A glycidyl ethers, ingredients in epoxy resins were studied fluorometrically. The commercial preparations Epikotes 824, 1001, 1003, reacted with guanosine at N-7 (Hemminki and Vainio 1980). Moreover, allyl, butyl, isopropyl and phenyl glycidyl ethers were shown to react with guanosine to make an N-7 alkyl product (Hemminki and Vainio 1980).

Glycidaldehyde, a bifunctional epoxide-derivative has been shown to react with guanosine in vitro to form a cyclic derivative between N-1 and N^2 (Goldschmidt et al. 1968). This type of adduct is unique, and is known to be formed only by another compound, chloroacetaldehyde.

Vinyl chloride, thought to be metabolized into reactive bifunctional species chloroethylene oxide and chloro-acetaldehyde, has been shown to produce three types of adducts with DNA in vivo (Osterman-Golkar et al. 1977, Green and Hathway 1978, Laib et al. 1981). They include N-7 alkyl guanine (Osterman-Golkar et al. 1977), etheno-adenine, a cyclic derivative between N-1 and N^6, and etheno-cytosine, a cyclic derivative between N-3 and N^4. The latter two products have been detected in a reaction between chloroacetaldehyde and DNA in vitro (Green and Hathway 1978) and in RNA in vivo (Laib and Bolt 1977, 1978). The cyclic structures formed resemble the one described above for the guanine adduct of glycidaldehyde.

Reactivity and Mutagenicity

We have previously carried out studies with epoxides and glycidyl ethers to correlate alkylation rates and mutagenicity to bacteria. Two types of alkylation tests were used: one applying 4-(p-nitrobenzyl)pyridine (Hemminki and Falck 1979) and other using deoxyguanosine in a fluorometric assay (Hemminki 1979). Mutagenicity was determined in a liquid incubation with E.coli WP2 uvrA in the absence of a metabolic activating system. 4-(p-nitrobenzyl)-pyridine reactions (Nelis et al. 1982), and mutagenicity of a number of epoxides have also been reported by other authors (Wade et al. 1978; Wade et al. 1979; Sugiura and Goto 1981). Induction of chromosome aberrations and SCE in cultured human lymphocytes by a series of epoxides was reported by Norppa et al. (1981).

Among the compounds tested by us trichloropropylene oxide was by far the most reactive and mutagenic, accounted for by the strong electronegativity of the chlorine atoms, making the epoxide a powerful nucleophile (Table 2). Epichlorohydrin and epibromohydrin were almost similar in their alkylation rates and mutagenicity, exceeding those of epifluorohydrin. The other epoxides tested, propylene oxide, butylene oxide and butadiene monooxide were markedly less reactive and mutagenic than the halogenated compounds. They were also less active than styrene oxide and glycidol.

Glycidyl ethers, including diglycidyl ethers of bisphenol A and aliphatic monoglycidyl ethers, were tested (Table 2). The reactivity of the bisphenol A derivatives decreased with increasing molecular weight. The same appeared to be true for mutagenicity, although the poor solubility of the high molecular weight epoxy resins interfered with assay. Anyway it was of interest that the epoxy resins were at least

Table 2. Alkylation rates and mutagenicity of epoxides and glycidyl ethers[1]

| Compound | Rate of Alkylation | | Mutagenicity |
	NBP %	dG %	WP2 uvrA %
Trichloropropylene oxide	610	830	460
Epichlorohydrin	100	100	100
Epibromohydrin	105	137	95
Epifluorohydrin	89	85	20
Styrene oxide	51	60[2]	30
Glycidol	38	60[2]	30
Propylene oxide	22	46	10
Butylene oxide	20	37	10
Butadiene monoxide	31	14	20
Epikote 828, mw. 370	57	49	149
Epikote 1001, mw. 1000	15	22	102
Epikote 1004, mw. 1300	5	10	-
Phenyl glycidyl ether	59	141	95
Butyl glycidyl ether	36	66	40
Allyl glycidyl ether	32	63	75
Isopropyl glycidyl ether	40	44	22

[1] As per cent in relation of epichlorohydrin.
[2] Guanosine alkylation.

Sources: Hemminki and Falck 1979; Hemminki 1979; Hemminki et al. 1980a; Hemminki and Vainio 1980.

as mutagenic on the molar basis as epichlorohydrin. The aliphatic glycidyl ethers were also quite active. Phenyl glycidyl ether was more active than the butyl, alkyl and isopropyl derivatives.

In general the studies with epoxides and glycidyl ethers show a good correlation with reactivity, as measured with NBP and deoxyguanosine, and mutagenicity. The presence of electron-attracting substituents, such as halogens, adjacent to the epoxide, increase the activity of the compound. By contrast, the presence of electron releasing groups, such as CH_3- or C_2H_3-, decrease the activity.

Carcinogenicity

Data on the carcinogenicity of epoxides has been surveyed by Van Duuren (1969) and by Ehrenberg and Hussain (1981). Carcinogenicity tests have been carried out by Van Duuren and his colleagues (1969), Weil et al. (1963) and Shimkin et al. (1966). Table 3 contains an updated list of the epoxides for which some positive animal data has been reported. The important industrial epoxides: ethylene oxide, propylene oxide, epichlorohydrin and styrene oxide have been found to be carcinogenic in one or more animal tests. For ethylene oxide human data has also been presented, suggesting that exposure during manufacturing may associate with an increased risk of leukaemia (Högstedt et al. 1979a, b).

Table 3. Epoxides and glycidyl ethers with some evidence on carcinogenicity

Compound	Route			Reference
	skin	subcutaneous	systemic	
Ethylene oxide		+		Dunkelberg 1979
Propylene oxide		+		Dunkelberg 1979
3,4-Epoxy-1-butene	+			Van Duuren 1969
1,2-Epoxyhexadecane	+			Van Duuren 1969
D,L-Diepoxybutane	+	+		Van Duuren 1969
meso-Diepoxybutane	+			Van Duuren 1969
Epichlorohydrin			+	Laskin et al. 1980
Styrene oxide			+	Maltoni et al. 1979

Table 3. Cont'd

Compound	Route			Reference
	skin	subcutaneous	systemic	
Glycidaldehyde	+	+		Van Duuren 1969
Bisphenol A diglycidyl ether	±			Holland et al. 1979
Diglycidyl ether	+			McCammon et al. 1957
Triethylene glycol diglycidyl ether			+	Shimkin et al. 1966

1,2-Epoxyhexadecane as well as D,L- and meso-diepoxybutane have also caused tumours in experimental animals. Data available on the carcinogenicity of glycidyl ethers is scanty. Bisphenol A diglycidyl ether (2,2-bis(p-glycidyloxy-phenol)propane and diglycidyl ether have caused skin tumors in mice (Holland et al. 1979; McCammon et al. 1957), although several negative results with the former have been reported by the chemical industry. Triethylene glycol diglycidyl ether caused lung tumor after an i.p. injection (Shimkin et al. 1966).

Conclusions

The presence of an epoxide group in a small molecular appears to confer ability to react covalently with cellular nucleophiles. The data surveyed here indicate that a number of simple epoxides are mutagenic and many of them have also been found to be carcinogenic in experimental animals. Ethylene oxide and epichlorohydrin have been shown to cause chromosomal aberrations in workers exposed occupationally, and limited data has been presented on the possible cancer risk on humans. Further experimental epidemiological studies are required on the health effects of epoxides and glycidyl ethers.

References

Brookes P, Lawley PD (1961). The alkylation of guanosine and guanylic acid. J Chem Soc 3923-3928.

Dunkelberg, H (1979). On the onkogenic activity of ethylene oxide and propylene oxide in mice. Br J Cancer 39:588.

Ehrenberg L, Hiesche KD, Osterman-Golkar S, Wennberg I (1974). Evaluation of genetic risk of alkylating agents: Tissue dose in the mouse from air contaminated with ethylene oxide. Mutation Res 24:83.

Ehrenberg L, Hussain S (1981). Genetic toxicity of some important epoxides. Mutation Res 86:1.

Goldschmidt BM, Biazej TP, Van Duuren BL (1968). The reaction of guanosine and deoxy-guanosine with glycid-aldehyde. Tetrahedron Lett 13:1583.

Green T, Hathway DE (1978). Interactions of vinyl chloride with rat-liver DNA in vivo. Chem-Biol Interact 22:211.

Hemminki K (1979). Fluorescence study of DNA alkylation by epoxides. Chem-Biol Interact 28:269.

Hemminki K (1980). Identification of guanine-adducts of carcinogens by their fluorescence. Carcinogenesis 1:311.

Hemminki K, Falck K (1979). Correlation of mutagenicity and 4-(p-nitrobenzyl)-pyridine alkylation by epoxides. Toxicol Lett 4:103.

Hemminki K, Falck K, Vainio H (1980a). Comparison of alkylation rates and mutagenicity of directly acting industrial and laboratory chemicals. Epoxides, glycidyl ethers, methylating and ethylating agents, halogenated hydrocarbons, hydrazine derivatives, aldehydes, thiuram and dithiocarbamate derivatives. Arch Toxicol 46: 277.

Hemminki K, Heinonen T, Vainio H (1981b). Alkylation of guanosine and 4-(p-nitrobenzyl)-pyridine by styrene oxide analogues in vitro. Arch Toxicol 49:35.

Hemminki K, Paasivirta J, Kurkirinne T, Virkki L (1980c). Alkylation products of DNA bases by simple epoxides. Chem-Biol Interact 30:259.

Hemminki K, Vainio H (1980). Alkylation of nucleic acid bases by epoxides and glycidyl ethers. In Holmstedt B, Lauwerys R, Mercier M, Roberfroid M (eds): "Mechanisms of toxicity and hazard evaluation," Amsterdam: Elsevier/North Holland, p 241.

Hogstedt C, Malmqvist N, Wadman B (1979a). Leukemia in workers exposed to ethylene oxide. JAMA 241:1132.

Hogstedt C, Rohlen O, Berndtsson BS, Axelson O, Ehrenberg L (1979b). A cohort study on mortality and cancer incidence among employees exposed to various chemicals in the production of ethylene oxide. Br J Ind Med 36:276.

Holland JM, Gosslee DG, Williams NJ (1979). Epidermal carcinogenicity of bis(2,3-epoxycyclopentyl)ether, 2,2-bis-(p-glycidyloxyphenyl)propane, and m-phenylenediamine in male and female C3H and C57BL=7 mice. Cancer Res 39:1718.

Laib RJ, Bolt HM (1977). Alkylation of RNA by vinyl chloride metabolites in vitro and in vivo: formation of 1-N^6-ethenoadenosine. Toxicology 8:185.

Laib RJ, Bolt HM (1978). Formation of 3,N^4-ethenocytidine moieties in RNA by vinyl chloride metabolites in vitro and in vivo. Arch Toxicol 39:235.

Laib RJ, Gwinner LM, Bolt HM (1981). DNA alkylation by vinyl chloride metabolites: etheno derivatives or 7-alkylation of guanine? Chem-Biol Interact 37:219.

Lawley PD (1976). Carcinogenesis by alkylation agents. In Searle CE (ed): "Chemical carcinogens," Monograph 173, Washington, D.C.: American Chemical Society, p 83.

Lawley PD, Jarman M (1972). Alkylation by propylene oxide of deoxyribonucleic acid, adenine, guanosine and deoxyguanylic acid. Biochem J 126:893.

Laskin S, Sellakumar AR, Kuschner M, Nelson N, LaMendola S, Rusch GM, Katz GV, Dulak NC, Albert RE (1980). Inhalation carcinogenicity of epichlorohydrin in noninbred Sprague-Dawley rats. J Natl Cancer Inst 65:751.

Lyle TA, Royer RE, Daub GH, Vander Jagt DL (1980). Reactivity-selectivity properties of reactions of carcinogenic electrophiles and nucleosides: influence of pH on site selectivity. Chem-Biol Interact 29:197.

Maltoni C, Failla G, Kassapidis G (1979). First experimental demonstration of carcinogenic effects of styrene oxide. Med Lav 5:358.

McCammon CJ, Kotin P, Falk HL (1957). The cancerogenic potency of certain diepoxides. Proc Am Assoc Cancer Res 2:229 (Abst).

Norppa H, Hemminki K, Sorsa M, Vainio H (1981). Effect of monosubstituted epoxides on chromosome aberrations and SCE in cultured human lymphocytes. Mutat Res 91:243.

Nelis HJCF, Airy SC, Sinsheimer JE (1982). Comparison of alkylation of nicotinamide and 4-(p-nitrobenzyl)-pyridine for the determination of aliphatic epoxides. Anal Chem 54:213.

Osterman-Golkar S, Hultmark D, Segerbäck D, Calleman CJ, Göthe R, Ehrenberg L, Wachtmeister CA (1977). Alkylation of DNA and proteins in mice exposed to vinyl chloride. Biochem Biophys Res Commun 76:259.

Paul JS, Pavelka MA (1971) Covalent binding of a carcinogenic epoxide to DNA in vivo. Fed Proc 30:448 (Abst).

Shimkin MB, Weisburger JH, Weisburger EK, Gubaref N, Suntzeff V (1966). Bio-assay of 29 alkylating chemicals by the pulmonary-tumor response in strain A mice, J Natl Cancer Inst 36:915.

Sugiura K, Goto M (1981). Mutagenicities of styrene oxide derivatives on bacterial test systems: relationship between mutagenic potencies and chemical reactivity. Chem-Biol Interact 35:71.

Wade DR, Airy SC, Sinsheimer JE (1978). Mutagenicity of aliphatic epoxides. Mutation Res 58:217.

Wade MJ, Moyer JW, Hine CH (1979). Mutation action of a series of epoxides. Mutation Res 66:367.

Walles S, Ehrenberg L (1968). Effects of α-methoxy ethylation on DNA in vitro. Acta Chem Scand 22:2727.

Van Duuren L (1969). Carcinogenic epoxides, lactones, and halo-ethers and their mode of action. Ann NY Acad Sci 163:633.

Weil CS, Condra N, C Haun, Striegel JA (1963). Experimental carcinogenicity and acute toxicity of representative epoxides. Am Ind Hyg Ass J 24:305.

Windmueller HG, Kaplan NO (1961). The preparation and properties of N-hydroxyethyl derivatives of adenosine, adenosine triphosphate, and nicotinamide adenine dinucleotide. J Biol Chem 236:2716.

Industrial Hazards of Plastics and Synthetic Elastomers, pages 385–396
© **1984 Alan R. Liss, Inc., 150 Fifth Ave., New York, NY 10011**

PREVENTING OF HAZARDS: EPOXY THERMOSETS

OLE SVANE AND NIELS KJÆRGAARD JØRGENSEN

DANISH LABOUR INSPECTION SERVICES

16-18, ROSENVÆNGETS ALLE, 2100 COPENHAVEN Ø

INTRODUCTION

Following a discussion in 1978, the Danish Labour Inspection
Service set into force a regulation on epoxy products, polyurethane and comparable substances and materials with similar
dangerous properties. In the following I will present the
contents of the Danish regulations concerning epoxy products,
and some of the difficulties that we have had both in elaborating the order concerning epoxy products and in controlling
whether people observed the rules. After presentation of the
regulation I will give you a short description of one of the
side products of the regulations, that is the registration
of substances and products covered by the order. Finally
follows a discussion of one of the clauses in the order which
contains a request for substitution, i.e. substituting a
dangerous product with a less dangerous product and/or substituting a dangerous process with a less dangerous process.

WHAT HAZARDS ARE WE PREVENTING?

As all constituents of the epoxy products were to be notified
to the work environment authority, we had the possibility of
regulating not only for the hazards related to epoxy resins
and curing agents. The epoxy products usually consist of
resins, curing agents and reactive diluents, e.i. short
chained epoxides with the ability of acting both as solvents
and as part of the cured polymer. Besides of the main constituents, usual products contain accelerators, stabilizers,
emullifiers, fillers and pigments. The public debate focused
on the resins and curing agents. However, the guidelines for

handling the products during the work procedure enabled us
to widen the scope of the prevention beyond the health hazards
of the two main constituents.

The main health hazards which were taken into consideration
were the skin allergenicity of the shorter epoxy resins and
the possible cancer risk of three epoxy resins, i.e. glycidyl
ethers (DGEBA) with MW 360, 720 and 960.

THE DANISH REGULATIONS

The order applies to epoxy products with all their constituents, including admixtures and any toxic, harmful, flammable
and explosive subsidiary agents used in connection with the
products. The order also applies to polyurethane products.
In a subsidiary clause it reads "comparable substances and
materials with similar dangerous properties shall also be
covered". This provision has never been put into use.

a. Registration.

One of the gains of the regulation is that the registration
of a dangerous product type could take place. Each product
has been registered with a total declaration.

b. Labeling and directions for use.

The supplier shall provide the substances with plain, legible
and indelible directions for use in Danish which shall be
glued to, or printed on the container or affixed in any other
way so that they remain with the container until the contents
have been fully used up.

c. Manufacture.

The regulation concerning manufacture is not different from
the general rules in the Danish regulations, i.e. the substances shall not be manufactured, prepared, used or processed unless measures have been taken to dispose of the
pollution of the air in the breathing zone, normally by mechanical ventilation.

d. Restriction of Application.

A new limitation is mentioned in the rules concerning spraying
with the products. Outside closed systems, spray-cabins or
spray-booths provided with effective ventilation spraying
is forbidden. This means that spraying with epoxy products
cannot take place in ship building. The clause explicitly
mentions the procedures which may be followed instead of
spraying, namely using filling knives, brushes, paint rol-
lers, joint-guns and other similar simple technical aids.
Actually this has had the consequence that any dispensation
from the mentioned rule must be discussed with the two
social partners in the actual work place, and the Labour Inspection. Dispensation has been given to a very limited extent.

e. Handling provisions.

The main point is that the work shall be organized and performed in such a manner, that any contact with skin is
avoided. This means that the person performing the work shall
wear suitable disposable gloves, under these fabric gloves,
and if needed a face protection screen and suitable working
clothes which fit tightly to the wrists and to the neck.
The consequence was that new types of disposable clothes
and gloves were manufactured. Very often they were made of
plastic and were more or less uncomfortable to wear. The
regulation requires that they are changed frequently and
always when they have been dirtied to such a degree that
there is a risk of skin contact. Even under these cirsumstan-
ces we envisaged that the personal protective equipment might
lead to physical and psychical strains as a consequence of
the nature of the work and temperature conditions. Therefore
a special clause allows the time spent by the employee at
work to be reduced for example by introducing special rest
periods in the course of the daily work.

f. Personal Hygienic Provisions.

As the contamination of the skin is considered a serious
danger, the provisions for good skin hygiene are quite strict.
There shall be unlimited access to handbasins with running
warm water at the place of work and, as far as possible,
in the very room where the work with the substances is being

carried out. The water taps shall be operated by the elbow
or the foot. In addition there shall be access to shower-
rooms with warm and cold water. In connection with the
washing facilities there shall be a suitable cleaning agent,
mild soap, soft dispensable towels, protective cream and fat
cream for application after washing and a waste container.
As quite often seen in Danish regulations persons who are
working with the substances shall have access to separate
changing rooms, where the ordinary clothes and the working
clothes shall be kept apart.

Access to the place where the substances are being handled
shall be limited, so that unauthorized employees or other
persons are kept out. That is the consequence of the mandatory training course.

g. Special Training Course.

Work may only be performed by persons who have undergone a
special training programme which consists of 16 lessons paid
by the employers. This is the first instance apart from the
pesticide area that courses are made mandatory for workers,
their own occupational health being the reason for the course.

j. Exclusion of Diseased Persons.

Persons suffering from eczema or confirmed epoxy allergy shall
not be allowed to work with epoxy products. Persons with
excessive perspiration of the hands (hyperhirdrosis manuum)
shall not be allowed to work with the substances either.

As an end result of the above mentioned preventive restric-
tions directions for use must be elaborated. It is the responsibility of the importer or the manufacturer to supply
the directions for use. The Labour Inspection can approve
the directions and certify that some particulars are mentioned, namely:

1) the designation of the substance and its fields of application
2) any safety measures to be observed
3) any risk in connection with the handling of the substance
4) notification of the substance to the Labour Inspection
   Service

5) name and address of the supplier

Having accepted the directions for use, the authorities will also be able to decide on labeling of the container. As a main rule the EEC-directives are followed, but in a few instances the Danish authorities have preferred to label more restrictively than described by the EEC-rules.

CONSIDERATIONS IN ELABORATING THE REGULATIONS - OR DIFFICULTIES.

The cause for the public discussion on epoxy thermosets was that a Danish group of researchers had confirmed a result on the mutagenicity of three of the most used epoxy resins. As described by Drs. Hemminki and Vainio the cancerogenicity in man of epoxy resins is still uncertain. During the work of elaborating the regulation it was made still more evident that other substances than the resins might be quite interesting from a toxicological point of view: First of all the so called reactive diluents, which means short-chain epoxides which act both as solvents and as part of the polymer network when the polymer has cured. The reactive diluents differ from the organic solvents in that they contain one or several epoxy groups and thereby form a chemical compound with the curing agent. By that means there is only slight evaporation of the product during drying or hardening and therefore only a small reduction in volume. This is of great importance in glue, joining material, and putty. About half of the reactive diluents were not classified according to EEC-rules, which meant that a considerable workload was placed in the Labour Inspection as the authority had to approve the directions for use, that is, to consider the fields of application and the possible exposure of skin and airways.

Other components, such as accellerators and stabilisors besides of solvents might also be as interesting as the original resin. Among the pigments several carcinogens were known, some of the emullifiers might also attract interest. In short: the question rose if it was possible to formulate a direction for use or subsequently to approve an importer's or producer's proposal of a direction for use. The solution has been to propose amendments to the proposals put forward by the producers and importers.

As a byproduct of the work with the directions of use some
earlier white spots on the map of regulations were discovered. For instance the rules of wearing masks with fresh air
supply were specified by a concession between the authority
and the two social partners: wearing masks with fresh air
supply during work which is strenuous was only allowed for
a total of four hours interspaced by at least one pause
after two hours. These rules have been taken over in other
areas.

It might be possible to find other forms of regulations
concerning the persons involved in the work, other than
training and exempting the few people with diagnosed eczema
and allergy. The usual way has earlier been to start a program of health surveillance. This was omitted as we in Denmark consider most health surveillance programs as an admit-
tance of the fact that the existing hygienic measures are
not sufficient. The question will be considered further in
respect of the isocyanate health hazards.

EXPERIENCES WITH THE REGULATIONS IN FORCE.

The impression is that all products imported to or produced
in Denmark are registered because epoxy thermosets are easily
identifiable, as they are a two component system which are
mixed just before use. We think that the declarations are
quite precise, but we also know that the declarations are
often changed as time goes, and that subsequent declarations
may not be notified. The number of new declarations per month
is so numerous that you may believe that you are dealing with
a normal - and full - number of notifications.

One of the most controversial points of the regulations has
been the mandatory training course of 16 hours, i.e. two
whole work-days, paid by the employer. The course is accepted by the "normal" employer, but also workers in computer
component production placing one drop of epoxy glue may themselves have difficulties in accepting a course.

A PRODUCTS REGISTER AT WORK.

One of the byproducts of the regulation has been that all
products imported and produced in Denmark are registered.

The registration has the following main points:

- identification of the name of the product, importer/produ-
  cer
- modes of application
- amount produced/imported
- declaration with specification
- warning, instruction for use
- toxicology: acute effects
                 irritation
                 allergens
                 long term effects
- disposal
- physical/chemical data

Apart from the purpose of simple control a register gives
the opportunity to analyse the total contents of the register
for a certain product such as epoxy thermosets. In the following I will present a few results on the reactive diluents.
Out of 600 declarations, 200 indicated that reactive diluents
were present. The group comprised of 26 different solvents
of which the 7 mostly used reactive solvents are presented
in figure 1. The material of this statement of account are
notifications received during the first year by the registry.
As seen from the figure the symbols A (allergen), M (mutagenic) appear in several of the most used reactive solvents.
Some of the most toxicologically active reactive solvents are
widely used in several products as seen in figure i. A common feature is that a fast acting reactive diluent is very
often used in epoxy thermosets which are dispersed in a
water base.

In another statement of account the contents of the registry
has been analysed in respect of the contents of organic solvents in epoxy and polyurethane products. As may be seen from
figure 2 the number of products vary from 971 (paints) to
20 (printing colours). 61 per cent of the printing colours
contain organic solvents, 51 per cent of the paints. 41 per
cent of the paints contain aromatic hydrocarbons, which is
the case for only 15 per cent of the printing colours.

A further analysis of all components of the declarations

may be pursued. After an initial mapping, it seems clear
.that defined purposes of analysing may render useful information. Some of the purposes may be: are there components
which are so unwanted that they should be discussed with
the producers with the purpose of eliminating them from the
market? Another purpose might be to compare products with
apparently similar applications, but with different toxicological properties.

SUBSTITUTION: DREAM OR REALITY?

A clause in the regulations runs: "The substances shall only
be used where experience has shown that the performance ob-
tainable by the use of another, less hazardous substance is
unsatisfactory".

This is an example of the principle of substitution which was
laid down in the Work Environment Act of 1977. A principle
which has been very much discussed during the years following
the enforcement of the law.

In simple words, two forms for substitution may occur:
Substitution of methods and substitution of substances.
Taking the actual example a substitution of methods might
mean that instead of using epoxy thermosets for joining
two metal parts you may prefer welding metal or you may prefer not to produce two metal parts but only one containing
both functions.

What is mentioned in the regulations on epoxy is the day to
day considerations:
Is it possible to use one product instead of another, obtai-
ning the same production results with less health risks?
After all the substitution of substances gives rise to the
following sub-questions:
- what is the technological goal of the process?
- which are the health risks of the process?
- what are the economic implications of the two?

The usual (and questionable) way of answering these questions is that you accept question number one in such a way
that the answer to the question is decided by technical parameters, for example the sellability, strength, durability
etc. of the product. If, however, the answer to the technological question is established the remaining problem is a

cost-effectiveness consideration. It means that the goal is
unquestionably defined but the cost may vary according to
the process chosen to obtain the results. The costs may be
of two kinds, namely: economic costs and health costs.

Looking back on the examples of constituents of certain epoxy
thermoset products, for instance in paints and in printing
colours, it seems obvious that specific considerations can
be taken. In the case of paints the possibility of exchan-
ging one product with another may be greater than in the
printing process. If two or three products are technologi-
cally acceptable, a toxicological evaluation will be the
means of giving the grounds for the choice of substitution.
Even if the amount of work seems insurmountable at the beginning, the substitution has been enforced with ordinary
paints in recent Danish regulations (3). In the regulation
of works with paints a simplified labeling system has been
ordered. Figure 3 shows an example of the labeling. Comparing the two labels shown in figure 3 the user gets the information by comparing the symbols before the hyphen, that
container A contains a fluid (solvent) which is more acceptable to his health than that in container B. By comparing
the symbols after the hyphen, the consumer is informed
that the epoxy compound in container A has the same accept-
ability from a health point of view as that in container B.
The consumer gets a very short piece of information - simple
and sometimes oversimplified. The labeling is in use today
and the choice is restricted, i.e.the more "healthy" product
for a certain production is mandatory. The substitution of
substances is a reality.

List of literature:
(1) Order concerning Polyurethane and Epoxy Products and
     comparable substances and materials with similar
     dangerous properties.
(2) Epoxy. Mange epoxy- og polyurethanprodukter indeholder
     organiske opløsningsmidler; Lisbet Seedorff and Arne
     Skov, Arbejdsmiljøinstituttet. Pas På! 10:81.
     Copenhagen.
(3) Order concerning Occupationally Related Painting.
     Danish Labour Inspection Service Order No.463 dated
     August 3rd, 1982 (in Danish: Bekendtgørelse om
     erhvervsmæssigt malearbejde).

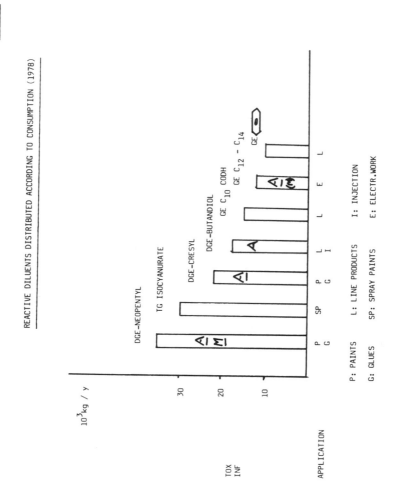

FIGURE 1

REACTIVE DILUENTS DISTRIBUTED ACCORDING TO CONSUMPTION (1978)

P: PAINTS   L: LINE PRODUCTS   I: INJECTION

G: GLUES   SP: SPRAY PAINTS   E: ELECTR.WORK

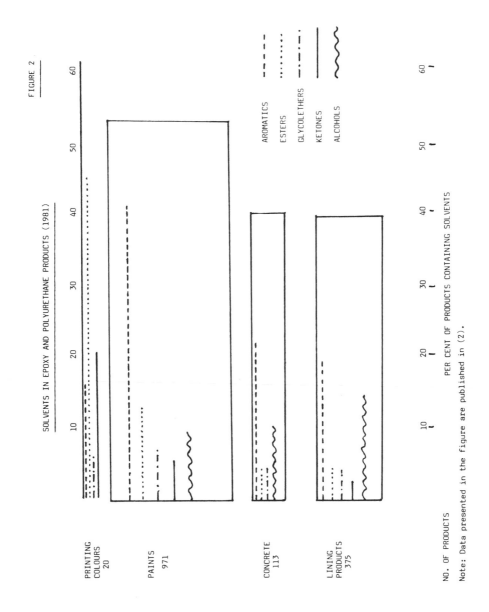

FIGURE 2

SOLVENTS IN EPOXY AND POLYURETHANE PRODUCTS (1981)

PRINTING
COLOURS
20

PAINTS
971

CONCRETE
113

LINING
PRODUCTS
375

NO. OF PRODUCTS

AROMATICS
ESTERS
GLYCOLETHERS
KETONES
ALCOHOLS

PER CENT OF PRODUCTS CONTAINING SOLVENTS

Note: Data presented in the figure are published in (2).

FIGURE 3

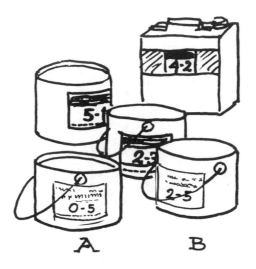

Industrial Hazards of Plastics and Synthetic Elastomers, pages 397–405
© 1984 Alan R. Liss, Inc., 150 Fifth Ave., New York, NY 10011

PRODUCTION AND PROCESSING OF SYNTHETIC ELASTOMERS

by E. Enwald, Industrial hygienist,

Nokia Corporation, Nokia Local Administration

Medical Department, PL 35 SF-37101 NOKIA

1. TYPES OF ELASTOMERS

Synthetic elastomers are mainly used in rubber industry together with or instead of natural rubber.

The most consumed types are

Abbreviation	Name
IR	Isoprene rubber, synthetic
BR	Butadiene rubber
CR	Chloroprene rubber
IIR	Isobutylene-isoprene 1. butyl rubber
NBR	Butadieneacrylonitrile 1. nitrile rubber
SBR	Styrene-butadiene rubber
EPM	Ethylene-propylene copolymer
EPDM	Ethylene-propylene-diene terpolymer
IM	Polyisobutylene
CSM	Chlorosulphonated polyeten
AU	Urethan rubber, ester-type
EU	Urethan rubber, ester-type
CO	Epichlorhydrin rubber, polychlormethyloxyran
ECO	Epichlorhydrin copolymer, ethyleneoxydchlormethyloxyran copolymer

Chemically cis-isoprene rubber resembles
natural rubber.

2. PRODUCTION

Synthetic elastomers are usually produced by
polymerization of monomeric basematerials. For
instance the needed   monomers for styrene-
butadiene rubber - production are styrene and
butadiene. They are producted from petroleum
sources. SBR may be produced by emulsion or by
solution polymerization. In emulsion made styrene-
butadiene rubber is produced by free radically
initiated emulsion polymerisation.

So called cold-SBRs are polymerized at about $5^{o}C$
using a redox initiating system. Polymerization
takes place in great cictern after that follows
decentation, coagulation - in this stage an
antioxidant is added -, water deduct, grinding,
drying, baling and packing. Solution polymerization is carried out in solution. Solvents
are hydrocarbons and lithium catalyst system
are used.

3. PROCESSING OF ELASTOMERS

Processing of elastomers means manufacturing of
rubber goods. The basic manufacturing operations
are quite similar in any rubber factory:

- storage of raw materials
- weighing of raw materials
- mixing of raw materials
- milling of compounds
- calendering or extruding of compounds
- cutting of components
- preparatory assembly
- assembly and building
- "curing" = vulcanizing
- inspecting and finishing
- packing
- storage and dispatch

The next sections include some data of measurements made in rubber factories in Finland on period from 1980-01-01 to 1982-06-30.

3.1. STORAGE OF RAW MATERIALS

Different kinds of rubber products, e.g. tyres, boots, rolls for paper machines, require different kinds of raw materials used in rubber compounding, but the basic components of the rubber mixture are:

Elastomers
- natural or synth. rubber
- "master batches" = some other raw materials (oil, carbon black, colours, sulpur) are precompounded with rubber

Peptizing agent
is added to the crude rubber in order to chemically break it down. It may be pentachlorthiophenol or its zinksalt.

Curing agents
are required in order to obtain crosslinking between the rubber molecules. Usually it is sulphur for diene containing elastomers. Saturated polymers are cured with peroxides.

Accelerators
promote a reaction between sulphur and rubber molecule, the reaction is usually quite slow. So accelerators increase the curing rate, but they also gain other advantages to rubber manufacturing. The most important types are:
aminederivates
aldehyde-amine condensation products
guanidinederivates
thiazoles
thiurams
dithiocarbamates
xantates and xantogenates

Activators
make accelerators more effective, they are mainly metal oxides: ZnO, MgO, PbO, Pb_3O_4. Some organic acids, such as stearic acid, are used together with metal oxides.

Antidegradants
- antioxidant is a commonly used collective name
for agents which decrease the deterioration of
rubber caused by heat, azon, flexing etc.
Chemically two main categories are:
amines and their derivates
phenols and their derivates
- antiozonants give a special protection against
ozon
- waxes and sun-checking agents, waxes mixed in
rubber blends are often acting as weathering
agents, but they also act as processing aid

Fillers
-inorganic reinforcing agents and fillers, the
most used are whiting $(CaCO_3)$, silicates, clays
and synthetic silica
- carbon blacks are made by partial combustion of
natural gas or oil. In rubber compounding they
are the most used reinforcing ingredients
- other organic fillers

Plasticizers and softeners
are usually asters e.g. adipates, phthalates,
selecates, stearates, phosphates or mineral oils

Tackifiers
- rosins and resins

Dyestuffs
- organic pigments
- TiO_2, iron pigments

Retardes, blowing agents, parfyms, flame
retarders, a.s.o.

Beside the compounding substances the other types
of materials are needed in manufacturing of rubber
products, e.g.
- antitack agents
- solvents
- bonding agents and adhesives
- mould release agents
- finishing materials

Problems mainly arise with dust.

Packages may be torn during transport. When
powderform materials are transferred to silos
and tanks, rather high amounts of dust may be
found on the breathing zone of the workers:
filling carbonblack silos up to 7 mg/m^3 and mineral
filler silos up to 14 mg/m^3. In the storehouses
for carbon blacks there may occur some amounts of
carbonmonoxide, if the ventilation is not adequate.

3.2. MIXING DEPARTMENT

The weighing and mixing procedures of raw materials
form nowadays an integrated whole, where materials
are handled quite automatically in enclosed
systems. Still there are plenty of old factories,
where parts of compounding materials have to be
weighed by hand. These substances are usually
accelerators, antioxidants and retarders, which
include the most hazardous substances of rubber
chemicals. A rather typical formulation may
look out like this:
- elastomers   100 parts
- fillers   3 "
- stearic oils   1 "
- zinc oxide   4 "
- plasticizing oils   10 "
- accelerators   2 "
- antioxidants   3 "
- resins   10 "
- sulphur   2 "

In the breathing zone of weighers the amounts of
dust have been between 2.1-4.4 mg/m^3, but in bad
working conditions even 11-25 mg/m^3. By weighing
mineral fillers the dust contents in br.z. have
been 2-21 mg/m^3.

Automatically or manually weighed matherials are
transferred into the chamber of mixing machine,
where all compounding ingredients are mixed up.
Usually two mixers are used. In the first one
so called premixtures are made, they may be
be stored for some time before the next stage.
In the second mixer the vulcanizing agents are
added, after which rubber compounds must be
processed on.

The machine-man may be exposed to the fumes
rising from mixer and to the dusts of the weighed
rubber chemicals, which he adds to mixer.
The fumes consist of the volative ingredients of
the rubber mixure: moisture, plasticizers, oils,
some antioxidants, monomeric impurities of
elastomers, a.s.o.

The dust concentrations may be high near mixers,
0.3-23 mg/m^3. The highest values are measured
near such machines, where preweighed chemicals
are added by pouring from open vessel.

The rubber compound is discharged from mixer into
a two-roll mill for sheeting off. In order to
prevent the uncured compound slabs adhering to
each other, the sheet belt is leaded through a
synthetic soap solution.

3.3. MILLING, EXTRUDING AND CALANDERING DEPARTMENTS

The rubber compounds are passed through one or
more milling machines - usually two-roll mills -,
where the even dispersion of the ingredients of
rubber compound occurs.

After milling warmed and softened compound
strips are transferred usually by conveyor belt
to calenders or to extrudes.

Multiple-roll calenders form the rubber compound
to thin sheet or apply it into textilefabric.
Calendered sheets may be wound off on to a roll,
textile or plastic sheets are used to prevent
the layers to stick each other, sometimes other
antitack agents e.g. talc or zincstearate are used.

In extruders rubber compounds are forced through
an appropriately shaped die, extruded rubber tape
is then cut into suitable lengths and set between
textile, plastic or metal sheets.

The main health hazards in these departments are
formed by fumes released from hot rubber compounds
and of rather high noise level near milling machines.

3.4. CUTTING OF COMPONENTS

The calendered rubber sheets and textile
reinforced rubber sheets are cut into peaces
sheet by sheet or in multilayer packs. The
cutting components may be done by sawing or
stamping with matrixis.

Textiles and some antitack agents can give rise
to substantial quantities of airborn dust. Also
noise levels may be high in some working places,
especially the impulse soundes of cutting strokes
may cause some hearing loss.

3.5. COMPONENT ASSEMBLY AND BUILDING

Rubber components are usually assembled manually.
Rubber parts which are transferred from foregoing
stages between textiles or plastic sheets can be
assembled to each other without any adhesive,
but usually surfaces are freshed with some solvent.

Nowadays the most used solvent is a mixture of
heptene based aliphatichydrocarbons, which
contains no aromatic hydrocarbon. Only in some
special cases, e.g. manufacturing oil resistant
rubber articles or when rubber to metal bonding
is needed, more dangerous solvents, such as
toluene, xylene, perchlorethylene or methylethyl-
ketone, are used in addhesives. Thus the rubber
products are assembled by sticking prepared parts
together. The greatest heath hazard is formed
of volatilised solvents. If petroleum spirits
with no or minimal benzene content are used in
large well-ventilated work areas with local
ventilation installations in some special working
place, the vapor concentrations can be kept on
satisfactory low levels.

3.6. VULCANIZING OR CURING

After assembling the rubber products have to be
vulcanized or cured before they are ready for use.
Usually heat is applied to the product under
pressure, during vulcanisation the vulcanising
agents react with rubber molecules and effect
crosslinking.

Some products are vulcanised in presses or molds
(e.g. tyres, tubes, vibration dampers), others
in autoclaves (e.g. rubber boots assembled on
metal lasts). Continuous curing systems are
to produce rubber tapes, profiles, hoses, a.s.o.

Heat conditions may be rather bad in the
vulcanizing departments and when the presses
or autoclaves are opened for removing curred
products, released fumes tend to create
environmental problems.

The composition of curing fumes is varying
greatly depending on the vulcanizing temperature
and presure, and the kind and amount of
defferent ingredients of rubber compounds used
to manufacture the rubber product.

Vapours may contain compounded ingredients as
such or impurities of raw materials. During
curing process many chemical reactions take
place producing also some volatile and toxic
substances.

The amount of different kinds of polycyclic
aromatic hyrdocarbons in curing fumes have
been investicated and e.g. bentso(a)pyrene
concentration were as follows:

10-80	ng/m^3	curing of	car tyres
4-170	"	"	butyltubes
24-32	"	"	catepillar belts
c. 17	"	"	rubber boots

PAH-samples were collected on glass fiber
filters from breathing zones of operators or
from working area very near of the sources of
vulcanizing fumes.

3.7. INSPECTION, FINISHING AND PACKING

Many of remaining procedures are made by hand,
so after vulcanisation hot rubber products are
usually cooled, hot articles may also still
involve fumes. Nowadays nearly every tyre is
going through x-ray inspection, operators are

usually well protected. Because the
finishing operations may be e.g. grinding,
trimming, repair, painting or cleaning , so
in this department workers may be exposed to
solvents, rubber dust and even to fumes of
pyrolysed rubber. In some packing departments
packing machines may cause high noise levels.

LITTERATURE

Hofman W (1980) Kautschuk-technologie. Gentner
Verlag Stuttgart.

Brydson JA (1978) Rubber Chemistry. Applied
Science Publichers Ltd, London.

WHO, IARC Monographs on the Evaluation of the
Carcinogenic Risk of Chemicals to Humans.
The Rubber Industry, Volume 28 IARC, Lyon,
Frence, April 1982.

Industrial Hazards of Plastics and Synthetic Elastomers, pages 407–419
© 1984 Alan R. Liss, Inc., 150 Fifth Ave., New York, NY 10011

GENOTOXIC EFFECTS OF ADDITIVES IN SYNTHETIC ELASTOMERS WITH
SPECIAL CONSIDERATION TO THE MECHANISM OF ACTION OF THIURA-
MES AND DITHIOCARBAMATES

Agneta Rannug, Ulf Rannug, and Claes Ramel

Division of Toxicological Genetics, Wallenberg
Laboratory, University of Stockholm, S-106 91
Stockholm, Sweden

INTRODUCTION

The widespread use of rubber and of additives to crude
rubber warrants the evaluation of possible health hazards in
their production and use. Many synthetic polymers used as
such or as co-polymers in rubber processing are associated
with mutagenic and carcinogenic effects in vitro and in vivo
(IARC, 1982). Butadiene is mutagenic towards Salmonella
typhimurium and recently also detected as a rodent carcinogen (Huff, 1982). Acrylonitrile is positive in animal carcinogenesis studies as well as in mutagenicity studies in
vitro (Fishbein, 1982). Chloroprene is embryotoxic, teratogenic, and mutagenic (IARC, 1979). Epichlorohydrin is positive in several in vitro test systems for mutagenicity and
have been found carcinogenic in mice (IARC, 1976). Styrene
the most commonly used polymer for rubber production also
poses a potential carcinogenic risk to man by its conversion
to styrene oxide, a substance that is mutagenic and which
also is carcinogenic in animals (Huff, 1982).

Usually each rubber product contains up to ten additives
beside the rubber polymer. The rubber additives are combined
with the polymers to give the products the desired properties, to facilitate processing, to extend the crude rubber,
and to give it a good finish. Approximately 14 classes of
additives are used for these purposes. In most cases little
is known of the toxicological properties of such additives
and in many cases they are not chemically well defined and
thus not to be found in the toxicological literature.

A toxicologic survey of rubber chemicals in Sweden by Holmberg and Sjöström (1977) indicated the use of at least 500 additives in Swedish manufacturing industries. 46 of these were selected for mutagenicity studies in a battery of short-term tests. Results from the recessive lethal tests on Drosophila, the point mutation tests on Chinese hamster V79-cells, the tests for induction of sister chromatid exchanges in Chinese hamster ovary cells and the induction of micronucleus in bone marrow cells of Chinese hamsters will be presented elsewhere (Donner et al., 1982). This paper will give the results from the Salmonella/microsome mutagenicity tests conducted on all 46 rubber additives.

Compounds were chosen for biological testing from four classes of additives, namely; curing agents, accelerators, antioxidants and retarders. All four classes contained mutagenic compounds. Most mutagens were found among accelerators and antioxidants.

Several compounds of the thiuram and dithiocarbamate types, with extensive use as accelerators, were found mutagenic at an early stage of these investigations (Hedenstedt et al., 1979). Tetraethylthiuram disulfide (TETD), an accelerator with many important other applications as well, in later experiments turned out to be a mutagen with the optimal mutagenicity found at low nontoxic concentrations . For these types of sulfurcontaining compounds the mechanism of biological and especially DNA-damaging action is complex and not fully understood. A more in detail study of their genotoxic effects has therefore been undertaken. A hypothesis explaining their mutagenic action via an indirect function on enzymes involved in the protection against oxygen radicals, is also presented.

The problem of identifying biologically active compounds in polymer processing industries is not solved simply by testing polymers and additives per se. Due to the formation of thermodegradation products during processing new biologically active substances emerge (Hedenstedt, 1981, Rannug--Hedenstedt and Östman, 1982, and Hoff et al., 1982).

MATERIALS AND METHODS

Samples of the additives used in this study were provided by Swedish and Finnish rubber factories and they were of technical purity. For the purpose of mechanism studies of tetramethylthiuram disulfide (TMTD), a recrystallized compound was used.

The mutagenesis assay was carried out as the plate-incorporation test according to Ames et al., 1975. The toxicity of each compound was determined in advance on a complete medium, and the doses to be tested for mutagenicity were chosen in regard to the toxic levels. S9 fractions were prepared from the pooled livers of eight male Sprague-Dawley rats that had been injected with Aroclor 1254 5 d before they were killed. The S9-mix contained 10% S9 fraction. The dependence on a NADPH-generating system was tested by exclusion of NADP from the S9-mix. His revertant colonies on four replicate plates per dose were counted after 48 h incubation at 37°C. When anaerobic treatment conditions were aimed at, jars equipped with a hydrogen-carbon dioxide generator, Pt catalyst and oxygen indicators were used to incubate the plates. Treatment in oxygen enriched atmosphere was carried out by exposing the bacteria in dessicators to a slow continuous streem of oxygen for three minutes prior to the incubation. The Salmonella strains used were originally provided by B. Ames, and the mutability of the strains and their genotypes were regularly checked by recommended procedures (Ames et al., 1975). Statistical analyses were performed with Student's t-test with the following significance levels: non-significant $P > 0.05$; $*$ $0.01 < P < 0.05$; $**$ $0.001 < P < 0.01$; $***$ $P \leqslant 0.001$.

RESULTS

Of 45 rubber additives tested for mutagenicity in the Salmonella microsome mutagenicity assay 16 compounds were not found to have any mutagenic properties when tested on strains TA1535, TA1537, TA1538, TA98 and TA100 with and without S9. These substances are indicated with - in table 1.

For another 15 compounds a negative response was also obtained in the tests. These substances were however not toxic for the Salmonella strains in higher concentrations and therefore the risk for a false negative test result, due to low solubility and insufficient penetration of the bacterial membranes, exists. In table 1 the result from such compounds are indicated with (-).

Table 1. Mutagenicity of rubber compounds. The lowest effective dose, ug per plate, is given in parenthesis.

CURING AGENTS

Dicumylperoxide	(-)	
2,5-Di-methyl-2,5-di(t-butylperoxi)hexane	(-)	
Dipentamethylenethiuram tetrasulfid (PTT)	-	
Dipentamethylenethiuram hexasulfid (PTH)	-	
4,4-Dithiodimorpholine	+	(100)

ACCELERATORS

N,N´-Diphenylguanidine (DPG)	-	
1,3-Diphenylthiocarbamide	-	
o-Totylbiguanide	(-)	
N,N´-Di-o-totylguanidine (DoTG)	(-)	
1,3-Diethyl-2-thiocarbamide (DETU)	(-)	
Zinc-2-mercaptobenzothiazole (Zn-MBT)	(-)	
2-Mercaptobenzothiazole (MBT)	(-)	
2,2-Dibenzothiazyldisulfide (MBTS)	(-)	
Tetramethylthiocarbamide	(-)	
Hexamethylenetetramine	(-)	
N-Cyclohexyl-2-benzothiazylsulfenamide (CBS)	(-)	
Tetramethylthiuramdisulfide (TMTD)	+	(5)
Tetraethylthiuramdisulfide (TETD)	+	(20)
Tetramethylthiurammonosulfide (TMTM)	+	(5)
Zinc dimethyldithiocarbamate (Ziram)	+	(5)
Copper dimethyldithiocarbamate	+	(1000)
Zinc diethyldithiocarbamate	+	(25)
Cadmium diethyldithiocarbamate	+	(10)
Tellurium diethyldithiocarbamate	-	
Zinc dibutyldithiocarbamate	-	
Nickel dibutyldithiocarbamate	-	
Zinc ethylphenyldithiocarbamate	+	(100)
Piperidine pentamethylenedithiocarbamate	-	
N,N´-Dimethyldiphenylthiuramdisulfid	(-)	
Dimethyldithiocarbamate Na$^+$salt (DTCA)	+	(5)

ANTIOXIDANTS

6-Ethoxy-1,2-dihydro-2,2,4-trimethylquinoline	+	(200)
Polymer. 2,2,4-Trimethyl-1,2-dihydroquinoline	(-)	
N,N´-Diphenyl-p-phenylenediamine (DPPD)	+	(10)
N-Phenyl-N´-isopropyl-p-phenylenediamine (IPPD)	-	
N-(1,3-Dimethylbutyl)-N´phenyl-PPD (6PPD)	-	
N,N´-Dicyclohexyl-PPD	+	(50)
N,N´-Bis(1,4-dimethylpentyl)-PPD	-	
2,6-Di-t-butyl-1-hydroxy-4-toluene (BHT)	(-)	
4,4´-Diaminodiphenylmethane	+	(50)
2-Mercaptobenzimidazole	-	
Aldol-α-naphtylamine	(-)	
Fenyl-α-naphtylamine	-	
Fenyl-β-naphtylamine	-	

RETARDERS

N-Nitrosodiphenylamine	-	
N-(Cyclohexylthio)-phthalimide	-	
N-Methyl-N-4-dinitrosoaniline	+	(5)

Curing agents

4',4-Dithiodimorpholine acted as a weak direct mutagen
of a base-pair substitution type. The mutagenicity on TA100
was diminished in the presence of S9 (Figure 1).

Antioxidants

Four antioxidants - 6-ethoxy-1,2-dihydro-2,2,4-tri-
methylquinoline, N,N'-diphenyl-p-phenylenediamine (DPPD),
N,N'-dicyclohexyl-p-phenylenediamine and 4,4'-diaminodiphe-
nylmethane - caused mutations in the presence of a metabolizing system plus cofactors on TA1538 and TA98 (Figure 1 and
2). They are relatively weak mutagens in this test system,
inducing revertants at or above the dose 50 μg per plate,
with the exeption of N,N'- diphenyl-p-phenylenediamine being
a little more active. These substances were all negative on
strain TA1535 and therefore the mutagenic action is of
frame-shift character although a certain effect was also
noticed with strain TA100 (data not shown).

Retarders

N-Methyl-N-4-dinitrosoaniline is a potent mutagen for
both base-pair substitution and frameshift sensitive strains
of Salmonella typhimurium including TA1535, TA1538, TA100,
and TA98. The dose response curve for TA1538 is shown in
figure 1. The mutagenicity was enhanced in the presence of a
complete metabolizing system. As shown in the figure the
mutagenicity was noticeably reduced by a metabolizing system
deficient of NADP.

Accelerators

Several accelerators were tested for mutagenicity. Only
those of the thiuram and dithiocarbamate types were found
positive in this study. For benzothiazoles and sulphenamides
the limited solubility in the water based test medium prevented a thorough investigation over a wide dose range.

The thiurames and dithiocarbamates gave mutagenic
effects at relatively low doses. In table 1 the lowest
effective doses are presented for 9 mutagenic accelerators
of this type. TMTD, one of the most active of the thiurames

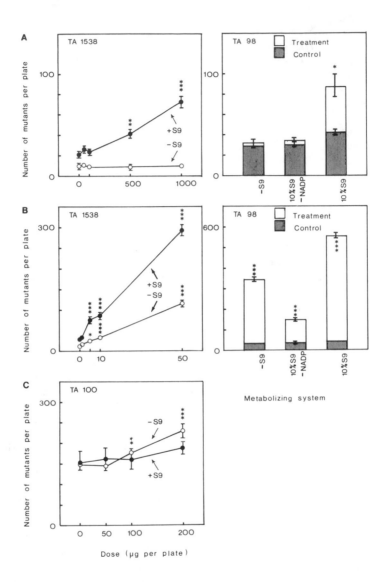

Figure 1. Mutagenicity of 4,4-diaminodiphenylmethane (A), N-methyl-N-4-dinitrosoaniline (B), and 4´,4-dithiodimorpholine (C) tested with and without S9. Test for cofactor dependency of the metabolic activation is shown to the right. The mean number of revertants from four plates are given ± S.E. Statistical evaluations described in "Methods".

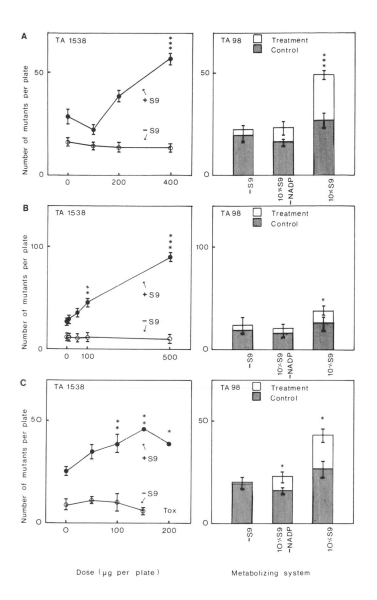

Figure 2. Mutagenicity of 6-ethoxy-1,2-dihydro-2,2,4-tri-
methylquinoline (A), N,N´-diphenyl-p-phenylenediamine (B),
and N,N´-dicyclohexyl-p-phenylenediamine (C) tested with and
without S9. For details see figure 1.

produced a direct effect on both base–pair substitution- and frameshift sensitive strains. The mutagenicity was elevated when a crude metabolizing system (S9) plus cofactors was added. This activation was also obtained with a microsomal preparation in the presence of an NADPH–generating system and a deactivation was noted which was independent of cofactors and was obtained with a cytosolic fraction of S9 (data not shown). The metabolism pattern described here is in common for the other thiurames and dithiocarbamates.

Oxygen sensitivity. The importance of oxygen in connection with the mutagenicity of TMTD was investigated in another experiment (figure 3). Two doses of TMTD, 5 and 50 µg per plate were tested on TA100 in the absence and presence of S9 with and without NADP. Incubation of the test series were performed in atmospheres containing different oxygen tensions.

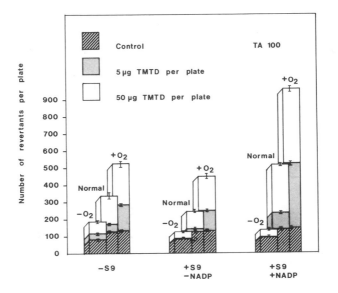

Figure 3. Mutagenicity of TMTD on Salmonella typhimurium TA100 after treatment in atmospheres with low ($-O_2$), normal, and high ($+O_2$) oxygen tensions. The presence of different metabolizing systems are indicated at the bottom of the figure.

Under standard conditions the direct effect was decreased in the presence of a S9—mix deficient of NADP and increased when a S9-mix with NADP was present. This overall pattern was also seen in test series incubated in an oxygen enriched environment. The mutagenic response was however doubled under these oxygen enriched conditions. Under anaerobic conditions the direct mutagenicity was reduced by one half of the value obtained under normal conditions. No metabolic activation was obtained in an oxygen deficient atmosphere.

DISCUSSION

The negative results will not be discussed in detail. The test method have some limitations for example in identifying poorly water soluble mutagens and mutagens causing oxidative damage to DNA (Ames et al., 1981). This fully explains the negative responses of curing agents of the peroxide type.

Positive results of some of the other compounds was to be expected from their carcinogenicity data. For example N-methyl-N,4-dinitrosoaniline and 4,4´-diaminodiphenyl-methane have both been associated with exess tumor production in animals (Holmberg and Sjöström, 1977) and have been withdrawn from use as rubber additives in Sweden.

The mutagenicity of a large number of accelerators of the thiuram and dithiocarbamate types, which also have antioxidant properties, were confusing. Mutagenicity due to alkylation of bacterial DNA is not plausible as there are no compounds among this group of additives that from an organic chemical point of wiev can be classified as alkylating. This was confirmed in tests for alkylation activity of thiuram and dithiocarbamate derivatives performed by Hemminki et al. (1980), using 4-(p-nitrobenzyl)-pyridine and deoxyguanosine as nucleophiles. They found no distinct alkylating activity of this type of compounds.

A number of investigations, however, have pointed to the DNA—damaging activity of thiurames and dithiocarbamates. Direct mutagenicity has been found in a number of short-term test systems for mutagenicity in bacteria and moulds (Hedenstedt et al., 1979, Zdzienicka et al., 1979, 1981, and Moriya, 1978) and Drosophila (Donner et al., 1982). Chromosomal damage have been reported (Pilinskaya, 1970 and 1971, and Donner et al., 1982) as well as aneuploidi (Upshall and Johnson, 1981).

Two properties of derivatives of dialkylthiuramic acid render them biologically active. These are for one thing their ability to form chelates with metals and for another their binding to SH-groups and the subsequent formation of mixed disulfides (Owens, 1969). Through enzyme inhibitions many fundamentally important cellular functions might be disturbed such as respiration, cell division and the protection against endogenous and exogenous oxidants.

Thiurames and dithiocarbamates probably constitutes a special class of mutagens as their inhibition of important enzymes may cause the mutagenic action. Among enzymes known to be inhibited by thiurames are catalase (Owens, 1969), superoxide dismutase (Forman et al.,1980), and glutathion peroxidase (Goldstein et al., 1979). These enzymes are all involved in the protection of the cell against oxidative damage induced by oxygen radicals and hydrogen peroxide, produced during the cellular reduction of oxygen to water (Fridovich, 1978). A steady state normally exists in respiring cells between the rate of production of reactive oxygen metabolites and their rate of inactivation. In case of an altered inactivation efficiency, oxygen radicals may exhibit a mutagenic action. In accordance with this, increased oxygen tension enhanced the mutagenicity of TMTD. Furthermore, oxygen at physiological concentrations have been found mutagenic towards oxygen sensitive mutants of S.typhimurium TA100 (Bruyninckx et al., 1978). The presumtion that thiurames and dithiocarbamates interfere with oxygen metabolism is further strengthened by the fact that oxygen toxicity towards rats is enhanced in the presence of diethyldithiocarbamate (Forman et al.,1980). In addition to the inhibitory action on enzymes, treatment with diethyldithiocarbamate has also been associated with production of hydrogen peroxide (Sinet et al., 1982).

A third property of these compounds, that may affect their genotoxic potentials is their reactivity towards nitrite under acidic conditions under formation of dialkylnitrosoamines.

The presence of biologically very reactive chemicals, e.g. some of the antioxidants and accelerators, must be seriously considered in regard to the risk for potentiation of genotoxic actions of other chemicals.

ACKNOWLEDGMENTS

The authors are indepted to Gislaved AB, Trelleborg AB, and Nokia AB for supplying the test substances. The technical assistance of Mrs Yvonne Eklund and Miss Ann Wiorek is gratefully acknowledged. This investigation was supported by grants from the Swedish Work Environment Fund and the Swedish Cancer Society.

REFERENSES

Ames B N, McCann J, Yamasaki E (1975). Methods for detecting carcinogens and mutagens with the Salmonella/mammalian-microsome mutagenicity test. Mutation Res 31:347.
Ames B N, Hollstein M C, Cathcart R (1981). Lipid peroxidation and oxidative damage to DNA. In Yagi K (ed): "Lipid peroxide in biology and medicine," New York: Academic Press, in press.
Bruyninckx W J, Mason H S, Morse S A (1978). Are physiological oxygen concentrations mutagenic? Nature 274: 10.
Donner M, Husgafvel-Pursiainen K, Jenssen D, Rannug A (1982). Mutagenicity of rubber additives and curing fumes: results from five short-term bioassays. Presented at the symposium "Genotoxic hazards in rubber industry", November 29-30, Helsinki, Finland.
Fishbein L (1982). Toxicity of the components of styrene polymers. This volyme.
Forman H J, York J L, Fischer A B (1980). Mechanism for the potentiation of oxygen toxicity by disulfiram. J Pharm Exp Therap 212:452.
Fridovich I (1978). The biology of oxygen radicals. Science 201:875.
Goldstein B D, Rozen M G, Quintavalla J C, Amoruso M A (1979). Decrease in mouse lung and liver gluthatione peroxdase activity and potentation of the lethal effects of ozone and paraquat by the superoxide dismutase inhibitor diethyldithiocarbamate. Biochem Pharmacol 28:27.
Hedenstedt A, Rannug U, Ramel C, Wachtmeister C A (1979). Mutagenicity and metabolism studies on 12 thiuram and dithiocarbamate compounds used as accelerators in the

Swedish rubber industry. Mutation Res 68:313.

Hedenstedt A, Ramel C, Wachtmeister C A (1981). Mutagenicity of rubber vulcanization gases in Salmonella typhimurium. J Toxicol Environ Health 8:805.

Hemminki K, Falck K, Vainio H (1980). Comparison of alkylation rates and mutagenicity of directly acting industrial and laboratory chemicals. Arch Toxicol 46:277.

Hoff A, Jacobsson S, Pfäffli P, Zitting A, Frostling H (1982). Degradation products of plastics. Scand J Work Environ Health 8 suppl 2.

Holmberg B, Sjöström B (1977). A toxicological survey of chemicals used in the Swedish rubber industry. Investiga- tion Report 1977:19. National Board of Occupational Safety and Health, Stockholm, Sweden.

Huff J E (1982). National toxicology programme and the carcinogenesis bioassays of the United States. This volume.

IARC (1976). IARC Monographs on the Evaluation of the Carcinogenic Risk of Chemicals to Humans, Vol. 11, Cadmium, nickel, some epoxides, miscellaneous industrial chemicals and general considerations on volatile anaesthetics. International Agency for Research on Cancer, Lyon, France, pp.131.

IARC (1979). IARC Monographs on the Evaluation of the Carcinogenic Risk of Chemicals to Humans, Vol. 19, Some Monomers, Plastics and Synthetic Elastomers and Acrolein. In- ternational Agency for Research on Cancer, Lyon, France, pp.131.

IARC (1982). IARC Monographs on the Evaluation of the Carcinogenic Risk of Chemicals to Humans, Vol. 28, The rubber industry. International Agency for Research on Cancer, Lyon, France, pp.154.

Moriya M, Kato K, Shirasu Y, Kada T(1978). Mutagenicity screening of pesticides in microbial systems. IV. Mutagenicity of dimethyldithiocarbamates and related fungicides. Abstract. Mutat Res 54:221.

Owens R G (1969). Organic sulfur compounds. In Torgeson D C (ed):"Fungicides II," New York and London: Academic Press, p 147.

Pilinskaya M A (1970). Chromosome aberrations in persons exposed to ziram during its production. Genetika 6:157.

Pilinskaya M A (1971). Cytogenetic effect of the fungicide ziram in an in vitro culture of human lymphocytes. Genetika 7:138.

Rannug-Hedenstedt A, Östman C (1981). Application of mutagenicity tests for detection and source assessment of genotoxic agents in the rubber work atmosphere. In

Sandhu S, Lewtas J, Nesnow S, Claxton L, Chernoff N
(eds): "Short-term Bioassays in the Analysis of Complex
Environmental Mixtures III," New York: Plenum, in press.
Sinet P-M, Garber P, Jerome H (1982). H_2O_2 production,
modification of the glutathione status and methemoglobin
formation in red blood cells exposed to
diethyldithiocarbamate in vitro. Biochem Pharmac 31:521.
Upshall A, Johson P E (1981). Thiuram-induced abnormal chromosome segregation in Aspergillus nidulans. Mutation Res
89:297.
Zdzienicka M, Zielenska M, Trojanowska M, Szymczyk T, Bigna-
mi M Carere A (1981). Microbial short-term assays with
thiram in vitro. Mutation Res 89:1.
Zdzienicka M, Zielenska M, Tudek B, Szymczyk T (1979). Mutagenic activity of thiram in Ames tester strains of
Salmonella typhimurium. Mutation Res 68:9.

Industrial Hazards of Plastics and Synthetic Elastomers, pages 421–427
© 1984 Alan R. Liss, Inc., 150 Fifth Ave., New York, NY 10011

CLOSING REMARKS

given at the
International Symposium On Occupational Hazards
Related To Plastics And Synthetic Elastomers
held at the Hanasaari Cultural Centre, Espoo, Finland
on 22-27 November 1982

J. E. Huff

National Toxicology Program
Research Triangle Park, North Carolina USA

Post hoc, ergo propter hoc

Fellow Students:

That salutation means of course that once a student always a student. The learning and sharing processes must never cease, for if we become comfortable in our meager storehouse of knowledge, then we will not only likely stagnate but our capabilities may eventually decrease.

Hence I like to think that we should attempt to remain always as children -- that is, never to discontinue exploration, never to stop asking questions, never to be satisfied. For as stated by Thomas Stearns Eliot in 1935, nearly fifty years ago, as part of his classic *Four Quartets (Little Gidding, V)*:

> *We shall not cease from exploration*
> *And the end of all our exploring*
> *Will be to arrive where we started*
> *And know the place for the first time*

You may recall from my lecture on the carcinogenicity of styrene (Huff, 1983) I stated that Bonastre first discovered styrene in 1831 by the distillation of the resin storax (IARC, 1979; Windholz, 1976). This alone is an interesting bit of information. However, as a small example of our pursuit of knowledge, I learned that storax (from the Latin styrax) is a balsam obtained from the trunk (inner bark) of

CLOSING REMARKS

Liquidambar orientalis Miller, known as Levant Storax, or of *Liquidambar styraciflua* Linne, known as American Storax (USP 1979). This storax resin, also called sweet oriental gum, is used as an incense, in perfumery, as an expectorant, and as a topical protectant. Each of us during this Symposium has learned about some intriguing truth that has stimulated us to pursue further.

The organizing committee asked me to give these closing remarks. I can only guess that they were confident I would not exercise the option of saying no. Joking aside, I am indeed honored by this invitation, and appreciate very much the opportunity to say a few concluding words at the end of such a satisfying -- yet work intensive -- Symposium. Thus, you see me before you once again.

Firstly and without exception all of us join as one in extending our sincere thanks and appreciation to Dr. Jorma Jarvisalo, the course leader; to Dr. Pirkko Pfaffli; to Dr. Harri Vainio; and to Ms. Outi Teperi for organizing an excellent and timely conference, for a comfortable working environment, and for maintaining a large reserve of patience with us. These four people, and I am sure others from the Institute of Occupational Health as well as from the Nordic Institute, should be recognized for their foresight in holding these important courses on the most relevant and current problems in the area of occupational health and safety.

So, what have we learned during this week? For one thing, we learned that we are a long distance from knowing every-thing we should know about monomers and polymers. Yet we must not neglect or discount the collected volumes of knowledge already gained.

On this Saturday morning we have just heard the distilled ideas and thoughts from the representatives of each of the four working groups: Hygienic Aspects, Toxicology, Occupa-tional Medicine and Health Care, and Predictive Studies. These volunteer speakers and their respective groups are to be congratulated. Their task was not easy. The outcome as evidenced by these short summaries was admirable. To combine in *ad hoc* working groups biologists, chemists, physicians, health administrators, engineers -- many with different native languages -- into coherent productive collectives, each preparing useful and scholarly endproducts,

CLOSING REMARKS

in such a short time must be considered close to heroic. With more time and thought these precis could well become valuable state-of-the-art concepts for future endeavors in the area of plastics.

From this International Symposium on Occupational Hazards Related to Plastics and Synthetic Elastomers we have heard and absorbed a tremendous volume of facts and figures. Some broad categorical examples include production processes and methodologies; and starting chemicals, added chemicals, reacted chemicals, altered chemicals, final chemicals, and degraded chemicals. We heard about available *in vitro* and *in vivo* toxicology methods and data; occupational hazards; identifying and monitoring these potential and real health decrements; and preventative measures.

Through the week these major chemical groupings were discussed and dealt with: polyvinyl and related polymers, styrene containing polymers, polyethylene and polypropylene, other thermoplastics, polyurethanes, epoxy thermosets, and synthetic elastomers or rubbers. An avalanche of information. You even survived learning from me about the Thanksgiving Day celebration and hearing about the United States National Toxicology Program.

Additionally we have gained from the experience and the wisdom of Sidney Weinhouse (1982) and of Takashi Sugimura (1982). Their advice honed over many years gives us considerable insight for establishing reasonable public health protective measures.

Right this moment before ending our six day Symposium we must all feel at least a little weary and a bit tired. Yet, when we return to our agencies, the knowledge we have all hidden away during these days will surface when needed, and surely this will make us better physicians, better inspectors, better workers, better scientists, better technicians, and, in my opinion, better people as well.

In these troubled times, we should consider ourselves fortunate indeed to have lived together these days. Gathered here in Hanasaari, an isolated hamlet near Helsinki, has been for me a unique and rewarding experience. Nothing at all like Richard Adams lamented in his sensitive book (1972) about rabbits:

CLOSING REMARKS

There is nothing that cuts you down to size
like coming to some strange and marvelous
place where no one even stops to notice
that you stare about you.

Much information, thoughts, ideas, and philosophies have been exchanged and shared. And friendships were established. Perhaps some time in the near future we shall be able to meet again and continue what began during this course.

After all, the message is clear: all of us are striving to make our own little area of the world a preferentially better and safer place in which to live. How one accomplishes or approaches this supreme goal is certainly unimportant, only that the peaceful and continuous effort is made.

About 2500 years ago Heraclitus predicted our yearning, our inquisitiveness, and, yes, even our search for new and changing knowledge:

There is nothing permanent except change.
No one can step into the same river twice,
because the river is never the same on two
separate occasions.

Thank you, and I look forward to seeing each of you again soon.

NOTES AND REFERENCES

1. *Post hoc, ergo propter hoc* translates to "after this, therefore because of this", and signals, to me at least, that much of our future efforts must be influenced by not only our most recent encounters and interactions (This Symposium), but to some extent by the composite of ALL our activities.

2. Quotations from T S. Eliot, R. Adams, and Heraclitus were added for emphasis while editing my hastily prepared remarks.

3. Huff, J.E. (1983). Carcinogenicity of Styrene, Styrene Oxide, Polystyrene, B-Nitrostyrene and Styrene, and Acrylonitrile. These Proceedings.

CLOSING REMARKS

4. IARC (1979). IARC Monographs on the Evaluation of the Carcinogenic Risk of Chemicals to Humans. Volume 19. Some Monomers, Plastics and Synthetic Elastomers, and Acrolein. 513 pages. International Agency for Research on Cancer, Lyon, France.

5. Windholz, M. Editor (1976). The Merck Index. An Encyclopedia of Chemicals and Drugs. Styrene, Monograph number 8657, page 1146. Merck & Co., Inc. Rahway, NJ.

6. USP (1979). The United States Pharmacopeia, Twentieth Revision and The National Formulary, Fifteenth Edition, both official from July 1, 1980; Article on Storax, page 742.

7. Weinhouse, S. (1982). Prometheus and Pandora - Cancer Research on our Diamond Anniversary: Presidential Address. Cancer Research 42: 3471-3474.

Dr. Weinhouse during his Presidential Address made these remarks on 29 April 1982 at the Seventy-Third Annual Meeting of the American Association for Cancer Research held in St. Louis, Missouri. These comments have been excerpted by me from page 3472 of his paper.

"The cancer researcher at this time ... is in the cross fire of a raging controversy ... dragged from his ivory tower into the rough-and-tumble arena of political debate and political action. He or she now is expected by an anxious public to have 'answers'."

"Much of the general public, and many of the vested interests, is either oblivious to or chooses to ignore the methodological and statistical principles that govern the testing for carcinogenicity. They also do not recognize the difficulties and pitfalls of the gathering and interpreting of experimental and epidemiological data. The expert who has to conduct and interpret experiments in animals is up against a pervasive attitude of the lay public that such tests are meaningless, and this view is often fostered by vested interests whose chief weapon is ridicule."

"In the tempest of these conflicting views and public misconceptions, the cancer researcher, using the principles of ethics and sound science ... must state as an article of faith, on which our whole science of biology depends, that if a substance is carcinogenic to animals

CLOSING REMARKS

it is a potential carcinogenic hazard to some people. This viewpoint, although buttressed by a compelling body of sound scientific evidence, is the focus of great controversy and misunderstanding."

8.   Sugimura, T. (1982). Mutagens, Carcinogens, and Tumor Promoters in our Daily Food. Cancer 49(10): 1970-1984.

In his 15 May 1982 paper Dr. Sugimura summarized these twelve items for cancer prevention "in a form easily under- standable by the general public". He emphasizes however that "these are, at best, very tentative recommendations for reducing exposure to carcinogens to prevent cancer, but currently this is the best advice we can give." He closes his article with the observation that "the most urgent problem facing researchers today is preventing cancer".

Cancer Prevention -- Tentative Recommendations for Reducing
Exposure to Carcinogens

  i.   Keep your diet well balanced, in terms of both taste and nutrition.
  ii.   Do not eat the same foods repeatedly and exclu- sively. Also exercise caution in taking the same medication over long periods.
  iii. Avoid excessive eating.
  iv.   Avoid drinking too much alcohol.
  v.   Refrain from excessive smoking.
  vi.   Take optimal daily doses of vitamins A, C, and E. Include a moderate amount of fibrous food ("roughage") in your diet.
  vii. Avoid excessive intake of salty food and do not drink too hot water, tea or coffee.
  viii. Avoid eating too many burnt parts of food, such as you find in charcoal-grilled meat and fish.
  ix.   Avoid moldy food which is not intentionally moldy, such as cheese.
  x.   Avoid excessive exposure to the sun.
  xi.   Avoid overwork so that you do not lower your resistance to disease.
  xii. Bathe or shower frequently.

9.   The contents in Notes 7 and 8 were presented at the Symposium in my lecture on the National Toxicology Program and long-term Carcinogenesis studies. Rather than include these comments and advice in that manuscript, their place- ment better fits here.

CLOSING REMARKS

10. In striving to accommodate the words of advice from Dr. Sugimura, Dr. Weinhouse, as well as other concerned scientific sages, we must remain mindful that the most righteous pathway to embrace may often be the least attractive, most disruptive, and perilously enigmatic since one will never know if the preventive measures adopted were correct, are protective of the public health, or reduces presumptive (or predictive) chemically-induced diseases. Dr. David Rall, Director of the National Institutes of Environmental Health Sciences and of the National Toxicology Program, emphasized these difficulties at the 1981 New York Academy of Sciences *Symposium on the Management of Assessed Risk for Carcinogens* These words should be read and remembered, for all those working in the area of public health need these guiding and enlightening thoughts:

"A risk can be imagined or it can be real; it can be immediate or distant. There are risks that an individual can control, and those over which he or she has no power. There are risks that have already resulted from past exposures and there are those that are predicted from exposures which have not taken place. Sensitivity to these distinctions is crucial not only for assessing and managing risks, but also for diagnosing and treating disease.

"Recognizing the problem is not enough, however. We must all make a commitment to deal with it. We need ways to achieve our goals.

Be aware that a commitment to deal with this problem buys us into a highly imprecise world of inadequate data and conflicting values. Not only is the information incomplete, but if we do not anticipate hazards and thereby prevent future disease, we will never know that we are right. In fact the naysayers will tell us that we cannot demonstrate that we were right, precisely because we are not willing to allow the evidence of disease to accumulate. In short, the price we pay for being right is that we cannot prove that doing nothing was wrong."

Rall, D.P. (1981). Issues in the Determination of Acceptable Risk. In: Management of Assessed Risk for Carcinogenic, Nicholson WJ, Editor, Ann NY Acad Sci 363: 139-144.

Index

PROGRESS IN CLINICAL AND BIOLOGICAL RESEARCH

Series Editors

Nathan Back
George J. Brewer
Vincent P. Eijsvoogel
Robert Grover

Kurt Hirschhorn
Seymour S. Kety
Sidney Udenfriend
Jonathan W. Uhr